Lecture Notes in Artificial Intelligence 5249

Edited by R. Goebel, J. Siekmann, and W. Wahlster

Subseries of Lecture Notes in Computer Science

Gerson Zaverucha
Augusto Loureiro da Costa (Eds.)

Advances in Artificial Intelligence – SBIA 2008

19th Brazilian Symposium on Artificial Intelligence
Savador, Brazil, October 26-30, 2008
Proceedings

Springer

Series Editors

Randy Goebel, University of Alberta, Edmonton, Canada
Jörg Siekmann, University of Saarland, Saarbrücken, Germany
Wolfgang Wahlster, DFKI and University of Saarland, Saarbrücken, Germany

Volume Editors

Gerson Zaverucha
Federal University of Rio de Janeiro
Caixa Postal 68511
Rio de Janeiro - 21941-972
Brazil
E-mail: gerson@cos.ufrj.br

Augusto Loureiro da Costa
Federal University of Santa Catarina
Department of Automation and Systems
Av. Ademar de Barros
Salvador, Bahia, CEP 88.040-900
Brazil
E-mail: augusto.loureiro@ufba.br

Library of Congress Control Number: Applied for

CR Subject Classification (1998): I.2, F.4.1, F.1.1, H.2.8

LNCS Sublibrary: SL 7 – Artificial Intelligence

ISSN 0302-9743
ISBN-10 3-540-88189-1 Springer Berlin Heidelberg New York
ISBN-13 978-3-540-88189-6 Springer Berlin Heidelberg New York

Springer is a part of Springer Science+Business Media

springer.com

Printed in Germany

Typesetting: Camera-ready by author, data conversion by Scientific Publishing Services, Chennai, India
Printed on acid-free paper SPIN: 12539758 06/3180 5 4 3 2 1 0

Preface

The 19th Brazilian Symposium on Artificial Intelligence (SBIA 2008) was held in Salvador, Bahia, Brazil, during October 26–30, 2008. It was hosted by the Federal University of Bahia (UFBa) and, as has occurred since 2002, was collocated with the Brazilian Symposium on Artificial Neural Networks (SBRN), now in its tenth edition. This year the two events took place jointly also with the Intelligent Robotics Journey (JRI).

SBIA, supported by the Brazilian Computer Society (SBC), is the leading conference in Brazil for the presentation of research and applications results in artificial intelligence. Since 1995, SBIA has become an international conference, with papers written exclusively in English, an international Program Committee (PC) and keynote speakers, and proceedings published in the *Lecture Notes in Artificial Intelligence* series of Springer. Since 1996, SBIA has been a biennial event.

The SBIA 2008 program included keynote talks/tutorials by some of the most distinguished researchers in the area, five workshops and a thesis and dissertation contest. SBIA 2008 continued the tradition of high selectivity for published papers and double-blind reviewing. A total of 142 submissions from 15 countries were received, of which only 27 were accepted for publication in this volume, yielding an acceptance rate of 19%. Each submission was reviewed by three PC members, and a recommendation was provided for each paper based on discussion among the reviewers moderated by the Program Chair.

We would like to thank all researches who submitted their papers to SBIA 2008. We are indebted to the 150 PC members, and also to the other reviewers, for their excellent work: thorough and timely reviews, and participation in the discussions. SBIA 2008 had the privilege and honor of having this PC and invited speakers. We gratefully acknowledge everyone in the Organizing Committee for their invaluable support, Aline Paes for helping with the preparation of this volume, along with the agencies CNPq and CAPES for the financial support. Finally, we would like to express our gratitude to Alfred Hoffman and his staff at Springer for their continuing support of SBIA.

October 2008

Gerson Zaverucha
Augusto Loureiro da Costa

Organization

Organizing Committee

General Chair

Augusto Loureiro da Costa	Universidade Federal da Bahia

General Co-chair

Evandro de Barros Costa	Universidade Federal de Alagoas

Program Chair

Gerson Zaverucha	Universidade Federal do Rio de Janeiro

Workshop Chair

Evandro de Barros Costa	Universidade Federal de Alagoas

Tutorial Chair

Gerson Zaverucha	Universidade Federal do Rio de Janeiro

Steering Committee

Fabio Gagliardi Cozman	Universidade de São Paulo
Jomi Fred Hübner	Universidade Regional de Blumenau
Fernando Osório	Universidade de São Paulo-São Carlos
Solange Rezende	Universidade de São Paulo-São Carlos
Jacques Wainer	Universidade Estadual de Campinas

Local Organizing Committee

Isamara Carvalho Alves	Universidade Federal da Bahia
Fabiana Cristina Bertoni	Universidade Estadual de Feira de Santana
Niraldo Roberto Ferreira	Universidade Federal da Bahia
Frederico Luiz G. de Freitas	Universidade Federal de Pernambuco
Allan Edgard Silva Freitas	Centro Federal de Educação Tecnológica da Bahia
Jorge da Costa Leite Jr.	Centro Federal de Educação Tecnológica da Bahia
Marcelo Santos Linder	Universidade Federal do Vale de São Francisco
Angelo Conrado Loula	Universidade Estadual de Feira de Santana

Program Committee

James Cussens	University of York, UK
Walter Daelemans	University of Antwerp, Belgium
Jesse Davis	University of Washington, USA
Luc De Raedt	Katholieke Universiteit Leuven, Belgium
Frank Dignum	Utrecht University, The Netherlands
Juergen Dix	Clausthal University of Technology, Germany
Marco Dorigo	Université Libre de Bruxelles, Belgium
Kurt Driessens	Katholieke Universiteit Leuven, Belgium
Lucia Drummond	Universidade Federal Fluminense, Brazil
Inês de Castro Dutra	Universidade do Porto, Portugal
Nelson Ebecken	Universidade Federal do Rio de Janeiro, Brazil
Ulle Endriss	University of Amsterdam, The Netherlands
Alexandre Evsukoff	Universidade Federal do Rio de Janeiro, Brazil
Stefano Ferilli	University of Bari, Italy
Peter Flach	University of Bristol, UK
Marcelo Finger	Universidade de São Paulo, Brazil
Felipe França	Universidade Federal do Rio de Janeiro, Brazil
Eibe Frank	University of Waikato, New Zealand
Alex Freitas	University of Kent, UK
Renata de Freitas	Universidade Federal Fluminense, Brazil
Johannes Fürnkranz	Technische Universität Darmstadt, Germany
Dov Gabbay	King's College London, UK
João Gama	Universidade do Porto, Portugal
Artur Garcez	City University, UK
Ana Cristina Bicharra Garcia	Universidade Federal Fluminense, Brazil
Thomas Gärtner	Fraunhofer IAIS, Germany
Lluis Godo	Instituto de Investigación en Inteligencia Artificial, Spain
Siome Goldenstein	Universidade Estadual de Campinas, Brazil
Moises Goldszmidt	Microsoft Research, USA
Fernando Gomide	Universidade Estadual de Campinas, Brazil
Marco Gori	University of Siena, Italy
Joseph Halpern	Cornell University, USA
Barbara Hammer	Clausthal University of Technology, Germany
Edward Hermann Haeusler	Pontifícia Universidade Católica do Rio de Janeiro, Brazil
Emilio Hernandez	Universidade de São Paulo, Brazil
Melanie Hilario	University of Geneva, Switzerland
Pascal Hitzler	University of Karlsruhe, Germany
Wiebe van der Hoek	University of Liverpool, UK
Steffen Hölldobler	TU Dresden, Germany
Jomi Hubner	Universidade Regional de Blumenau, Brazil
Katsumi Inoue	National Institute of Informatics, Japan

Catholijn Jonker	Delft University of Technology, The Netherlands
Alipio Jorge	Universidade do Porto, Portugal
Nikola Kasabov	Auckland University of Technology, New Zealand
Kristian Kersting	MIT, USA
Irwin King	The Chinese University of Hong Kong, Hong Kong
Sven Koenig	University of Southern California, USA
Stefan Kramer	TU München, Germany
Luis Lamb	Universidade Federal do Rio Grande do Sul, Brazil
Nada Lavrac	Jozef Stefan Institute, Slovenia
Zhao Liang	Universidade de São Paulo-São Carlos, Brazil
Vera Strube de Lima	Pontifícia Universidade Católica Rio Grande do Sul, Brazil
John Lloyd	Australian National University, Australia
Alessio Lomuscio	Imperial College London, UK
Michael Luck	King's College London, UK
Teresa Ludermir	Universidade Federal de Pernambuco, Brazil
Donato Malerba	University of Bari, Italy
Ramon de Mantaras	Institut d'Investigación en Intelligència Artificial, Spain
Ana Teresa Martins	Universidade Federal do Ceará, Brazil
Stan Matwin	University of Ottawa, Canada
Wagner Meira Jr.	Universidade Federal de Minas Gerais, Brazil
Maria Carolina Monard	Universidade de São Paulo-São Carlos, Brazil
Luiza Mourelle	Universidade Estadual do Rio de Janeiro, Brazil
Hiroshi Motoda	US Air Force Research Laboratory, Japan
Maria das Graças V. Nunes	Universidade de São Paulo-São Carlos, Brazil
Luiz Satoru Ochi	Universidade Federal Fluminense, Brazil
Eugenio Oliveira	Universidade do Porto, Portugal
Irene Ong	University of Wisconsin Madison, USA
Fernando Osório	Universidade de São Paulo-São Carlos, Brazil
Ramon Otero	Universidad da Coruña, Spain
David Page	University of Wisconsin Madison, USA
Tarcisio Pequeno	Universidade Federal do Ceará, Brazil
Bernhard Pfahringer	University of Waikato, New Zealand
Aurora Pozo	Universidade Federal do Paraná, Brazil
Foster Provost	New York University, USA
Jan Ramon	Katholieke Universiteit Leuven, Belgium
Soumya Ray	Oregon State University, USA
Solange Rezende	Universidade de São Paulo-São Carlos, Brazil
Carlos Ribeiro	Instituto Tecnológico da Aeronáutica, Brazil

Dan Roth	University of Illinois at Urbana-Champaign, USA
Celine Rouveirol	Université de Paris XI, France
Jeffrey Rosenchein	Hebrew University of Jerusalem, Israel
Fariba Sadri	Imperial College London, UK
Sandra Sandri	Institut d'Investigación en Intelligència Artificial, Spain
Paulo Santos	Centro Universitário da FEI, Brazil
Lorenza Saitta	Università del Piemonte Orientale, Italy
Torsten Schaub	University of Potsdam, Germany
Jude Shavlik	University of Wisconsin Madison, USA
Jaime Sichman	Universidade de São Paulo, Brazil
Carles Sierra	Institut d'Investigación en Intelligència Artificial, Spain
Alexandre Silva	Universidade Federal do Rio de Janeiro, Brazil
Ricardo Silva	University of Cambridge, UK
Flávio Soares C da Silva	Universidade de São Paulo, Brazil
Guillermo Simari	Universidad Nacional del Sur, Argentina
Marcilio de Souto	Universidade Federal do Rio Grande do Norte, Brazil
Marc Schoenauer	Université Paris Sud, France
Jan Struyf	Katholieke Universiteit Leuven, Belgium
Prasad Tadepalli	Oregon State University, USA
Carlos Thomaz	Centro Universitario da FEI, Brazil
Michael Thielscher	TU Dresden, Germany
Francesca Toni	Imperial College London, UK
Flavio Tonidandel	FEI, Brazil
Luis Torgo	Universidade do Porto, Portugal
Jacques Wainer	Universidade Estadual de Campinas, Brazil
Toby Walsh	NICTA and University of New South Wales, Australia
Lipo Wang	Nanyang Technological University, Singapore
Renata Wassermann	Universidade de São Paulo, Brazil
Li Weigang	Universidade de Brasilia, Brazil
Michael Wooldridge	University of Liverpool, UK
Wamberto Vasconcelos	University of Aberdeen, UK
Sheila Veloso	Universidade Estadual do Rio de Janeiro, Brazil
Marley Vellasco	Pontifícia Universidade Católica do Rio de Janeiro, Brazil
Renata Vieira	Pontifícia Universidade Católica Rio Grande do Sul, Brazil
Aline Villavicencio	Universidade Federal do Rio Grande do Sul, Brazil
Bianca Zadrozny	Universidade Federal Fluminense, Brazil
Filip Železný	Czech Technical University in Prague, Czech Republic

Additional Reviewers

Prasanna Balaprakash
Guilherme Bastos
Tristan Behrens
João Bertini
Reinaldo Bianchi
Inacio Lanari Bo
Anarosa Brandao
Nicolas Bredeche
Fabricio Breve
Nils Bulling
Philippe Caillou
Sara Casare
Paulo Castro
George Cavalcanti
Ivan Costa
Marc Esteva
Katti Faceli
Andrea Giovanucci
Maury Gouvêa Jr.
Renato Krohling
Omar Lengerke
Priscila M. V. Lima
Ana Lorena
Felipe Meneguzzi
Olana Missura
Mariá Cristina Nascimento
Peter Novak
Luis Nunes
Marco Montes de Oca
Thiago Pardo
Artur Alves Pessoa
Gauthier Picard
Cláudio Policastro
Ronaldo Prati
Jakob Puchinger
Luis Paulo Reis
Tsang Ing Ren
Pedro Pereira Rodrigues
Raquel Ros
André Rossi
André Renato da Silva
Cassia Santos
Haroldo Santos
Antonio Selvatici
Marcone Jamilson Souza
Hirokazu Taki
Satoshi Tojo
Eduardo Uchoa
Sebastian Alberto Urrutia
Germano Vasconcelos
Laurent Vercouter
Nicolau Werneck

Supporting Scientific Society

SBC	Sociedade Brasileira de Computação

Sponsoring Institutions

CNPq	Conselho Nacional de Desenvolvimento Científico e Tecnológico
CAPES	Coordenação de Aperfeiçoamento de Pessoal de Nível Superior

Table of Contents

Invited Papers

Tutorials

Computer Vision and Pattern Recognition

Distributed AI: Autonomous Agents, Multi-Agent Systems and Game Theory

Knowledge Representation and Reasoning

Machine Learning and Data Mining

Natural Language Processing

Robotics

Logical and Relational Learning

Luc De Raedt

Department of Computer Science, Katholieke Universiteit Leuven
Celestijnenlaan 200 A B-3001, Heverlee, Belgium
luc.deraedt@cs.kuleuven.be

I use the term logical and relational learning (LRL) to refer to the subfield of machine learning and data mining that is concerned with learning in expressive logical or relational representations. It is the union of inductive logic programming, (statistical) relational learning and multi-relational data mining and constitutes a general class of techniques and methodology for learning from structured data (such as graphs, networks, relational databases) and background knowledge. During the course of its existence, logical and relational learning has changed dramatically. Whereas early work was mainly concerned with logical issues (and even program synthesis from examples), in the 90s its focus was on the discovery of new and interpretable knowledge from structured data, often in the form of rules or patterns. Since then the range of tasks to which logical and relational learning has been applied has significantly broadened and now covers almost all machine learning problems and settings. Today, there exist logical and relational learning methods for reinforcement learning, statistical learning, distance- and kernel-based learning in addition to traditional symbolic machine learning approaches. At the same time, logical and relational learning problems are appearing everywhere. Advances in intelligent systems are enabling the generation of high-level symbolic and structured data in a wide variety of domains, including the semantic web, robotics, vision, social networks, and the life sciences, which in turn raises new challenges and opportunities for logical and relational learning. These developments have led to a new view on logical and relational learning and its role in machine learning and artificial intelligence. In this talk, I shall reflect on this view by identifying some of the lessons learned in logical and relational learning and formulating some challenges for future developments.

Reference

1. De Raedt, L.: Logical and Relational Learning. Springer, Heidelberg (in press)

G. Zaverucha and A. Loureiro da Costa (Eds.): SBIA 2008, LNAI 5249, p. 1, 2008.

Transfer Learning by Mapping and Revising Relational Knowledge

Raymond J. Mooney

Department of Computer Sciences, University of Texas at Austin
1 University Station C0500, Austin, TX 78712-0233, USA
mooney@cs.utexas.edu

1 Transfer Learning (TL)

Traditional machine learning algorithms operate under the assumption that learning for each new task starts from scratch, thus disregarding knowledge acquired in previous domains. Naturally, if the domains encountered during learning are related, this *tabula rasa* approach wastes both data and computational resources in developing hypotheses that could have potentially been recovered by simply slightly modifying previously acquired knowledge. The field of *transfer learning* (TL), which has witnessed substantial growth in recent years, develops methods that attempt to utilize previously acquired knowledge in a *source* domain in order to improve the efficiency and accuracy of learning in a new, but related, *target* domain [7,6,1].

2 Statistical Relational Learning (SRL)

Traditional machine learning methods also assume that examples are represented by fixed-length feature vectors and are *independently and identically distributed* (i.i.d). *Statistical relational learning* (SRL), studies techniques that combine the strengths of relational learning (e.g. *inductive logic programming*) and probabilistic learning of graphical models (e.g. *Bayesian networks* and *Markov networks*). By combining the power of logic and probability, such methods can perform robust and accurate reasoning and learning about complex relational data [2]. Also, SRL frequently violates the i.i.d. assumption since examples are not independent, in which case, inferences about examples must be made in unison (i.e. collective classification).

3 TL for SRL

Most TL research addresses supervised feature-vector classification or reinforcement learning. In contrast, our research has focussed on developing TL methods for SRL. *Markov logic networks* (MLNs) are an expressive SRL formalism that represents knowledge in the form of a set of weighted clauses in first-order predicate logic [5]. We have developed an initial MLN transfer system, TAMAR, that

G. Zaverucha and A. Loureiro da Costa (Eds.): SBIA 2008, LNAI 5249, pp. 2–3, 2008.

first autonomously maps the predicates in the source MLN to the target domain and then revises the mapped structure to further improve its accuracy [3]. Our results on transfer learning between three real-world data sets demonstrate that our approach successfully reduces the amount of computational time and training data needed to learn an accurate model of a target domain compared to learning from scratch.

We view transferring an MLN to a new domain as consisting of two subtasks: *predicate mapping* and *theory refinement*. In general, the set of predicates used to describe data in the source and target domains may be partially or completely distinct. Therefore, the first transfer task is to establish a mapping from predicates in the source domain to predicates in the target domain. For example, the predicate `Professor` in an academic source domain may map to `Director` in a target movie domain. TAMAR searches the space of type-consistent mappings and determines the mapping that results in an MLN that best fits the available data in the target domain. Once a mapping is established, clauses from the source domain are translated to the target domain. However, these clauses may not be completely accurate and may need to be revised, augmented, and re-weighted in order to properly model the target data. This revision step uses methods similar to those developed for theory refinement [4], except the theory to be revised is *learned* in a previous domain rather than manually constructed for the target domain by a human expert.

Acknowledgements

I would like to thank Lilyana Mihalkova and Tuyen Huynh for their significant contributions to this research, which is partially supported by the U.S. Defense Advanced Research Projects Agency under contract FA8750-05-2-0283.

References

1. Banerjee, B., Liu, Y., Youngblood, G.M. (eds.): Proceedings of the ICML 2006 Workshop on Structural Knowledge Transfer for Machine Learning, Pittsburgh, PA (2006)
2. Getoor, L., Taskar, B. (eds.): Introduction to Statistical Relational Learning. MIT Press, Cambridge (2007)
3. Mihalkova, L., Huynh, T., Mooney, R.J.: Mapping and revising Markov logic networks for transfer learning. In: Proceedings of the Twenty-Second Conference on Artificial Intelligence (AAAI-2007), Vancouver, BC, pp. 608–614 (July 2007)
4. Richards, B.L., Mooney, R.J.: Automated refinement of first-order Horn-clause domain theories. Machine Learning 19(2), 95–131 (1995)
5. Richardson, M., Domingos, P.: Markov logic networks. Machine Learning 62, 107–136 (2006)
6. Silver, D., Bakir, G., Bennett, K., Caruana, R., Pontil, M., Russell, S., Tadepalli, P. (eds.): Proceedings of NIPS-2005 Workshop on Inductive Transfer: 10 Years Later (2005)
7. Thrun, S., Pratt, L. (eds.): Learning to Learn. Kluwer Academic Publishers, Boston (1998)

Developing Robust Synthetic Biology Designs Using a Microfluidic Robot Scientist

Stephen Muggleton

Department of Computing, Imperial College London
180 Queen's Gate, London SW7 2BZ, UK
shm@doc.ic.ac.uk

Synthetic Biology is an emerging discipline that is providing a conceptual framework for biological engineering based on principles of standardisation, modularity and abstraction. For this approach to achieve the ends of becoming a widely applicable engineering discipline it is critical that the resulting biological devices are capable of functioning according to a given specification in a robust fashion. In this talk we will describe the development of techniques for experimental validation and revision based on the development of a microfluidic robot scientist to support the empirical testing and automatic revision of robust component and device-level designs. The approach is based on probabilistic and logical hypotheses [1] generated by active machine learning. Previous papers [2,3] based on the author's design of a Robot Scientist appeared in Nature and was widely reported in the press. The new techniques will extend those in the speaker's previous publications in which it was demonstrated that the scientific cycle of hypothesis formation, choice of low-expected cost experiments and the conducting of biological experiments could be implemented in a fully automated closed-loop. In the present work we are developing the use of Chemical Turing machines based on micro-fluidic technology, to allow high-speed (sub-second) turnaround in the cycle of hypothesis formation and testing. If successful such an approach should allow a speed-up of several orders of magnitude compared to the previous technique (previously 24 hour experimental cycle).

References

1. De Raedt, L., Frasconi, P., Kersting, K., Muggleton, S.H. (eds.): Probabilistic Inductive Logic Programming. LNCS (LNAI), vol. 4911, pp. 1–27. Springer, Heidelberg (2008)
2. Muggleton, S.H.: Exceeding human limits. Nature 440(7083), 409–410 (2006)
3. King, R.D., Whelan, K.E., Jones, F.M., Reiser, P.K.G., Bryant, C.H., Muggleton, S.H., Kell, D.B., Oliver, S.G.: Functional genomic hypothesis generation and experimentation by a robot scientist. Nature 427, 247–252 (2004)

G. Zaverucha and A. Loureiro da Costa (Eds.): SBIA 2008, LNAI 5249, p. 4, 2008.

Logic, Probability and Learning, or an Introduction to Statistical Relational Learning

Luc De Raedt

Department of Computer Science, Katholieke Universiteit Leuven
Celestijnenlaan 200 A B-3001, Heverlee, Belgium
luc.deraedt@cs.kuleuven.be

Probabilistic inductive logic programming (PILP), sometimes also called statistical relational learning, addresses one of the central questions of artificial intelligence: the integration of probabilistic reasoning with first order logic representations and machine learning. A rich variety of different formalisms and learning techniques have been developed and they are being applied on applications in network analysis, robotics, bio-informatics, intelligent agents, etc. This tutorial starts with an introduction to probabilistic representations and machine learning, and then continues with an overview of the state-of-the-art in statistical relational learning. We start from classical settings for logic learning (or inductive logic programming) namely learning from entailment, learning from interpretations, and learning from proofs, and show how they can be extended with probabilistic methods. While doing so, we review state-of-the-art statistical relational learning approaches and show how they fit the discussed learning settings for probabilistic inductive logic programming.

This tutorial is based on joint work with Dr. Kristian Kersting.

References

1. De Raedt, L.: Logical and Relational Learning. Springer, Heidelberg (in press, 2008)
2. De Raedt, L.: Kristian Kersting: Probabilistic logic learning. SIGKDD Explorations 5(1), 31–48 (2003)
3. De Raedt, L., Kersting, K.: Probabilistic Inductive Logic Programming. In: Ben-David, S., Case, J., Maruoka, A. (eds.) ALT 2004. LNCS (LNAI), vol. 3244, pp. 19–36. Springer, Heidelberg (2004)

G. Zaverucha and A. Loureiro da Costa (Eds.): SBIA 2008, LNAI 5249, p. 5, 2008.

Text Mining

Raymond J. Mooney

Department of Computer Sciences, University of Texas at Austin
1 University Station C0500, Austin, TX 78712-0233, USA
mooney@cs.utexas.edu

Most data mining methods assume that the data to be mined is represented in a structured relational database. However, in many applications, available electronic information is in the form of unstructured natural-language documents rather than structured databases. This tutorial will review machine learning methods for text mining. First, we will review standard classification and clustering methods for text which assume a vector-space or "bag of words" representation of documents that ignores the order of words in text. We will discuss naive Bayes, Rocchio, nearest neighbor, and SVMs for classifying texts and hierarchical agglomerative, spherical k-means and Expectation Maximization (EM) methods for clustering texts. Next we will review information extraction (IE) methods that use sequence information to identify entities and relations in documents. We will discuss hidden Markov models (HMMs) and conditional random fields (CRFs) for sequence labeling and IE. We will motivate the methods discussed with applications in spam filtering, information retrieval, recommendation systems, and bioinformatics.

G. Zaverucha and A. Loureiro da Costa (Eds.): SBIA 2008, LNAI 5249, p. 6, 2008.

From ILP to PILP

Stephen Muggleton

Department of Computing, Imperial College London
180 Queen's Gate, London SW7 2BZ, UK
shm@doc.ic.ac.uk

Inductive Logic Programming (ILP) is the area of Computer Science which deals with the induction of hypothesised predicate definitions from examples and background knowledge. Probabilistic ILP (PILP) extends the ILP framework by making use of probabilistic variants of logic programs to capture background and hypothesised knowledge. ILP and PILP are differentiated from most other forms of Machine Learning (ML) both by their use of an expressive representation language and their ability to make use of logically encoded background knowledge. This has allowed successful applications in areas such as Systems Biology, computational chemistry and Natural Language Processing. The problem of learning a set of logical clauses from examples and background knowledge has been studied since Reynold's and Plotkin's work in the late 1960's. The research area of ILP has been studied intensively since the early 1990s, while PILP has received increasing amounts of interest over the last decade. This talk will provide an overview of results for learning logic programs within the paradigms of learning-in-the-limit, PAC-learning and Bayesian learning. These results will be related to various settings, implementations and applications used in ILP. It will be argued that the Bayes' setting has a number of distinct advantages for both ILP and PILP. Bayes' average case results are easier to compare with empirical machine learning performance than results from either PAC or learning-in-the-limit. Broad classes of logic programs are learnable in polynomial time in a Bayes' setting, while corresponding PAC results are largely negative. Bayes' can be used to derive and analyse algorithms for learning from positive only examples for classes of logic program which are unlearnable within both the PAC and learning-in-the-limit framework. It will be shown how a Bayesian approach can be used to analyse the relevance of background knowledge when learning. General results will also be discussed for expected error given a k-bit bounded incompatibility between the teacher's target distribution and the learner's prior.

G. Zaverucha and A. Loureiro da Costa (Eds.): SBIA 2008, LNAI 5249, p. 7, 2008.

Density of Closed Balls in Real-Valued and Autometrized Boolean Spaces for Clustering Applications

C.G. González, W. Bonventi Jr., and A.L. Vieira Rodrigues*

Technology Center-University of Sorocaba-SP-Brazil
{carlos.gonzalez,waldemar.bonventi,andrea.rodrigues}@uniso.br

Abstract. The use of real-valued distances between bit vectors is customary in clustering applications. However, there is another, rarely used, kind of distances on bit vector spaces: the autometrized Boolean-valued distances, taking values in the same Boolean algebra, instead of $\mathbb{R}$. In this paper we use the topological concept of closed ball to define density in regions of the bit vector space and then introduce two algorithms to compare these different sorts of distances. A few, initial experiments using public databases, are consistent with the hypothesis that Boolean distances can yield a better classification, but more experiments are necessary to confirm it.

1 Introduction

Data clustering or unsupervised classification is an important field of research in computer science (see [1,2]), with a number of applications in computer programs. In clustering, the distance concepts play a fundamental role in algorithm design. To classify objects we can deal with two types of features: those which can take values from a continuous range (a subset of $\mathbb{R}$) or from a discrete set. If this discrete set has only two elements, we have binary features or properties. Furthermore, in the most cases, finite discrete or symbolic features can be easily represented by binary values, which are represented by bit vectors. Hence, a number of distance concepts have been used since the early clustering works to deal with binary features [3]. Because of the metric spaces are generally defined as taken values in $\mathbb{R}$, the usual distances in clustering (and the distances on bit vector spaces) are real-valued. But in [4,5] we find another kind of metric spaces, called "autometrized", taking values in the same algebra, instead of $\mathbb{R}$. These autometrized distances are rarely used in clustering and the authors do not know any work comparing real-valued and autometrized distances in clustering.

The purpose of this paper is to confront real-valued and autometrized Boolean-valued distances. Since both kind of distances are very different, we choose a theoretical framework in which similar concepts, techniques and algorithms are used:

* The first author wish to express his gratitude to Prof. Marcelo Finger for valuable comments on some early ideas related to this paper. Thanks are also given to the referees, specially referee 3, for suggestions improving the quality of this paper. This work was supported by the Fapesp-SP-Brazil.

G. Zaverucha and A. Loureiro da Costa (Eds.): SBIA 2008, LNAI 5249, pp. 8–22, 2008.

we employ the topological concept of closed ball and the same density definition for both kind of metrics. Thus the comparison can be carried out in a simple way.

The organization of this paper is as follows. Section 2 presents the basic concepts of ultrametric and Boolean valued spaces and the appropriated concepts of closed ball for these spaces. Section 3 develops the theoretic framework: explains the concepts of Boolean algebra of bit vectors, closed ball and least closed ball in real-valued and autometrized Boolean-valued spaces. Section 4 carried out a discussion about the application of Boolean-valued distances in clustering. Finally, sections 5, 6 and 7 introduce two algorithms, show preliminary experimental results and get the final conclusions of this work.

2 Ultrametric and Boolean-Valued Autometrized Spaces

2.1 Ultrametric Spaces

The definition of ultrametric space had been created from the standard definition of metric space by replacing the sum with maximum operation in the triangle inequality.

Definition 1. *Let S be a non void set and let d be a map: $d : S \times S \longrightarrow \mathbb{R}$. We said that $\langle S, d\rangle$ is a* ultrametric space *if, $\forall x, y, z \in S$, it holds:*
a) $d(x,x) = 0$;
b) $d(x,y) = d(y,x)$;
c) If $x \neq y$, *then* $d(x,y) \neq 0$;
d) $d(x,z) \leq max(d(x,y), d(y,z))$.
The map d is called an ultrametric.

In [6] Krasner has introduced the term "ultrametrique" and has explained the most important properties of ultrametric spaces: every point inside a ball is itself at the center of the ball, two balls are either disjoint or contained one within the other, all triangles are either equilateral or isosceles, and so on (see [7] for other explanation of basic properties).

Some properties, for instance, previously named on equilateral and isosceles triangles, are derived from the linearity of $\mathbb{R}$. From this, consider condition d in definition 1: in any triangle each edge must be lower or equal to the largest one. Then, a scalene triangle falsify condition d. For an equilateral triangle the equality holds. Furthermore, an isosceles, non equilateral, triangle satisfies condition d if and only if one edge is lower than each other. The existence of the maximum and the comparability of the edges follow from the linearity of $\mathbb{R}$.

The well-known topological definition of closed ball can be extended to ultrametric spaces (for topological concepts, see [8]). Despite that the definition is the same that in metric spaces, the geometrical sense of closed ball is quite different in ultrametric spaces.

Definition 2. *Let $\langle S, d\rangle$ be an ultrametric space. For $o \in S, r \in \mathbb{R}$,* the closed ball of center o and radius r, $C_r(o)$, *is:*

$$C_r(o) = \{x \in S : d(o, x) \leq r\}$$

A set $D \subseteq S$ *is called a* closed ball *if there exists* r *and* o *as above, such that* $D = C_r(o)$.

The named property that any point in a closed ball can be considered as center is the following proposition.

Proposition 1. *Let* $\langle S, d\rangle$ *be an ultrametric space,* $o \in S, r \in \mathbb{R}$ *and* $x \in C_r(o)$. *Then* $C_r(x) = C_r(o)$.

Proof. We will show that $C_r(x) \subseteq C_r(o)$. Let $y \in C_r(x)$. Then $d(x, y) \leq r$. Since $x \in C_r(o)$, we have $d(o, x) \leq r$. From condition d in definition 1 and the property of the maximum operation, we obtain $d(o, y) \leq \max(d(o, x), d(x, y)) \leq r$. The other side is analogous. □

From the proof of proposition 1 we see that two closed balls of the same radius are disjoint or identical. Hence, if we fix an ultrametric and consider a radius r, then we partition the entire space into closed balls of radius r. In clustering terminology, the radius r is a classifier. It is easy to show that for any pair of closed ball there exists only two possibilities: one is included within the other or they are disjoint, but it is necessary to use that the radius are comparable, i.e. the linearity of $\mathbb{R}$.

When we want to specify a closed ball in a metric space, we need to find the center and the radius. However, in a ultrametric space this is reduced to find the radius, since any point in the closed ball can be considered as a center.

2.2 Autometrized Boolean-Valued Spaces

We recall that a Boolean algebra B is defined as a complemented, distributive lattice with minimum ($\mathbb{0}$) and maximum ($\mathbb{1}$) [9,10], denoting with $\leq$ the partial order of B and with $\wedge$ and $\vee$ the meet or infimum and the join or supremum operations. Furthermore, we denote by $-$ and $\triangle$ the complement and symmetric difference operations of the Boolean algebra.

Despite that standard metrics can be defined on a Boolean algebra (for instance, the Hamming distance), Blumenthal in 1952 introduced another modification of the standard definition of metric space in eliminating $\mathbb{R}$ of the definition, using a Boolean algebra instead. The word "autometrized" is used in these cases, i.e. for metric spaces that take values in the same structure. A number of autometrized structures are known: Brouwerian Algebras, Newman Algebras, commutative l-groups, etc. (see [5]). In a Boolean algebra, a two elements set can not have a maximum. Hence, Blumenthal replaced the maximum of definition 1 with the supremum operation ($\vee$). The main concept of this paper is called autometrized Boolean-valued space:

Definition 3. *Let* B *be a Boolean algebra and let* $\mathbb{0}$ *the minimum of* B. *Then, for* $d : B \times B \longrightarrow B$ *and* $x, y, z \in B$, $\langle B, d\rangle$ *is a* autometrized Boolean-valued space (BVS) *if:*

a) $d(x, x) = \mathbb{0}$;
b) $d(x, y) = d(y, x)$;
c) If $x \neq y$, *then* $d(x, y) \neq \mathbb{0}$;
d) $d(x, z) \leq d(x, y) \vee d(y, z)$.
The map d is called a Boolean-valued distance.

Blumenthal has used the symbol + instead of the $\vee$, thus his definition looks as the usual for metric spaces. However, this modification is very relevant. From the theoretic viewpoint, $\mathbb{R}$ has a well defined structure: is continuous, have a countable dense subset, is linearly ordered, has no first no last element, etc. The Boolean structures are discrete, partially ordered, with maximum and minimum, etc. The set $\mathbb{R}$ has so many operations defined on it, much more than a Boolean algebra. The most usual operations of the Boolean algebra, meet and join, are idempotent, differently from the standard operations on $\mathbb{R}$. The consequences of this different structure in clustering applications are discussed in Section 4.

In [4] the distance $d(x, y)$ in autometrized Boolean algebras is defined as the symmetric difference (XOR in bit vectors)

$$d(x, y) = x \bigtriangleup y = (x \wedge -y) \vee (-x \wedge y)\,.$$

In [4] are also proved the conditions in Definition 3 and some geometric properties of this distance.

By making a little modification of closed ball standard definition in metric spaces, we obtain the following.

Definition 4. *Let* $\langle B, d\rangle$ *be a BVS. For* $o, r \in B$, the closed ball of center o and radius r, $C_r(o)$, *is:*

$$C_r(o) = \{x \in B : \ d(o, x) \leq r\}$$

A set $D \subseteq B$ *is called a* closed ball *if there exists* r *and* o *as above, such that* $D = C_r(o)$.

Since the supremum is lower than any other upper bound, we obtain $d(o, x) \vee d(x, y) \leq r$ from $d(y, x) \leq r$ and $d(o, x) \leq r$ and an analogous of proposition 1 holds in BVS.

3 Spaces on Bit Vectors

Let k be a positive integer and let S be the space of all bit vectors v of length k. Since S has the structure of a Boolean algebra, we use the expression *a Boolean algebra of bit vectors*, omitting the fixed k, as well in the cases that the Boolean algebra structure is not relevant. For a bit vector v, we denote v_i its i-th coordinate (the i-th place), omitting that $1 \leq i \leq k$.

3.1 Real-Valued Metrics: Hamming and Jaccard

Let us fix a Boolean algebra of bit vectors. If we denote by $qb(v)$ the number of bits equaling 1 in the bit vector v, then the well-known Hamming metric $h(v, w)$

between two bit vectors is defined by $h(v, w) = qb(v \text{ XOR } w)$. The proof that this distance is a metric is straightforward.

Another widely used distance is Jaccard j:

$$j(a,b) = \begin{cases} 0 & \text{if } a = b \\ 1 - \frac{qb(a \wedge b)}{qb(a \vee b)} = \frac{qb(a \triangle b)}{qb(a \vee b)} & \text{otherwise} \end{cases}$$

with $qb(v)$ as above. Notice that this definition avoids the problem of dividing by zero (divide overflow), as usual in definitions of Jaccard coefficient and Jaccard distance found in the literature, for example: $jaccardcoef(A, B) = \frac{|A \cap B|}{|A \cup B|}$ (see http://en.wikipedia.org/wiki/Jaccard_index). To prove that Jaccard distance is a metric, the only non trivial condition is the triangle inequality. A proof can be found in [11].

The Euclidean distance also can be defined on a Boolean algebra of bit vectors in the usual way. Then, another example of metric space is $\langle B, g \rangle$ with B a Boolean algebra of bit vectors and g the Euclidean distance.

Example 1. Let v and w be the bit vectors $v = \langle 0,0,1,0,1,1,0,1 \rangle$ and $w = \langle 0,1,1,0,0,1,1,0 \rangle$. With the notation above we have:

Euclidean	Hamming	Jaccard
$g(v, w) = 2$	$h(v, w) = 4$	$j(v, w) = \frac{1}{3}$

Considering bit vector algebras, the Euclidean, the Hamming and the Jaccard distances are not ultrametric. In this sense, counterexamples of condition d in definition 1 can be easily found for these distance concepts: let $x = \langle 1, 0 \rangle$, $y = \langle 1, 1 \rangle$ and $z = \langle 0, 1 \rangle$. Hence, let us see a well-known example of an ultrametric space onto a bit vector algebra: for bit vectors v and w, let $d(v, w)$ the largest coordinate such that v and w differ. When $v = w$, we can consider that these sequences differ until the 0-th coordinate, or to simply define $d(v, v) = 0$. Such g is an ultrametric.

3.2 Autometrized Boolean-Valued Spaces of Bit Vectors

The order and the operations of the Boolean algebra are defined in the standard way. The order is defined point-to-point (i.e. the bit-to-bit order). Thus, $v = \langle v_1, v_2, \ldots, v_k \rangle$. If v and w are two bit vectors of length k, we define the point-to-point order in the usual way:

$$v \leq w \quad \text{if and only if} \quad v_i \leq w_i, \quad \text{for all } i$$

and the meet and the join operations also are defined point-to-point too:

$$v \wedge w = \langle v_1 \wedge w_1, v_2 \wedge w_2, \ldots, v_k \wedge w_k \rangle$$
$$v \vee w = \langle v_1 \vee w_1, v_2 \vee w_2, \ldots, v_k \vee w_k \rangle$$

A well-known theorem states that two finite Boolean algebras of the same cardinality are isomorphic [10]. Then, we can think any finite Boolean algebra as a algebra of bit vectors with $\mathbb{0}$ and $\mathbb{1}$ for the bit vectors of the all zeros and of all ones, respectively.

The symmetric difference $\triangle$ is the more common Boolean-valued distance, represented by the XOR operation on bit vectors. By using the v and w of subsection 3.1, we have $v \triangle w = \langle 0,1,0,0,1,0,1,1 \rangle$. Two Boolean distances can be incomparable: let $t = \langle 1,0,1,0,0,1,1,0 \rangle$ and thus $v \triangle t = \langle 1,0,0,0,1,0,1,1 \rangle$. Hence, we have $v \triangle w \nleq v \triangle t$ and $v \triangle t \nleq v \triangle w$ by the bit-to-bit order. The strange situation is that the distance between v and w is not the same that between v and t. However neither t is closer of v than w, nor w is closer of v than t.

3.3 Examples of Closed Balls in Bit Vector Spaces and Minimal Closed Balls

Let us see some examples. We will use the Boolean algebra of the bit vectors of length 4. In Jaccard space, the closed ball of radius 0.5 and center $\langle 1,1,0,0 \rangle$ (figure 1) is:
$C_{0.5}(\langle 1,1,0,0 \rangle) = \{\langle 1,1,1,1 \rangle, \langle 1,1,1,0 \rangle, \langle 1,1,0,1 \rangle, \langle 1,1,0,0 \rangle, \langle 1,0,0,0 \rangle, \langle 0,1,0,0 \rangle\}$

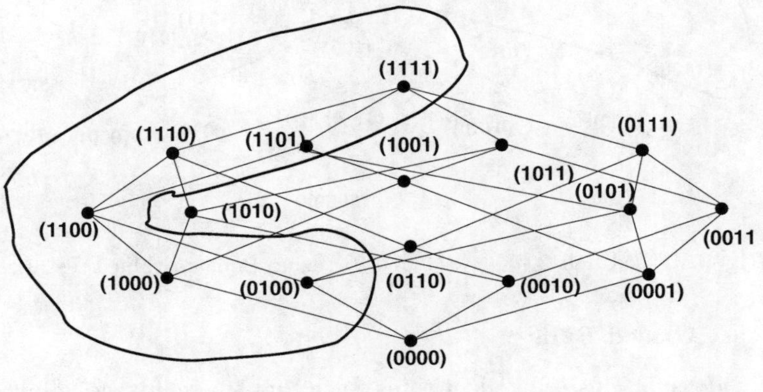

Fig. 1. Closed ball $C_{0.5}(\langle 1,1,0,0 \rangle)$ using Jaccard distance

In Hamming space, the closed ball of radius 1 and center $\langle 1,1,0,0 \rangle$ (see figure 2) is:
$C_1(\langle 1,1,0,0 \rangle) = \{\langle 1,1,1,0 \rangle, \langle 1,1,0,1 \rangle, \langle 1,1,0,0 \rangle, \langle 1,0,0,0 \rangle, \langle 0,1,0,0 \rangle\}$
In the XOR autometrized BVS, the radius r and the distance between two elements are no more real numbers, but elements of the same Boolean algebra, i.e. bit vectors of length four. Let us see the closed ball of radius $\langle 0,0,1,1 \rangle$ and center $\langle 1,1,0,0 \rangle$: $C_{\langle 0,0,1,1 \rangle}(\langle 1,1,0,0 \rangle)$. The distance between the center $\langle 1,1,0,0 \rangle$ and the element $\langle 1,1,1,1 \rangle$ is $\langle 1,1,0,0 \rangle \triangle \langle 1,1,1,1 \rangle = \langle 0,0,1,1 \rangle$ and $\langle 0,0,1,1 \rangle \leq \langle 0,0,1,1 \rangle = r$ (bit-to-bit). Hence, by definition 4, we have $\langle 1,1,1,1 \rangle \in C_{\langle 0,0,1,1 \rangle}(\langle 1,1,0,0 \rangle)$. Thus, this closed ball (see figure 3) is:
$C_{\langle 0,0,1,1 \rangle}(\langle 1,1,0,0 \rangle) = \{\langle 1,1,1,1 \rangle, \langle 1,1,0,1 \rangle, \langle 1,1,1,0 \rangle, \langle 1,1,0,0 \rangle\}$

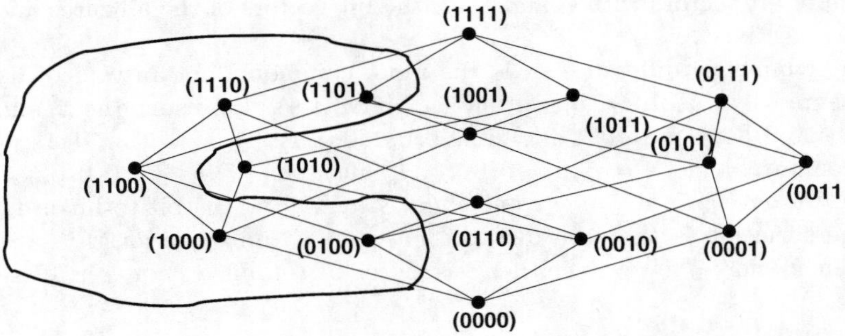

Fig. 2. Closed ball $C_1\left(\langle 1,1,0,0\rangle\right)$ using Hamming distance

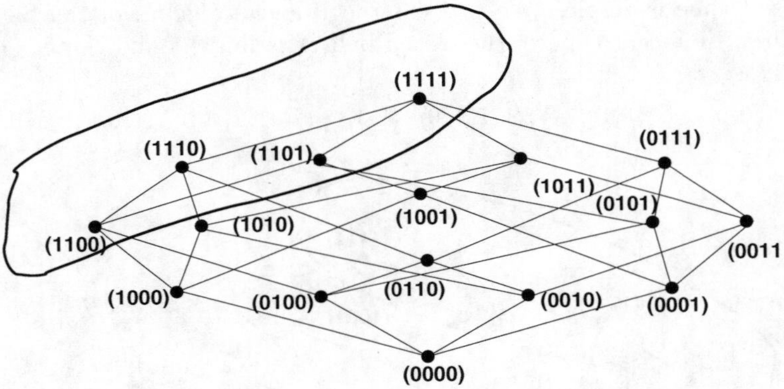

Fig. 3. Closed ball $C_{\langle 0,0,1,1\rangle}\left(\langle 1,1,0,0\rangle\right)$ using Boolean-valued distance

3.4 Least Closed Balls

The early ideas for clustering algorithms, such that single linkage, minimal span tree and UPGMA (see [12]) use operations and the linear order of $\mathbb{R}$ and can not be directly implemented with autometrized Boolean-valued distances. Hence, in order to implement, validate and compare both distances, we present two density-based clustering algorithms. First, we discuss the density concept and its relationship with the topological concept of closed ball.

Data clustering is frequently looking for denser regions of the feature space to select the elements which belong to these regions into clusters. In some sense, it can be said that data clustering has as a target the dense regions of the feature space. In the sequel, we aim to specify the density concept to be used later in clustering algorithms.

The simplest concept of density for finite spaces is to consider a set of elements related to the total of nodes of some region of the space including this set. Frequently, given a set A of points, we need a way to determinate a region of space which includes A. A possible answer to this problem is to consider the least closed ball which includes A, the so-called spherical clusters, where "least closed ball" means the ball with lowest radius.

Definition 5. *Let $\langle B, d\rangle$ be a a metric, ultrametric or Boolean-valued space and let $A \subseteq B, A \neq \varnothing$. Furthermore, let $C_r(c)$ a closed ball in $\langle B, d\rangle$. Then $C_r(c)$ is called a* minimal closed ball including A *if $A \subseteq C_r(c)$ and there is not an $r' < r$ such that $A \subseteq C_{r'}(c)$.*

But this approach based on the concept of least closed ball including a given set of elements yields a number of problems. For example, in certain spaces can exist two different minimal closed balls including a certain set A. For example, in the Hamming space of the bit vectors of length 4, let $A = \{\langle 1,1,1,0\rangle, \langle 1,0,0,0\rangle\}$: the centers $\langle 1,1,0,0\rangle$ and $\langle 1,0,1,0\rangle$ yield closed balls which include A, have radius 1 and have 5 nodes: $C_1(\langle 1,1,0,0\rangle) = \{\langle 1,1,1,0\rangle, \langle 1,1,0,1\rangle, \langle 1,1,0,0\rangle, \langle 1,0,0,0\rangle, \langle 0,1,0,0\rangle\}$ and $C_1(\langle 1,0,1,0\rangle) = \{\langle 1,1,1,0\rangle, \langle 1,0,1,1\rangle, \langle 1,0,1,0\rangle, \langle 1,0,0,0\rangle, \langle 0,1,0,0\rangle\}$ (some features on closed balls in Hamming spaces can be found in [13]).

But, even the question of how to find a minimal closed ball which includes a given set is far away from triviality. We will look for both a center and a radius, but in certain spaces it can be hard to find both of them. Of course, we can use a *brute force* procedure. For example, let B be a Boolean algebra of bit vectors and let $A \subseteq B$. For each element $x \in B$, find the maximum of the distances with the elements of A, call it $m(x)$. Then, let $m(c) = \min\{m(x) : x \in B\}$ (the minimum of this set) and set c as center and $m(c)$ as radius. With little spaces we can use *brute force* procedures, but these are very limited procedures, since the number of elements of a Boolean algebra grows exponentially. In certain spaces, as Euclidean or Hamming, we can idealize best procedures to find a minimal closed ball. Certain clustering algorithms do not look for a center, but for a centroid, to simplify the search.

3.5 Closed Balls in Jaccard Space

Let B be a Boolean algebra of bit vectors. In the sequel, let us analyze the closed balls including a given set of elements $A \subseteq B$ in Jaccard space. In this space there is always such a closed ball, because the whole space B is one of them. Furthermore, the radiuses of closed balls are real-valued and hence they are linearly ordered. Since the set of closed balls is finite, the set of radiuses is too. Since every set of radiuses has a least element, there exists a closed ball including A with least radius.

Proposition 2. *Let B be a Boolean algebra of bit vectors, $\langle B, j\rangle$ a Jaccard space and let $A \subseteq B, A \neq \varnothing$. There is always a minimal closed ball C in $\langle B, j\rangle$ such that $A \subseteq C$.* □

However, the existence of least closed balls can fail in Jaccard space.

Example 2. Consider the Boolean algebra of the bit vectors of length 4. There are two different minimal closed balls including $\{\langle 1,1,0,0\rangle, \langle 1,0,0,0\rangle\}$: $C_{0.5}$ $(\langle 1,0,0,0\rangle)$ and $C_{0.5}(\langle 1,1,0,0\rangle)$.

Given a set $A \neq \varnothing$. It is possible that $c \notin A$, for each minimal closed ball $C_r(c)$ such that $A \subseteq C_r(c)$, i.e. no element of A is the center of some minimal closed ball including A. Then, the seek for a center can be a excessively time-consuming task if the whole space must be analyzed. Hence, the following propositions are relevant in practice (see http://www.geocities.com/autometrized22/mathematical.zip for the proof).

Proposition 3. *Let B be a Boolean algebra of bit vectors and let $A \subseteq B, A \neq \varnothing$. Furthermore, let $C_1(c)$ a closed ball of radius 1. Then, there exists a $c' \in A$ such that $C_1(c') = C_1(c)$, $c' \leq \bigvee A$ and $c' \geq \bigwedge A$.* □

Proposition 4. *Let B be a Boolean algebra of bit vectors and let $A \subseteq B, A \neq \varnothing$. Furthermore, let $C_r(c)$ a minimal closed ball of radius $r < 1$. Then $c \leq \bigvee A$ and $c \geq \bigwedge A$.*

3.6 Least Closed Balls in Autometrized BVS

Fixed a autometrized BVS $\langle B, d\rangle$. Let $A \subseteq B$. We will analyze the question again: is there a least closed ball in B including A? Note that the B itself is a closed ball including A. Since the set of radius is finite and partially ordered, there is a minimal closed ball including A. But the question of the existence of the least closed ball including A is hard to solve. We present a construction using the symmetric difference as distance.

Let $\langle B, \triangle\rangle$ the autometrized BVS of bit vectors with the symmetric difference as distance. Furthermore, let $A \subseteq B$, $A \neq \varnothing$, $A = \{a_1, a_2, \ldots, a_n\}$. To define the desired open ball, we need the center and the radius. Because of the ultrametric properties of $\langle B, \triangle\rangle$, any element of the closed ball can be taken as center. Then, we set a_1, the first element of A, as center. The radius is defined by:

$$r = \bigvee_{i=1}^{n} (a_1 \triangle a_i)$$

In http://www.geocities.com/autometrized22/mathematical.zip a proof of theorem 1 can be found.

Theorem 1. *$C_r(a_1)$ is the least closed ball such that $A \subseteq C_r(a_1) \subseteq B$.*

Now, the density, is computed by:

$$\delta = \frac{|A|}{2^{qb(r)}}$$

where $|A|$ is the cardinal number of A and $qb(r)$ is the number of bits equaling 1 in vector r. Note that $2^{qb(r)} = |C_r(a_1)|$ (this has a straightforward inductive proof on the length of the vector r).

4 The Application of Boolean-Valued Distances in Clustering

The difference of structure and operations between $\mathbb{R}$ and the Boolean algebras has several consequences in clustering applications. Let us see some of these.

Given $x, y \in \mathbb{R}$ we can take the least number, but in a Boolean algebra B two different elements can be incomparable. The fact that there is no total order on Boolean algebras yields a problem when we try to find the least distance: it may not exist. Some algorithms use common operations on $\mathbb{R}$. For example, we can compute the average of a set of distances. But the average of a set of elements has no sense in a Boolean algebra. Recall that Euclidean distance uses the operations of sum, square and square root on $\mathbb{R}$. On the other hand, the operations of supremum and infimum in a Boolean algebra are idempotent. Completely different structure and operations imply completely different clustering techniques.

In a more general way, when we consider the application of Boolean distances in clustering, the following questions immediately arise:

Question 1. How Boolean-valued concepts can be implemented in actual programs?
Question 2. How to use the Boolean-valued distance concepts to yield clustering algorithms?

When a scientist changes his research framework, immediately he feels the tools which he lost, but do not yet see the tools which he gets. Perhaps the main loss is the linear order of the set of values of the distance map: for an element x, it can occur that we can not determine what of two other elements, say y and z, is closer of x, because of the lack of comparability of the Boolean-valued distances: $d(x,y) \not\leq d(x,z)$ and $d(x,z) \not\leq d(x,y)$. This is a serious problem and in some of our techniques we need to use the Hamming distance together the Boolean distance to circumvent it. We also lost the rich mathematics of $\mathbb{R}$: we are not able anymore to accomplish simple operations, for example, the average of a set of values.

But if we are able to apply the Boolean distances in clustering, still remains the following question:
$*$ Is the autometrized Boolean-valued space a suitable structure for clustering applications?

More specifically:

Question 3. Can the Boolean-valued space clustering algorithms yield a good classification?
Question 4. Can these algorithms be validated?
Question 5. Are these algorithms efficient?

4.1 Clustering Implementation of Autometrized BVS

First, we will answer question #1. We work on a Boolean algebra of bit vectors. An unsigned or integer type can be used to represent the bit vectors in the usual way. For example, suppose that we work in the space of vectors of 8 bits. Then the vector $\langle 0,1,0,0,1,0,1,1\rangle$ represents an object that has the first, second, fourth and seventh properties (using the right to left order in bit vectors) and is encoded as the integer 75 (see the programs in http://www.geocities.com/autometrized22/programs.zip for details). The AND, OR, XOR and NOT operations are implemented bit-to-bit in most processors and are available in programming languages. For instance, the C language uses `&`,`|`,`^` and `~` for these. The bit-to-bit $\leq$ can not be used directly, i.e. `V <= W`, since it compares the integer values and not the bit-to-bit ones. By using a well-known equivalence in Boolean algebras, we can write $(x \wedge y) = x$ (`x & y == x` in C language) for $x \leq y$.

5 Algorithms

In this section we describe density-based clustering algorithms and use them to confront Jaccard and Boolean-valued distances. When we use Jaccard distance, the search for the center is a time-consuming procedure. Hence, we replace the seek for a center, by a faster procedure seeking for a centroid. In the Boolean case, any element of the closed ball can be used as a center and thus there is no seek for a center. In http://www.geocities.com/autometrized22/programs.zip can be found the programs that implement these algorithms. Complexity analysis of the algorithms presented here shows that it is around $O\left(n^3\right)$. By using dynamic programing and suitable structures (a heap, for example), this complexity can be lowered.

5.1 Encoding

The encoding of binary properties using bit vectors is done in the usual way, taking the 1 value if the object has the property, 0 otherwise. For an example of an usual encoding, let us consider a market basket database over items $1,2,3,4,\ldots,k$. Each transaction is encoded as a bit vector of length k: $\langle v_1, v_2, \ldots, v_k\rangle$, where v_i has the 1 value if the transaction has the item i, and 0 otherwise.

The so-called symbolic data, when they take a finite number of values, can be encoded into binary taking each possible value of the attribute as a bit. Thus, in the mushroom data base in [14], the attribute cap-color can take the values brown, buff, cinnamon, gray, green, pink, purple, red, white or yellow and we use one bit for brown, one for buff and so on.

For continuous data, a number of discretization techniques can be used. But this is a complex discussion, out of the scope of this paper.

5.2 The Big Ball Density Algorithm

This is a hierarchical agglomerative algorithm (see [2]). In the main step, two clusters are selected if the least closed ball including these cluster has a maximal density. To compute this density, we also consider the elements inside the ball which do not belong to these clusters, if they exist.

```
for each element e
  make_unitary_cluster( e );
end_for;
number_of_clusters <-- number_of_elements;
while number_of_clusters > 1 do
  maximal_density <-- 0;
  for each cluster pair {x,y} do
    find a minimal closed ball b which includes x,y
    compute the density d of b
    if d > maximal_density then
      cl1 <-- x;  cl2 <-- y;  maximal_density <-- d;
    end_if;
  end_for;
  create a new cluster with the elements of cl1 and cl2;
  delete( cl1 ); delete( cl2 );
  number_of_clusters <-- number_of_clusters - 1;
end_while;
```

5.3 Ball in the Middle Algorithms

This is a hierarchical agglomerative algorithm too. For each pair of clusters, we make a list of the nearest elements from a cluster to another. Since Boolean-valued distances can be incomparable, we use the Hamming distance in the Boolean case. Then, we find the least closed ball which includes that list and compute the density. Select the pair of clusters with the larger density.

```
for each element e
  make_unitary_cluster( e );
end_for;
number_of_clusters <-- number_of_elements;
while number_of_clusters > 1 do
  maximal_density <-- 0;  least_distance <-- INFINITE;
  for each cluster pair {x,y} do
    compute the distance d between the nearest elements of x,y
    if d > least_distance then
      continue;
    least_distance <-- d;
    make the list_of_nearest;
    find a minimal closed ball b which includes list_of_nearest;
    compute the density d of b
    if d > maximal_density then
      cl1 <-- x;  cl2 <-- y; maximal_density <-- d;
      end_if;
    end_for;
    create a new cluster with the elements of cl1 and cl2;
    delete( cl1 );  delete( cl2 );
    number_of_clusters <-- number_of_clusters - 1;
  end_while;
```

6 Experimental Results

We have used three databases in [14]. Despite these databases are primarily used for supervised machine learning, they were also used several times in the literature to test unsupervised classification. In this sense, we are looking for "natural structure" of data (which can exist or not), without using the supplied target classes. Then, we compare the "natural structure of data" founded by the algorithm with the target classes, to evaluate the classification. This is a standard procedure in several papers in the literature.

The comparison was carried out by using the same algorithms with Jaccard distance and autometrized Boolean XOR. We also used a procedure to eliminate outliers. The discussion about outliers elimination is out the scope of this paper (see [2] for such question). For each data base and algorithm, the same outliers elimination was used for both distances, and thus this elimination does not influence the results.

The programs in http://www.geocities.com/autometrized22/programs.zip implement the algorithms. These programs were used to yield the results showed in Tables 1, 2 and 3. When using Jaccard distances, we use a centroid of the closed ball. The centroid found by our method is not the center of a minimal closed ball in nearly 6% of the cases, with average error of 5% in the radius. In most of cases (>96%), the true center and the centroid yields the same clustering.

6.1 Results Using the Congress, Hepatitis and Zoo Databases

We used the database congress in [14]: this data set includes votes for each of the U.S. House of Representatives Congressmen on the 16 key votes in 1984. Usually, the bit vector coordinate v_i takes the 1 value if the vote was "yes", 0 otherwise. The target classes are republicans and democrats and they were not used in the clustering procedure. We used a strong outliers elimination to find some structure in data. Table 2 shows, Boolean distances were successfully to find these structures, but Jaccard has failed. We also include data from [16] corresponding to the ROCK algorithm and a "traditional hierarchical clustering algorithm". At this date, [16] has 507 citations named in http://scholar.google.com.

The hepatitis database in [14] contains the medical and personal data from patients with hepatitis. The target classes are alive or death. Both Boolean and Jaccard distances fails in finding some structure of the data consistent with the target classes (see table 2).

Table 1. Results using the Congress databases

		Big ball		Middle ball			ROCK		Traditional	
		Class 1	Class 2	Class 1	Class 2		Class 1	Class 2	Class 1	Class 2
Jaccard	Cluster 1	1	1	7	1	Cluster 1	22	144	52	157
	Cluster 2	131	106	84	87	Cluster 2	201	5	215	11
Boolean	Cluster 1	15	105	0	88					
	Cluster 2	117	2	91	0					

Table 2. Results using the Hepatitis databases

		HEPATITIS			
		Big ball		Middle ball	
		Class 1	Class 2	Class 1	Class 2
Jaccard	Cluster 1	7	10	22	42
	Cluster 2	24	105	9	73
Boolean	Cluster 1	0	6	7	7
	Cluster 2	31	109	24	108

Table 3. Results using the zoo database

	Jaccard														**Boolean**														**Results in [15]**						
	Big ball							**Middle ball**							**Big ball**							**Middle ball**													
Class	1	2	3	4	5	6	7	1	2	3	4	5	6	7	1	2	3	4	5	6	7	1	2	3	4	5	6	7	1	2	3	4	5	6	7
Cluster 1	33	0	4	0	4	0	0	16	0	1	9	0	0	0	40	0	0	0	0	0	0	37	0	0	0	0	0	0	41	0	0	0	0	0	0
Cluster 2	8	0	0	0	0	0	0	15	0	0	0	2	4	5	1	20	1	0	0	0	0	0	20	0	0	0	0	0	0	20	1	0	0	0	0
Cluster 3	0	20	0	0	0	0	0	9	0	1	0	1	0	0	0	0	4	13	4	0	0	4	0	3	13	4	0	0	0	0	4	13	4	0	1
Cluster 4	0	0	1	13	0	0	7	1	7	1	0	0	1	3	0	0	0	0	0	8	3	0	0	0	0	0	8	6	0	0	0	0	0	8	9
Cluster 5	0	0	0	0	0	8	3	0	13	0	4	1	3	0	0	0	0	0	0	0	7	0	0	0	0	0	0	2							

The zoo database is an artificial data set in [14]. It has seven target classes, but [15] affirms that the natural structure has four classes. We set to five the number of clusters. In [15] is used a hierarchical agglomerative algorithm, based on a distance concept called "weighted dissimilarities". The results in [15] using the zoo database are showed in the last column of table 3, together with results using the ball in the middle and the big ball algorithms with Boolean and Jaccard metrics. Again, autometrized Boolean-valued distances present best results that Jaccard distance.

7 Conclusions

Despite of very different features of real-valued and Boolean-valued distances, our method of comparison, based on similar concepts and algorithms suitable for both kind of distances, yields preliminary experimental results consistent with the hypothesis that autometrized distances are suitable to find the structure of databases. But the evidence presented here does not suffice for conclusive results. It is necessary more experiments with natural and artificial databases, as well as to use another autometrized and real-valued distances and the study of geometrical and topological concepts which can be applied in a general way on both kind of distances.

Our future work is the test of programs to reply questions #3, #4 and #5 of section 4 in a conclusive manner.

References

1. Jain, A.K., Murty, M.P., Flynn, P.J.: Data clustering: A review. ACM Computing Surveys 31(3), 264–322 (1999)
2. Theodoridis, S., Koutroumbas, K.: Pattern Recognition. Academic Press, San Diego (1998)
3. Zhang, B., Srihari, S.N.: Properties of binary vector dissimilarity measures. Technical report, State University of New York at Buffalo (2003), `http://ftp.cedar.buffalo.edu/papers/articles/CVPRIP03_propbina.pdf`
4. Blumenthal, L.M.: Boolean geometry. I. Rend. Circ. Mat. Palermo II Ser. 1, 343–360 (1952)
5. Swamy, K.L.N.: A general theory of autometrized algebras. Mathematischen Annalen 157, 65–74 (1964)
6. Krasner, M.: Nombres semi-reels et espaces ultrametriques. Compte Rendus d'Academie des Sciences de Paris. Tome II 219, 433 (1944)
7. Ramal, R., Toulouse, G., Virasoro, M.: Ultrametricity for physicists. Reviews of Modern Physics 58(3), 765–788 (1986)
8. Zomorodian, A.J.: Topology for Computing. Cambridge University Press, Cambridge (2005)
9. Sikorski, R.: Boolean Algebras. Springer, Heidelberg (1969)
10. Monk, J.D.: A brief introduction to Boolean algebras (2004), `http://www.colorado.edu/math/courses/monkd`
11. Ramon, J.: Clustering and Instance Based Learning in First Order Logic. PhD thesis, Katholieke Universiteit Leuven, Belgium, Department of Computer Science (2002)
12. Sneath, P.H.A., Sokal, R.R.: Numerical Taxonomy: the principles and pratice of numerical classification. W.H. Freeman and Company, San Francisco (1974)
13. Moraglio, A., Poli, R.: Topological interpretation of crossover. In: Deb, K., et al. (eds.) GECCO 2004. LNCS, vol. 3102, pp. 1377–1388. Springer, Heidelberg (2004)
14. Blake, C., Merz, C.: UCI repository of machine learning databases. University of California, Irvine, Dept. of Information and Computer Sciences (1998)
15. Mali, K., Mitra, S.: Clustering and its validation in a symbolic framework. Pattern Recognition Letters 24(14), 2367–2376 (2003)
16. Guha, S., Rastogi, R., Shim, K.: ROCK: A robust clustering algorithm for categorical attributes. Information Systems 25(5), 345–366 (2000)

Multi-Dimensional Dynamic Time Warping for Image Texture Similarity

Rodrigo Fernandes de Mello[1] and Iker Gondra[2]

[1] University of São Paulo
Institute of Mathematics and Computer Science
São Carlos, SP, Brazil
mello@icmc.usp.br
[2] St. Francis Xavier University
Department of Mathematics, Statistics and Computer Science
Antigonish, NS, Canada
igondra@stfx.ca

Abstract. Modern content-based image retrieval systems use different features to represent properties (e.g., color, shape, texture) of the visual content of an image. Retrieval is performed by example where a query image is given as input and an appropriate metric is used to find the best matches in the corresponding feature space. Both selecting the features and the distance metric continue to be active areas of research. In this paper, we propose a new approach, based on the recently proposed Multidimensional Dynamic Time Warping (MD-DTW) distance [1], for assessing the texture similarity of images with structured textures. The MD-DTW allows the detection and comparison of arbitrarily shifted patterns between multi-dimensional series, such as those found in structured textures. Chaos theory tools are used as a preprocessing step to uncover and characterize regularities in structured textures. The main advantage of the proposed approach is that explicit selection and extraction of texture features is not required (i.e., similarity comparisons are performed directly on the raw pixel data alone). The method proposed in this preliminary investigation is shown to be valid by proving that it creates a statistically significant image texture similarity measure.

Keywords: Content-Based Image Retrieval, Texture, Dynamic Time Warping, Similarity Measure, Distance Measure, Chaos Theory.

1 Introduction

In recent years, the rapid development of information technologies and the advent of the Web have accelerated the growth of digital media and, in particular, image collections. As a result, new mechanisms to search on large image databases have been proposed. One of the first approaches was keyword-matching, which uses a textual representation and is based on the manual annotation of images with descriptive keywords. This approach is not only subjective and error-prone but also very time-consuming and cumbersome for large databases.

G. Zaverucha and A. Loureiro da Costa (Eds.): SBIA 2008, LNAI 5249, pp. 23–32, 2008.

Recently, automatic image labeling approaches [2,3,4] have been proposed as an attempt to improve the manual annotation of images. In [2], image recognition techniques are employed to automatically assign a limited number of descriptive keywords. This approach is limited by the fact that current image recognition methods are not completely reliable and, as a consequence, the assigned keywords have to be verified by a person. Other works such as [3] consider the textual context of images, in web pages, to automatically extract descriptive keywords (such as those that appear in captions). The performance of those approaches is lower than the one obtained by using manual annotation. Furthermore, their applicability is limited in situations where there is no textual context (such as in photo albums). Those textual description approaches can only obtain part of the richness and complexity of an image's visual content.

To overcome these problems, content-based image retrieval (CBIR) [5] was proposed in the early 1990's. The basic idea is to directly use the visual content when determining image similarity. Retrieval is performed by using a query image as input, which has a feature set extracted to represent its visual content (e.g., color, shape, texture). Afterwards, an appropriate metric is applied to find the best matches in the corresponding feature-set space. In this context, texture is one of the most important visual characteristics when defining similarity among images.

Texture is defined as the repetition of a certain atomic pattern (or texton) residing in a region. For a low-level image analysis, texture features play a very important role in distinguishing textured regions from one another based on the measurement of optical homogeneity of surfaces. In the 1970's, Haralick et al. [6] proposed the first systematic analysis of texture by using a co-occurrence matrix, which describes the distribution of co-occurring pixel values at a given offset. Based on such matrix, several texture features (e.g., contrast, entropy), which explore the spatial dependence of pixel values, can be extracted. Other researchers carried on investigating texture. For example, Tamura et al. [7] developed a set of texture features designed to measure the visual properties of coarseness, contrast, directionality, line-likeness, regularity, and roughness which, based on conducted psychological experiments, are thought to dominate human visual perception of texture. Research into other techniques, such as the use of wavelet-based texture features, has also been very active.

When considering texture similarity, the selection of both a set of features and a distance metric continues to be the most critical decision. The selection of features requires the application of methods to extract the most relevant visual characteristics, which are then used to compare to other textures. The distance metric is responsible for the comparison of the feature values of different textures.

The complexity of feature extraction and the observation that structured texture contains regular repeated patterns has motivated this work to investigate the suitability of the Dynamic Time Warping (DTW) distance [8] as a feature-independent (i.e., based on the raw pixel values only) measure of image texture similarity. DTW was designed to find the minimal distance between two series considering their synchronization through shifts. In this context, we consider raw pixel data to compute texture similarity. All the relevant features of a texture are, in some way, "hidden" in the raw information. Besides that, any feature extraction technique has its limitations because, as of yet, no one has completely figured out and implemented a technique that is able to obtain the same characterization quality as a person. Thus, instead of attempting to determine texture similarity based on a small set of (probably incomplete) features, we find that it may be advantageous to use a similarity measure that is based on the raw data itself.

The objective of this research is to obtain some preliminary evidence as to whether a DTW-based feature-independent texture similarity measure can actually result in a good, and statistically significant, retrieval performance. This paper is organized as follows: Section 2 introduces the chaos theory concepts of Embedded and Separation dimensions, which are used as a preprocessing step to obtain a descriptive representation of textures. In Section 3 we introduce the traditional and multi-dimensional DTW. Our approach (and its evolution) is described in section 4. Experimental results with a real data set are presented in Section 5. Finally, concluding remarks are given in Section 6.

2 Embedded and Separation Dimensions

The occurrence of repeated patterns, in structured textures, has encouraged us to investigate a similarity measure which considers textures as series of events. From that, we decided to consider DTW, a technique that can measure the minimal distance between two series, by considering possible synchronization points. Although, during our studies, we observed that the original image representation may not be the most suitable when uncovering the regularities and patterns of textures. Based on such conclusion, we decided to consider chaos theory tools to unfold and reorganize data textures according to possible hidden regularities. The concepts considered in this work are presented next.

Chaos theory is defined as the qualitative study of unstable aperiodic behavior in deterministic nonlinear dynamical systems [9,10]. Attempts to understand the behavior of such systems have led to the development of several works (e.g., [10,11]) which aim at characterizing possible internal regularities. In order to understand how the internal regularities can be obtained, consider the logistic equation (1) with the initial conditions $t \in [0, 4000]$, $b = 3.8$ and $x_0 = 0.5$.

$$x_{t+1} = b \times x_t \times (1.0 - x_t) \quad (1)$$

Figure 1(a) presents the outputs for the logistic function. If we directly apply a similarity measure comparing this series to another one, we may obtain bad results. A better solution comes by applying chaos theory tools to unfold the full series behavior through two dimensional analysis: the embedded and the separation dimensions. By obtaining those two dimensions, we can understand how information has to be represented in the space, before using it in further comparisons (in the case of texture retrieval).

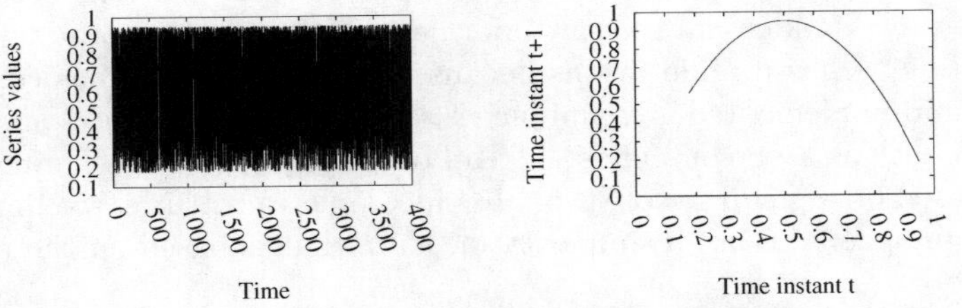

(a) Outputs (b) Reconstructed with $m = 2, \tau = 1$

Fig. 1. Logistic Function

Whitney [12] proposed that a series $x_0, x_1, ..., x_{n-1}$ could be reconstructed into a multidimensional space $x_n(m, \tau) = (x_n, x_{n+\tau}, ..., x_{n+(m-1)\tau})$ where m is the embedded dimension and τ is a fixed time delay. According to this study, each series can be reconstructed and, consequently, simplified to be understood and compared. To better understand those dimensions, consider the same logistic function outputs reconstructed in a multidimensional space where $m = 2$ and $\tau = 1$. This reconstruction, which is basically the plot of x_t versus x_{t+1}, is presented in figure 1(b). Now the behavior of the logistic function, which was apparently a random walk (figure 1(a)), can be understood and modeled in an easier way.

Basically, the embedded dimension defines the number of axis that we will plot the series to unfold its full behavior. Some series can only be understood when using more than two dimensions. Besides the embedded, there is the separation dimension which helps to extract the periodic behavior of a series. This basically tells the number of points we may look back in the history to detect regular behavior (or patterns, this is also known as the seasonability of the series). Next we discuss how to determine good values for both dimensions.

According to Abarbanel [13] we have to apply the autocorrelation function (equation (2), where $E[.]$ is the expected value, μ is the average, k is the time shift size and σ^2 is the variance) on a series and use its first minimum as the separation dimension. The autocorrelation measures how well the series matches itself considering a time separation. This is useful to find repeated patterns in a series. However, this technique is formulated for linear series, and, consequently, it may not present good results for non-linear and chaotic ones.

$$ACF(k) = \frac{E[(X_i - \mu)(X_{i+k} - \mu)]}{\sigma^2} \tag{2}$$

Fraser and Swinney [14] studied and confirmed that the Auto Mutual Information (AMI) technique presents better results to estimate the separation dimension. This technique is not linear-dependent, and consequently is more interesting for this work. To use this technique we must calculate it for different time shifts and adopt the first minimum of the function. The average mutual information is defined in equation (3) where X and Y have, respectively, the probability distribution functions P_X, P_Y and X and Y occur in pairs with the joint distribution P_{XY} [15].

$$I(X;Y) = \int dxdy P_{XY}(x,y) \log_2 \frac{P_{XY}(x,y)}{P_X(x)P_Y(y)} \quad (3)$$

After defining the separation dimension we must find the embedded dimension. Takens [16] and Mañé [17] studied and confirmed that the upper limit of the embedded dimension D_e, an integer value, can be defined using the fractal dimension D_f according to $D_e > 2.0 \times D_f$. Although this dimension is usually larger than necessary. For instance, the fractal dimension of the Lorenz attractor is 2.06 [18], consequently the upper limit of the embedded dimension would be $D_e > 2.0 \times 2.06$, then it is 5. Although, according to [19] the attractor can be represented in $D_e = 3$. The important point here is that we can represent the attractor with a number of dimensions smaller than 5, which reduces complexity. From a mathematical point of view, using either 3 or 5 results in no difference, because once the attractor is unfolded the analysis can be conducted. Although, if we unnecessarily work with more dimensions, we add complexity and execution time to obtain solutions [19]. This conclusion motivated works on how to define a good embedded dimension for different series.

The traditional approach to obtain the minimum embedding dimension is by calculating any system invariant (such as the Lyapunov exponent) to different embedded dimension values and observe when it saturates. The complexity of this approach motivated Kennel *et al.* [19] to propose the False Nearest Neighbors (FNN) method. In FNN, the nearest neighbors for each point, in the space, are calculated, initially with the embedded dimension equal to 1. Then, the Euclidean distance from the point to its nearest neighbor is calculated. Afterwards, a new dimension is added and the distance of the point to its nearest neighbor obtained. If this distance increases, those two points are considered false neighbors. This happens because the attractor being modeled needs more dimensions to be unfolded and studied.

Kennel *et al.* [19] consider a embedded dimension d where the rth nearest neighbor of $y(n)$ is $y^{(r)}(n)$. The Euclidean distance between the point $y(n)$ and its rth nearest neighbor is obtained by equation (4). Adding a new dimension, we go for $d+1$ and add the coordinate $(d+1)^{th}$ in each vector $y(n)$. The new coordinate is $x(n+Td)$ which is included in the new Euclidean distance equation (5).

$$R_d^2(n,r) = \sum_{k=0}^{d-1} (x(n+kT) - x^{(r)}(n+kT))^2 \quad (4)$$

$$R_{d+1}^2(n,r) = R_d^2(n,r) + (x(n+dT) - x^{(r)}(n+dT))^2 \quad (5)$$

Then, the criterion is to measure the distance variation when adding the new dimension as presented in equation (6).

$$V_{n,r} = \sqrt{\frac{R_{d+1}^2(n,r) - R_d^2(n,r)}{R_d^2(n,r)}} = \frac{|x(n+Td) - x^{(n)}(n+Td)|}{R_d^2(n,r)} \quad (6)$$

The authors indicate that, if $V_{n,r} > R_{tol}$, then the points are considered false neighbors, where R_{tol} is a threshold. They conclude that $R_{tol} \geq 10.0$ is enough to generate good results. This reconstruction, using the embedded and separation dimensions, unfolds the attractor and can be applied to any series. After unfolding, we can better and more easily study the behavior of a series. Then, we use the new dataset (the reorganized representation of the image) to compare against others. This new representation has less complexity to understand the image regularities and, consequently, to model it.

3 Multi-Dimensional Dynamic Time Warping

The distance between two hypothetical series can be quantified using different measures, one of them is Dynamic Time Warping (DTW) [8]. This technique aligns two series to find the ideal warp (the best synchronization point) in order to minimize the distance between them.

In order to understand it, consider two series $S = s_0, s_1, ..., s_{m-1}$ and $T = t_0, t_1, ..., t_{n-1}$ of length m and n, respectively. Firstly, DTW (algorithm 1) creates an m-by-n matrix d where each element (ith, jth) represents the distance $d(S_i, T_j) = (S_i - T_j)^2$ between each pair of points S_i and T_j. Afterwards, DTW creates a new matrix D to accumulate the total distance between each possible pair of points of the two series. This step fills out the matrix D where the elements represent all possible alignments (synchronizations) of the two series and their distances.

After calculating the matrix, the DTW distance is computed through the summing of the shortest possible path that starts at the right bottom of the matrix and goes up to the left-top element. This path represents the best synchronization between the two series and the sum of all of its matrix elements is the DTW distance. DTW was designed for one-dimensional series. However, there are many applications in which calculating an optimal alignment requires the use of multi-dimensional series. Holt *et al.* [1] proposed the Multi-Dimensional Dynamic Time Warping (MD-DTW), an approach to calculate the DTW by synchronizing multi-dimensional series, which is basically an extension of the original DTW, where the matrix D is created by computing the distance between k-dimensional points (where, differently from the original approach, k can be larger than 1). This approach preprocesses the multi-dimensional series, which must have the same number of dimensions, according to algorithm 2. The last step of this algorithm is the execution of the traditional DTW (algorithm 1) considering the matrix D as the result of the preprocessing phase.

4 Proposed Approach

The first investigations we conducted, using DTW as a similarity measure to retrieve images, generated better results than random retrieval. Firstly, each

Algorithm 1. Dynamic Time Warping Algorithm

Let m and n be the length of the series S and T, respectively.
Let d be the matrix which computes the distance of pairs of values of S and T.
for $i = 0$ to $m-1$ **do**
  for $j = 0$ to $n-1$ **do**
    $d[i][j] = (S[i] - T[j])^2$;
  end for
end for
Let D be the matrix with the DTW distance among pairs of elements of series S and T.
$D[0][0] = d[0][0]$;
for $i = 1$ to $m-1$ **do**
  $D[i][0] = d[i][0] + D[i-1][0]$;
end for
for $j = 1$ to $n-1$ **do**
  $D[0][j] = d[0][j] + D[0][j-1]$;
end for
for $i = 1$ to $m-1$ **do**
  for $j = 1$ to $n-1$ **do**
    $D[i][j] = min(D[i-1][j], D[i-1][j-1], D[i][j-1]) + d[i][j]$;
  end for
end for
The total DTW distance between the two series is stored at the matrix element $D[m-1][n-1]$.

Algorithm 2. Multi-dimensional Dynamic Time Warping Algorithm

Let S, T be two series of dimension K and length n and m, respectively.
Normalize each dimension of S and T separately to a zero mean and unit variance.
Fill the matrix D according to:
$D(i,j) = \sum_{k=1}^{K} |S(i,k) - T(j,k)|$
Consider the matrix D to compute the traditional DTW algorithm (instead of the matrix D of the traditional approach).

RGB color image was converted to the corresponding grayscale image. Afterwards, we applied a Laplace edge detection algorithm to the grayscale images. Then, before applying DTW as the similarity measure, we organized the pixel values of the preprocessed images in two different ways. To better understand them, let $P = \{p_{0,0}, p_{0,1}, ..., p_{1,0}, p_{1,1}, ..., p_{r-1,c-1}\}$, where $0 \leq p_{i,j} \leq 255$ is the grayscale value of the pixel at the intersection of the ith row and jth column of an image with r rows and c columns, be the matrix representation of an image. In the first way, we organized the series using the pixels in the following order $\{p_{0,0}, p_{0,1}, ..., p_{1,0}, p_{1,1}, ..., p_{r-1,c-1}\}$; in the second approach, the pixels were organized as $\{p_{0,0}, p_{1,0}, ..., p_{0,1}, p_{1,1}, ..., p_{r-1,c-1}\}$.

After organizing the pixels, we conducted experiments comparing images by using the two different orders but the results were inconsistent (i.e., most query images resulted in very different retrieval performances when using the two orders). Thus, the first conclusion was that neither one of this one-dimensional orders was a good representation of image textures. That is, the original texture representation may not be the most suitable for uncovering existing regularities or patterns. This motivated us to study and apply chaos theory tools to perform optimal data reorganization.

For each image, the embedded and the separation dimensions, were first computed. Then, each image was reconstructed into a multidimensional space $x_n(m,\tau) = (x_n, x_{n+\tau}, ..., x_{n+(m-1)\tau})$ where m is the embedded dimension and

τ is the separation dimension. The resulting multi-dimensional series were then used as the input for MD-DTW-based similarity comparisons.

5 Experimental Results

Consider a database consisting of a set of images $\mathcal{D}$. Let x be a query image and $\mathcal{A} \subset \mathcal{D}$ be the subset of images in $\mathcal{D}$ that are relevant to x. After processing x, the image retrieval method generates $\mathcal{R} \subset \mathcal{D}$ as the retrieval set. Then, $\mathcal{R}^+ = \mathcal{R} \cap \mathcal{A}$ is the set of relevant images to x that appear in $\mathcal{R}$. Users want the database images to be ranked according to their relevance to x and then be presented with only the k most relevant images so that $|\mathcal{R}| = k < |\mathcal{D}|$. Thus, images are ranked by their distance to the query image and, in order to account for the quality of image rankings, precision at a cut-off point (e.g., k) is commonly used. Thus, the performance of the image retrieval method is commonly measured by *precision*, which quantifies the ability to retrieve only relevant images and is defined as $precision := \frac{|\mathcal{R}^+|}{|\mathcal{R}|}$.

The objective of our experiments was to obtain preliminary evidence as to whether the proposed approach actually creates a statistically significant image texture similarity measurement. Therefore, we tested its performance against a uniform random retrieval to select the images in $\mathcal{R}$. The *Texture* data set, obtained from MIT Media Lab [20] was used for evaluation. There are 40 different texture images that are manually classified into 15 classes. Each of those images is then cut into 16 non-overlapping images of size 128x128. Thus, there are 640 images in the database. Sample images are shown in figure 2.

Fig. 2. Sample images from *Texture* data set

Each image was used as a query and the precision and recall of a retrieval set of $k \in [1, 640]$ nearest images was measured. The results of our approach and the random retrieval are presented in figure 3, which shows the degradation of precision as k increases. That is, attempting to increase recall results in the introduction of more non-relevant images into $\mathcal{R}$. Thus, precision-recall graphs have a classical concave shape. In order to increase both precision and recall, the curve should move up and to the right (as observed after the initial decrease (i.e., at the middle) of the curve in figure 3) so that both recall and precision are higher at every point along the curve.

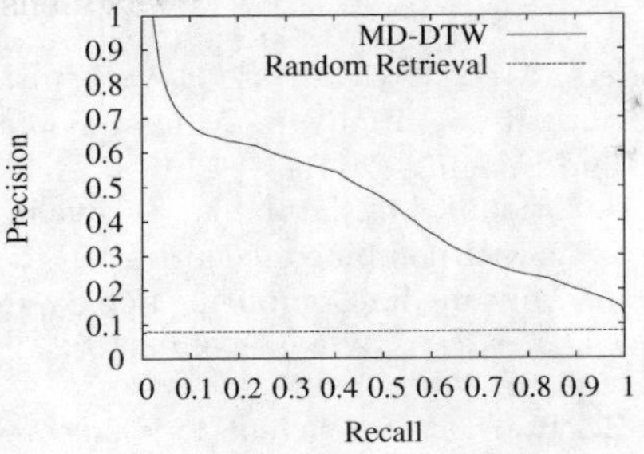

Fig. 3. Precision and recall

It is common to compare retrieval methods by using a fixed retrieval set size. A reasonable (and commonly used) value for the size of the retrieval set $k = 20$. The average precision over the 640 queries with $k = 20$ for the proposed approach and for random retrieval was 0.594 and 0.079 respectively. As can be observed by these results and figure 3, the proposed approach performed surprisingly well and is obviously statistically different than uniform random retrieval. Table 1 shows the average precision for each of the texture classes shown in Figure 2.

Table 1. Average Precision for each of the 15 texture classes shown in figure 2

Class	Average Precision	Class	Average Precision	Class	Average Precision
1	0.974	2	0.970	3	0.797
4	0.998	5	0.422	6	1
7	0.134	8	0.573	9	0.786
10	0.263	11	0.190	12	0.472
13	0.238	14	0.194	15	0.979

6 Conclusions

Based on the observation that structured image textures contain repeated patterns, we investigated the possibility of using the MD-DTW [1], which allows for the comparison of arbitrarily shifted patterns in multi-dimensional series, as a measure of image texture similarity. Chaos theory tools were used in the preprocessing step in order to allow the identification, characterization and unfolding of regularities in the raw data of structured textures. The main advantage of the proposed approach is that explicit selection and extraction of texture features is not required (i.e., similarity comparisons are performed directly on the raw pixel data alone). The proposed approach performed surprisingly well when compared against uniform random retrieval thus proving that it creates a statistically significant image texture similarity measure. This is an encouraging preliminary result that motivates us to continue to work on the MD-DTW as a feature-independent measure of texture similarity.

References

1. ten Holt, G.A., Reinders, M.J.T., Hendriks, E.A.: Multi-Dimensional Dynamic Time Warping for Gesture Recognition. In: Annual Conference of the Advanced School for Computing and Imaging (2007)
2. Ono, A., Amano, M., Hakaridani, M., Satoh, T., Sakauchi, M.: A flexible content-based image retrieval system with combined scene description keywords. In: IEEE International Conference on Multimedia Computing and Systems, pp. 201–208 (1996)
3. Shen, H.T., Ooi, B.C., Tan, K.L.: Giving meanings to WWW images. In: ACM Multimedia, pp. 39–48 (2000)
4. Srihari, R.K., Zhang, Z., Rao, A.: Intelligent indexing and semantic retrieval of multimodal documents. Information Retrieval 2, 245–275 (2000)
5. Smeulders, A.W.M., Worring, M., Santini, S., Gupta, A., Jain, R.: Content-based image retrieval at the end of the early years. IEEE Transactions on Pattern Analysis and Machine Intelligence 22(12), 1349–1380 (2000)
6. Haralick, R.M., Shanmugam, K., Dinstein, I.: Texture features for image classification. IEEE Transactions on Systems, Man, and Cybernetics 3(6), 610–621 (1973)
7. Tamura, H.: Textural features corresponding to visual perception. IEEE Transactions on Systems, Man, and Cybernetics 8(6), 460–473 (1978)
8. Kruskall, J., Liverman, M.: The symmetric time warping problem: from continuous to discrete. In: Time Warps, String Edits and Macro-molecules: The Theory and Practice of Sequence Comparison, pp. 125–161. Addison-Wesley, Reading (1983)
9. Edmonds, A.N.: Time Series Prediction Using Supervised Learning and Tools from Chaos Theory. PhD Thesis, University of Luton (1996)
10. Casdagli, M.: Nonlinear prediction of chaotic time series. Physica D Nonlinear Phenomena 35, 335–356 (1989)
11. Gu, S., Wang, Z., Chen, J.: The fractal research and predicating on the times series of sunspot relative number. Applied Mathematics and Mechanics 20(1), 84–89 (1999)
12. Jackson, E.A.: Perspectives of Nonlinear Dynamics. Cambridge University Press, Cambridge (1989)
13. Abarbanel, H.D.I., Brown, R., Sidorowich, J.J., Tsimring, L.S.: The analysis of observed chaotic data in physical systems. Reviews of Modern Physics 65, 1331–1392 (1993)
14. Fraser, A.M., Swinney, H.L.: Independent coordinates for strange attractors from mutual information. Phys. Rev. A 33(2), 1134–1140 (1986)
15. The Multiple-Dimensions Mutual Information Program (Matt Kennel), `http://www-ncsl.postech.ac.kr/en/softwares/archives/mmi.tar.Z`
16. Takens, F.: Detecting strange attractors in turbulence. In: Dynamical Systems and Turbulence, pp. 366–381. Springer, Heidelberg (1980)
17. Mañé, R.: On the dimension of the compact invariant sets of certain nonlinear maps. In: Dynamical Systems and Turbulence, pp. 230–242. Springer, Heidelberg (1980)
18. Medio, A., Gallo, G.: Chaotic Dynamics: Theory and Applications to Economics. Cambridge University Press, Cambridge (1993)
19. Kennel, M.B., Brown, R., Abarbanel, H.D.I.: Determining embedding dimension for phase-space reconstruction using a geometrical construction. Phys. Rev. A 45(6), 3403–3411 (1992)
20. Picard, R., Graczyk, C., Mann, S., Wachman, J., Picard, L., Campbell, L.: MIT media lab: Vision texture database, `http://vismod.media.mit.edu/vismod/imagery/VisionTexture/`

Audio-to-Visual Conversion Via HMM Inversion for Speech-Driven Facial Animation

Lucas D. Terissi* and Juan Carlos Gómez

Laboratory for System Dynamics and Signal Processing
FCEIA, Universidad Nacional de Rosario
CIFASIS, CONICET
Riobamba 245bis, 2000, Rosario, Argentina
{lterissi,jcgomez}@fceia.unr.edu.ar

Abstract. In this paper, the inversion of a joint Audio-Visual Hidden Markov Model is proposed to estimate the visual information from speech data in a speech driven MPEG-4 compliant facial animation system. The inversion algorithm is derived for the general case of considering full covariance matrices for the audio-visual observations. The system performance is evaluated for the cases of full and diagonal covariance matrices. Experimental results show that full covariance matrices are preferable since similar, to the case of using diagonal matrices, performance can be achieved using a less complex model. The experiments are carried out using audio-visual databases compiled by the authors.

Keywords: Hidden Markov Models, Audio-Visual Speech Processing, Facial Animation.

1 Introduction

Speech driven animation of virtual characters is playing an increasingly important role due to the widespread use of multimedia applications such as computer games, online virtual characters, video telephony, and other interactive human-machine interfaces. Among the different approaches proposed in the literature to model audio-visual data, the ones based on Hidden Markov Models (HMM) have proved to yield more realistic results when used in applications of speech driven facial animation.

Earlier approaches for speech-driven facial animation systems, such as the works in [1], [2], [3] and [4], resort to different HMM structures and require the use of Viterbi optimization algorithm [5] in the training or synthesis stages. This leads to video predictions of limited quality due to the high noise sensitivity of Viterbi algorithm. To address this limitation, Choi *et al* [6] have proposed a Hidden Markov Model Inversion (HMMI) method for audio-visual conversion. HMMI was originally introduced in [7] in the context of robust speech recognition. In HMMI, the visual output is generated directly from the given audio

* Corresponding author.

G. Zaverucha and A. Loureiro da Costa (Eds.): SBIA 2008, LNAI 5249, pp. 33–42, 2008.

input and the trained HMM by means of an expectation-maximization (EM) iteration, thus avoiding the use of the Viterbi sequence and improving the performance of the estimation [8]. Recently, Xie *et al* [9] proposed a coupled HMM approach and derived an expectation maximization (EM)-based A/V conversion algorithm for the CHMMs, which converts acoustic speech into reasonably good facial animation parameters.

In this paper, a speech driven MPEG-4 compliant facial animation system is proposed. A joint audio-visual Hidden Markov Model (AV-HMM) is trained using audio-visual data and then Hidden Markov Model inversion is used to estimate the animation parameters from speech data. The feature vector corresponding to the visual information during the training is obtained via Independent Component Analysis (ICA). Previous approaches based on HMMs consider diagonal covariance matrices for the audio-visual observation, invoking reasons of computational complexity. In this paper, the use of full covariance matrices is investigated. Simulation results show that the use of full covariance matrices leads to an accurate estimation of the visual parameters, yielding a performance similar to that of using diagonal covariance matrices, but with a less complex model and without affecting significantly the computational load.

The rest of the paper is organized as follows. An overview of the speech driven facial animation system is presented in section 2. The AV-HMM is introduced in section 3, where an HMMI algorithm for the general case of considering full covariance matrices for the audio-visual observations is also derived. In section 4, the proposed algorithm for feature extraction is described. The MPEG-4 compliant facial animation technique is presented in section 5. Experimental results and some concluding remarks are included in sections 6 and 7, respectively.

2 Speech Driven Facial Animation System Overview

A block diagram of the proposed speech driven animation system is depicted in Fig. 1. An audiovisual database is used to estimate the parameters of a joint AV-HMM. This database consists of videos of a talking person with reference marks in the region around the mouth, see Fig. 2(a).

In a first training stage, feature parameters of the audiovisual data are extracted. The audio part of the feature vector consists of mel-cepstral coefficients, while the visual part are the coefficients in a ICA representation of the above mentioned set of reference marks. In a second training stage, the audio part of the AV-HMM is re-trained using audio data from a speech-only database. Retraining only the audio part of the model allows to obtain a more robust model against inter-speaker variability, avoiding the need to record videos of speakers with the reference marks on their faces.

For the speech driven animation, speech data is used to estimate the visual features by inversion of the AV-HMM using the technique described in section 3. From these data, Facial Animation Parameters (FAPs) of the MPEG-4 [10] standard are computed to generate the facial animation.

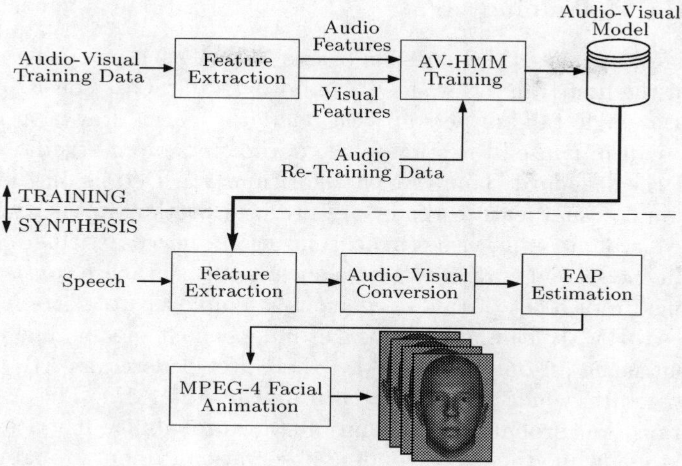

Fig. 1. Schematic representation of the speech driven animation system

3 Audio Visual Model

In this paper, a joint AV-HMM is used to represent the correlation between the speech and facial movements. The AV-HMM, denoted as λ_{av}, is characterized by three probability measures, namely, the state transition probability distribution matrix (A), the observation symbol probability distribution (B) and the initial state distribution (π), and a set of N states $S = (s_1, s_2, \dots, s_N)$, and audiovisual observation sequence $O_{av} = \{o_{av1}, \dots, o_{avT}\}$. In addition, the observation symbol probability distribution at state j and time t, $b_j(o_{avt})$, is considered a continuous distribution which is represented by a mixture of M Gaussian distributions

$$b_j(o_{avt}) = \sum_{m=1}^{M} c_{jm} \mathcal{N}(o_{at}, o_{vt}, \mu_{jm}, \Sigma_{jm}) , \qquad (1)$$

where c_{jm} is the mixture coefficient for the m-th mixture at state j and $\mathcal{N}(o_{at}, o_{vt}, \mu_{jm}, \Sigma_{jm})$ is a Gaussian density with mean μ_{jm} and covariance Σ_{jm}. The audio-visual observation o_{avt} is partitioned as $o_{avt} \triangleq \left[o_{at}^T, o_{vt}^T\right]^T$, where o_{at} and o_{vt} are the audio and visual observation vectors, respectively.

A single ergodic (that is one in which transitions among all the states are allowed) HMM is proposed to represent the audiovisual data. An alternative to an ergodic model, would be a set of left-to-right HMMs representing the different phonemes (with associated visemes) of the particular language. These models have been used in the context of speech modeling by several authors, see for instance [9]. An ergodic model provides a more compact representation of the audiovisual data, without the need of phoneme segmentation, which is required when left-to-right models are used. In addition, this has the advantage of making the system adaptable to any language.

3.1 AV-HMM Training

The training of the AV-HMM consists of two stages, each one using a different database. In the first training stage, an audiovisual database consisting of a set of videos of a single talking person with reference marks drawn on the region around the mouth, is used to estimate the parameters of an ergodic AV-HMM, resorting to the standard Baum-Welch algorithm [11]. Details on the composition of the audiovisual feature vector are given in Section 4, where procedures to take into account audio-visual synchronization and co-articulation are also described. In the second training stage, a speech-only database consisting of audio recordings from a set of talking persons is employed to re-train the audio part of the AV-HMM, leading to a speaker independent model. The re-training is carried out using an only audio HMM (hereafter denoted as A-HMM), with the same structure, which is constructed from the AV-HMM. The A-HMM has the same transition probability and initial state probability matrices obtained in the first stage, while the corresponding observation symbol probability distribution is re-estimated from the speech-only database. The observation symbol probability distribution is parameterized by μ_{jm}, Σ_{jm} and c_{jm}, see equation (1). To emphasize the mix composition of the AV-HMM, the mean and covariance parameters can be partitioned as

$$\mu_{jm} = \begin{bmatrix} \mu^a_{jm} \\ \mu^v_{jm} \end{bmatrix} \quad , \quad \Sigma_{jm} = \begin{bmatrix} \Sigma^a_{jm} & \Sigma^{av}_{jm} \\ \Sigma^{va}_{jm} & \Sigma^v_{jm} \end{bmatrix} \quad , \tag{2}$$

where the superscript a and v denote the audio and visual parts, respectively. During the second training stage, only μ^a_{jm} and Σ^a_{jm} are re-estimated using speech-only data. Finally, the re-estimated parameters are fed back into the AV-HMM.

3.2 Audio-to-Visual Conversion

Hidden Markov Model Inversion (HMMI) was originally proposed in [7] in the context of robust speech recognition. Choi and co-authors [6] used this technique to estimate the visual features associated to audio features for the purposes of speech driven facial animation. Typically, it is assumed [7], [6], [9] a diagonal structure for the covariance matrices of the Gaussian mixtures, invoking reasons of computational complexity. This assumption is relaxed in this paper allowing for full covariance matrices. This leads to more general expressions for the visual feature estimates.

The idea of HMMI for audio-to-visual conversion is to estimate the visual features based on the trained AV-HMM, in such a way that the probability that the whole audiovisual observation has been generated by the model is maximized. It has been proved [11] that this optimization problem is equivalent to the maximization of the auxiliary function

$$Q(\lambda_{av};\lambda_{av},O_a,O_v,O'_v) \triangleq \sum_{j=1}^{N}\sum_{m=1}^{M} P(O_a,O_v,j,m \mid \lambda_{av}) \ \log \ P(O_a,O'_v,j,m \mid \lambda_{av})$$

$$= \sum_{j=1}^{N}\sum_{m=1}^{M} P(O_a,O_v,j,m \mid \lambda_{av}) \ \Bigg[\log \ \pi_{j_0} + \sum_{t=1}^{T} \log \ a_{j_{t-1}j_t} +$$

$$+ \sum_{t=1}^{T} \log \ \mathcal{N}(o_{at},o'_{vt},\mu_{j_t m_t},\Sigma_{j_t m_t}) + \sum_{t=1}^{T} \log \ c_{j_t m_t}\Bigg] \ , \tag{3}$$

that is

$$O'_v = \arg\max_{O'_v} \{Q(\lambda_{av};\lambda_{av},O_a,O_v,O'_v)\} \ , \tag{4}$$

where O_a, O_v and O'_v denote the matrices containing the audio, visual and estimated visual sequences from $t = 1,\ldots,T$, respectively, π_{j_0} denotes the initial probability for state j and $a_{j_{t-1}j_t}$ denotes the state transition probability from state j_{t-1} to state j_t.

The solution to the optimization problem in (4) can be computed by equating to zero the derivative of Q with respect to o'_{vt}. Considering that the only term that depends on o'_{vt} is the one involving the Gaussians, this derivative can be written as

$$\frac{\partial Q(\lambda_{av};\lambda_{av},O_a,O_v,O'_v)}{\partial o'_{vt}} = \sum_{j=1}^{N}\sum_{m=1}^{M} P(O_a,O_v,j,m \mid \lambda_{av})\times$$

$$\times \frac{\partial}{\partial o'_{vt}} \left[\sum_{t=1}^{T} \log \ \mathcal{N}(o_{at},o'_{vt},\mu_{j_t m_t},\Sigma_{j_t m_t})\right] = 0 \ . \tag{5}$$

Considering that

$$\log \ \mathcal{N}(o_{at},o'_{vt},\mu_{j_t m_t},\Sigma_{j_t m_t}) = \log \frac{1}{(2\pi)^{d/2}\sqrt{|\Sigma_{j_t m_t}|}} -$$

$$-\frac{1}{2}\begin{bmatrix} o_{at}-\mu^a_{j_t m_t} \\ o_{vt}-\mu^v_{j_t m_t}\end{bmatrix}^T \begin{bmatrix} \Phi^a_{j_t m_t} & \Phi^{av}_{j_t m_t} \\ \Phi^{v_t a_t}_{j_t m_t} & \Phi^v_{j_t m_t}\end{bmatrix} \begin{bmatrix} o_{at}-\mu^a_{j_t m_t} \\ o_{vt}-\mu^v_{j_t m_t}\end{bmatrix} \ , \tag{6}$$

where d is the dimension of o_{avt} and

$$\Sigma^{-1}_{j_t m_t} = \begin{bmatrix} \Phi^a_{j_t m_t} & \Phi^{av}_{j_t m_t} \\ \Phi^{va}_{j_t m_t} & \Phi^v_{j_t m_t}\end{bmatrix} ,$$

the estimated visual observation becomes

$$o'_{vt} = \left[\sum_{j=1}^{N}\sum_{m=1}^{M} P(o_a,o_v,j,m \mid \lambda_{av})\Phi^v_{jm}\right]^{-1} \times$$

$$\times \sum_{j=1}^{N}\sum_{m=1}^{M} P(o_a,o_v,j,m \mid \lambda_{av}) \left[\Phi^v_{jm}\mu^v_{jm} - \Phi^{va}_{jm}(o_{at}-\mu^a_{jm})\right] \ . \tag{7}$$

For the case of diagonal matrices, equation (7) reduces to

$$o'_{vt} = \left[\sum_{j=1}^{N} \sum_{m=1}^{M} P(o_a, o_v, j, m \mid \lambda_{av}) \Phi^v_{jm} \right]^{-1} \times \sum_{j=1}^{N} \sum_{m=1}^{M} P(o_a, o_v, j, m \mid \lambda_{av}) \Phi^v_{jm} \mu^v_{jm} , \tag{8}$$

which is equivalent to the equation derived in [6].

As is common in HMM training, the estimation algorithms (7) and (8) are implemented in a recursive way, initializing the visual observation randomly.

4 Feature Extraction

The audio signal is partitioned in frames with the same rate as the video frame rate. A number of Mel-Cepstral Coefficients in each frame (a_t) are used in the audio part of the feature vector. To take into account the audiovisual co-articulation, several frames are used to form the audio feature vector $o_{at} = \left[a^T_{t-t_c}, \ldots, a^T_{t-1}, a^T_t, a^T_{t+1}, \ldots, a^T_{t+t_c}\right]^T$ corresponding to the visual feature vector o_{vt}.

For the visual part, the coefficients in an Independent Component representation of the coordinates of marks in the region around the mouth of the speaking person are used, see Fig. 2(a). Let $\mathbf{F} = \{\mathbf{f}_1, \mathbf{f}_2, \ldots, \mathbf{f}_T\}$ represent the training data collected from videos. Each vector $\mathbf{f}_t = [x_1^{(t)}, x_2^{(t)}, \ldots, x_P^{(t)}, y_1^{(t)}, y_2^{(t)}, \ldots, y_P^{(t)}]^T$ contains the coordinates $(x_p^{(t)}, y_p^{(t)})$ of each mark ($p = 1, 2, \ldots, P$) for the t-th frame, $t = 1, 2, \ldots, T$.

Let $\mathbf{f}_0$ be the neutral facial expression, mainly defined as the expression with all face muscles relaxed and the mouth closed [10]. The relative facial deformation (with respect to the neutral expression) at each frame can be computed as $\mathbf{d}_t = \mathbf{f}_t - \mathbf{f}_0$, and a deformation matrix can then be defined as

$$\mathbf{D} = [\mathbf{d}_1, \mathbf{d}_2, \ldots, \mathbf{d}_T] . \tag{9}$$

The different facial expressions in the training data are represented by the columns of matrix $\mathbf{D}$. The idea is to represent any facial expression as the linear combination of a reduced number of independent vectors. The dimensionality reduction can be performed by Principal Component Analysis [12]. The PCA stage yields an uncorrelated set of vectors. It is desirable to have a statistically independent set of vector so that information contained in each vector will not provide information on any of the others. This is the main idea in ICA. Summarizing, ICA after PCA will be performed on the data matrix $\mathbf{D}$.

Several algorithms are available in the literature for ICA computation. The reader is referred to [12] and the references therein. In this paper, the symmetric decorrelation based FastICA algorithm as implemented in [13] was employed.

As a result of the ICA processing, any facial deformation can then be computed as

$$\mathbf{f}_t = \sum_{k=1}^{K} o_{vt_k} \mathbf{u}_k + \mathbf{f}_0 , \tag{10}$$

where $\{\mathbf{u}_k\}_{k=1}^{K}$ are the independent components from $\mathbf{D}$ and o_{vt_k} is the k-th component of the visual vector o_{vt}. The coefficients o_{vt_k} are computed in two stages. In the first stage, the mark locations are estimated using image processing techniques. In the second stage, the coefficients o_{vt_k} are computed in such a way that the facial expression is given by the linear combination of the ICs vectors that best match the mark estimation computed in the first stage. Details of this procedure can be found in [14].

5 Facial Animation

As already mentioned, the facial animation technique proposed in this paper is MPEG-4 compliant. The MPEG-4 standard defines 64 Facial Animation Parameters and 84 Feature Points (FPs) on a face model in its neutral state [15]. FAPs represent a complete set of basic facial actions such as head motion, and eye, cheeks and mouth control. FPs are used as reference points to perform the facial deformation.

Based on the estimated facial expression for each frame, the associated FAPs can be determined by computing the displacement of a set of marks from their corresponding position in the neutral facial expression. For instance, the marks encircled in red in Fig. 2(a) can be associated to FAP3 corresponding to jaw opening. Figure 2(b) shows the resulting expression after applying the estimated FAP3 to the neutral expression (several other FAPs, in addition to FAP3, have also been applied to produce the mouth opening and cheek movements). Similarly, several subsets of marks can be associated to the different FAPs.

6 Experimental Results

For the audio-visual training, videos of a talking person with reference marks on the region around the person's mouth were recorded at a rate of 30 frames per seconds, with (320×240) pixels resolution. The audio was recorded at 11025Hz synchronized with the video. The videos consist of sequences of the Spanish utterances corresponding to the digits zero to nine in random order. For the re-training of the audio part of the AV-HMM, an only-audio database consisting of recordings of sequences of the utterances corresponding to the digits zero to nine by 25 speakers (balance proportion of males and females) was collected.

Experiments were performed with AV-HMM with full and diagonal covariance matrices, different number of states and mixtures in the ranges $[3, 20]$ and $[2, 19]$, respectively, and different values of the co-articulation parameter t_c in the range $[2, 5]$. In the experiments, the audio feature vector a_t is composed by the first eleven non-DC Mel-Cepstral coefficients, while the visual feature vector o_v is of dimension two ($K = 2$ in equation (10)). The performances of the

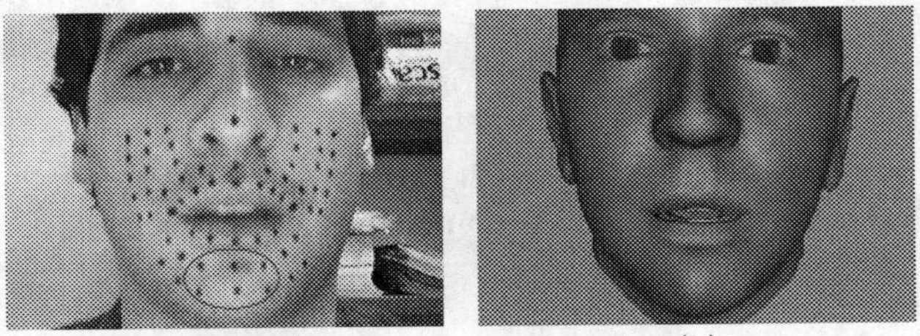

(a) (b)

Fig. 2. (a) Real person facial expression. Marks associated to FAP3 are encircled in red. (b) Synthesized facial expression.

different models were compared by computing the Average Mean Square Error (AMSE)(ϵ), and the Average Correlation Coefficient (ACC)(ρ) between the true and estimated visual parameters, defined as

$$\epsilon = \frac{1}{TK} \sum_{k=1}^{K} \frac{1}{\sigma_{v_k}^2} \sum_{t=1}^{T} \left[o'_{vt_k} - o_{vt_k} \right]^2 , \tag{11}$$

$$\rho = \frac{1}{TK} \sum_{t=1}^{T} \sum_{k=1}^{K} \frac{(o_{vt_k} - \mu_{v_k})(o'_{vt_k} - \mu'_{v_k})}{\sigma_{v_k} \sigma'_{v_k}} , \tag{12}$$

respectively, where μ_{v_k} and σ_{v_k} denote the mean and the variance of the true visual observation, respectively, and μ'_{v_k} and σ'_{v_k} denote the mean and variance of the estimated visual parameters, respectively.

For the quantification of the visual estimation accuracy, a separate audio-visual dataset, different from the training dataset, was employed. The following results correspond to a co-articulation parameter $t_c = 5$, which proves to be the optimal value in the given range. Fig. 3(a) and Fig. 3(b), show the AMSE and the ACC as a function of the number of states and the number of mixtures for an AV-HMM with full covariance matrix. In this case, equation (7) applies for the estimation of the visual observations o'_{vt}. As can be observed, as the number of states and the number of mixtures increase, the AMSE increases and the ACC decreases, indicating that the accuracy of the estimation deteriorates. This is probably due to the bias-variance tradeoff inherent to any estimation problem. The optimal values for the number of states and mixtures could be for this case $N = 4$ and $M = 2$, respectively, corresponding to $\epsilon = 0.47$ and $\rho = 0.75$.

Fig. 3(c) and Fig. 3(d), show the AMSE and the ACC as a function of the number of states and the number of mixtures for an AV-HMM with diagonal covariance matrix. In this case, equation (8) applies for the estimation of the visual observations o'_{vt}. As can be observed, to obtain a similar accuracy a more complex model (larger number of states or mixtures) is required. For this case, the optimal values are $N = 19$ and $M = 3$, corresponding to $\epsilon = 0.47$ and $\rho = 0.76$. The use of full covariance matrices affects the computational complexity during the training stage but, since this is carried out off-line, this does not represent a problem. During the synthesis stage (visual estimation through HMM

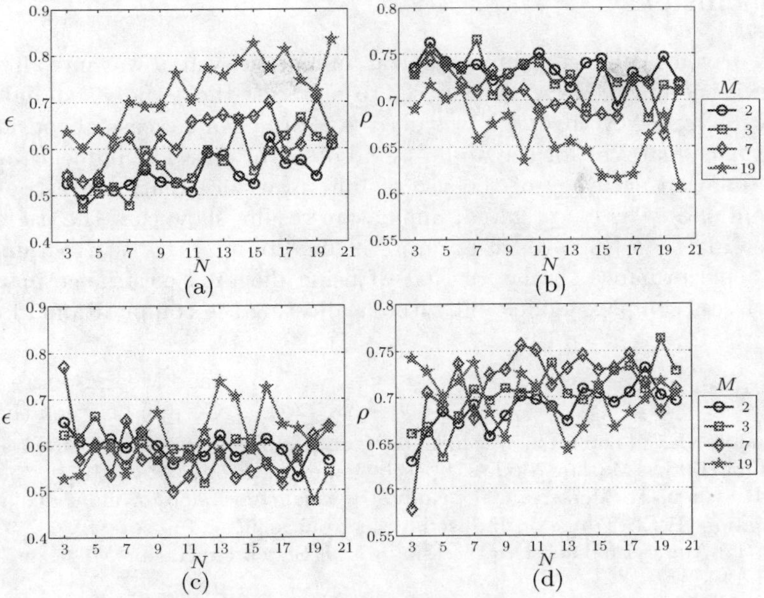

Fig. 3. *AMSE* (ϵ) and *ACC* (ρ) as a function of the number of states N and the number of mixtures M. Where (a) and (b) correspond to the case of full covariance matrices and, (c) and (d) correspond to the case of diagonal covariance matrices.

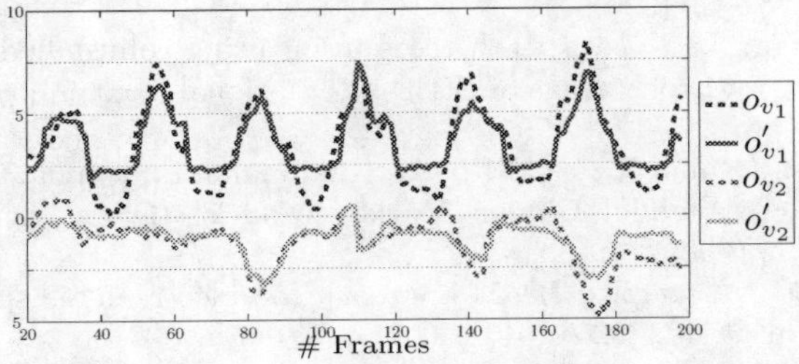

Fig. 4. True (dashed line) and estimated (solid line) visual observations

inversion), and due to the low dimension of the visual feature vector ($K = 2$), the computational load is similar to the case of using diagonal covariance matrices for the same number of states and mixtures.

The above arguments allow one to conclude that the use of full covariance matrices is preferable from the point of view of both computational complexity and accuracy.

The true and estimated visual parameters for the case of full covariance matrices with $N = 4$ states and $M = 2$ mixtures (optimal values) are represented in Fig. 4, where a good agrement can be observed.

7 Conclusions

A speech driven MPEG-4 compliant facial animation system was introduced in this paper. A joint AV-HMM is proposed to represent the audio-visual data and an algorithm for HMM inversion was derived for the general case of considering full covariance matrices for the audio-visual observations. The influence on the visual estimation accuracy of the use of full covariance matrices, as opposed to diagonal ones, was investigated. Simulation results show that the use of full covariance matrices leads to an accurate estimation of the visual parameters, yielding a performance similar to that of using diagonal covariance matrices, but with a less complex model and without affecting the computational load.

References

1. Yamamoto, E., Nakamura, S., Shikano, K.: Lip movement synthesis from speech based on Hidden Markov Models. Speech Communication 26(1-2), 105–115 (1998)
2. Rao, R., Chen, T., Mersereau, R.: Audio-to-visual conversion for multimedia communication. IEEE Trans. on Industrial Electronics 45(1), 15–22 (1998)
3. Chen, T.: Audiovisual speech processing. IEEE Signal Processing Magazine 18(1), 9–21 (2001)
4. Brand, M.: Voice puppetry. In: Proceedings of SIGGRAPH, Los Angeles, CA USA, pp. 21–28 (August 1999)
5. Viterbi, A.J.: Error bounds for convolutional codes and an asymptotically optimal decoding algorithm. IEEE Trans. on Information Theories 13, 260–269 (1967)
6. Choi, K., Luo, Y., Hwang, J.: Hidden Markov Model inversion for audio-to-visual conversion in an MPEG-4 facial animation system. Journal of VLSI Signal Processing 29(1-2), 51–61 (2001)
7. Moon, S., Hwang, J.: Noisy speech recognition using robust inversion of Hidden Markov Models. In: Proceedings of IEEE International Conf. Acoust., Speech, Signal Processing, pp. 145–148 (1995)
8. Fu, S., Gutierrez-Osuna, R., Esposito, A., Kakumanu, P., Garcia, O.: Audio/visual mapping with cross-modal Hidden Markov Models. IEEE Trans. on Multimedia 7(2), 243–252 (2005)
9. Xie, L., Liu, Z.Q.: A coupled HMM approach to video-realistic speech animation. Pattern Recognition 40, 2325–2340 (2007)
10. ISO/IEC IS 14496-2, Visual (1999)
11. Baum, L.E., Sell, G.R.: Growth functions for transformations on manifolds. Pacific Journal of Mathematics 27(2), 211–227 (1968)
12. Hyvärinen, A., Karhunen, J., Oja, E.: Independent Component Analysis. John Wiley & Sons, Inc., New York (2001)
13. Gävert, H., Hurri, J., Särelä, J., Hyvärinen, A.: FastICA package for MATLAB. Lab. of Computer and Information Science, Helsinki University of Technology
14. Terissi, L.D., Gómez, J.C.: Facial motion tracking and animation: An ICA-based approach. In: Proceedings of 15th European Signal Processing Conference, Poznań, Poland, September 3-7, pp. 292–296 (2007)
15. Ostermann, J.: Face Animation in MPEG-4. In: MPEG-4 Facial Animation - The Standard, Implementation and Applications, pp. 17–56. John Wiley & Sons, Chichester (2002)

Discriminant Eigenfaces: A New Ranking Method for Principal Components Analysis

Carlos Eduardo Thomaz[1] and Gilson Antonio Giraldi[2]

[1] Department of Electrical Engineering, FEI, São Paulo, Brazil
[2] Department of Computer Science, LNCC, Rio de Janeiro, Brazil

Abstract. Principal Component Analysis (PCA) is one of the most successful approaches to the problem of creating a low dimensional data representation and interpretation. However, since PCA explains the covariance structure of all the data, the first principal components with the largest eigenvalues do not necessarily represent important discriminant directions to separate sample groups. In this work, we investigate a new ranking method for the principal components. Instead of sorting the principal components in decreasing order of the corresponding eigenvalues, we propose the idea of using the discriminant weights given by separating hyperplanes to select among the principal components the most discriminant ones. Our experimental results have shown that the principal components selected by the separating hyperplanes are quite useful for understanding the differences between sample groups in face image analysis, allowing robust reconstruction and interpretation of the data as well as higher recognition rates using less linear features.

Keywords: Principal Components Analysis, Separating Hyperplanes, Small Sample Size Problems, EigenFaces.

1 Introduction

Principal Component Analysis (PCA) is one of the most successful approaches to the problem of creating a low dimensional data representation and interpretation. However, since PCA explains the covariance structure of all the data its most expressive components [1], that is, the first principal components with the largest eigenvalues, do not necessarily represent important discriminant directions to separate sample groups.

A common practice to identify the important linear directions for separating sample groups is to use Fisher's Linear Discriminant Analysis (LDA) [2,3,4] rather than PCA. However, when the dimension of the feature space is greater than the number of groups, LDA can find at most the number of groups - 1 meaningful discriminant directions [5,6,7]. In the case of two-class problems, a similar limitation can occur in the SVM classifier [9].

In this work, we propose a new ranking method for the principal components given by the group-differences extracted by separating hyperplanes obtained through LDA and SVM. Rather than sorting the principal components in decreasing order of the corresponding eigenvalues, we propose the idea of using

G. Zaverucha and A. Loureiro da Costa (Eds.): SBIA 2008, LNAI 5249, pp. 43–52, 2008.

the discriminant weights given by separating hyperplanes to select among the principal components the most discriminant ones. Such a set of principal components ranked in decreasing order of the discriminant weights is called here as the *discriminant principal components*.

To evaluate the discriminant principal components, the following two-group separation tasks have been performed using face images: female versus male (gender) experiments, and non-smiling versus smiling (expression) experiments. The experimental results carried out in this work show that the principal components selected by the separating hyperplanes are quite useful for understanding the differences between sample groups in face image analysis, allowing robust reconstruction and interpretation of the data as well as higher recognition rates using less linear features.

2 Principal Components Analysis (PCA)

Let an $N \times n$ training set matrix X be composed of N input face images with n pixels. This means that each column of matrix X represents the values of a particular pixel observed all over the N images. Let this data matrix X have covariance matrix S with respectively P and Λ eigenvector and eigenvalue matrices, that is,

$$P^T S P = \Lambda. \tag{1}$$

It is a proven result that the set of m $(m \leq n)$ eigenvectors of S, which corresponds to the m largest eigenvalues, minimizes the mean square reconstruction error over all choices of m orthonormal basis vectors [3]. Such a set of eigenvectors that defines a new uncorrelated coordinate system for the training set matrix X is known as the principal components. In the context of face recognition, those $P_{pca} = [p_1, p_2, ..., p_m]$ components are frequently called eigenfaces [8].

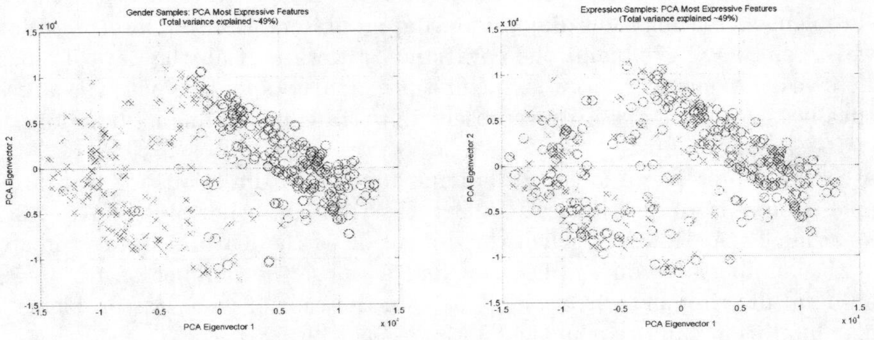

Fig. 1. Two-dimensional most expressive subspace identified by PCA for the gender (left) and expression (right) samples. In the corresponding scatter plots, male (non-smiling) samples are coded with a circle whereas female (smiling) samples are coded with a cross.

However, since PCA explains the covariance structure of all the data its most expressive components [1], that is, the first principal components with the largest eigenvalues, are not necessarily the most discriminant ones. Figure 1 shows the face samples used in this work to illustrate this problem. The left and right plots show the gender and expression samples, respectively, projected on the first two principal components. As can be seen, PCA clearly fails to recover the important features for separating smiling from non-smiling samples. The reason why PCA fails in this case is that the smiling and non-smiling changes are not as substantial as the male and female ones and, consequently, the first principal components that maximize the total scatter across all the images encode essentially variations due to the gender information.

3 Separating Hyperplanes

In this section we describe the main principles behind the statistical separating hyperplanes defined by LDA and SVM [2,3,9].

The primary purpose of LDA is to separate samples of distinct groups by maximizing their between-class separability while minimizing their within-class variability. Its main objective is to find a projection matrix W_{lda} that maximizes the Fisher's criterion:

$$W_{lda} = \arg\max_{W} \frac{|W^T S_b W|}{|W^T S_w W|}, \tag{2}$$

where S_b and S_w are respectively the between and within class scatter matrices [2,3]. The Fisher's criterion is maximized when the projection matrix W_{lda} is composed of the eigenvectors of $S_w^{-1} S_b$ with at most number of groups - 1 nonzero eigenvalues [2]. In the case of a two-class problem, the LDA projection matrix is in fact the leading eigenvector w_{lda} of $S_w^{-1} S_b$, assuming that S_w is invertible.

SVM [9] is primarily a two-class classifier that maximizes the width of the margin between classes, that is, the empty area around the separating hyperplane defined by the distance to the nearest training samples. It can be extended to multi-class problems by solving essentially several two-class problems.

Given a training set that consists of N pairs of $(x_1, y_1), (x_2, y_2) \ldots (x_N, y_N)$, where x_i denote the n-dimensional training observations and $y_i \in \{-1, 1\}$ the corresponding classification labels. The SVM method [9] seeks to find the hyperplane defined by

$$f(x) = (x \cdot w) + b = 0, \tag{3}$$

which separates positive and negative observations with the maximum margin. It can be shown that the solution vector w_{svm} is defined in terms of a linear combination of the training observations, that is,

$$w_{svm} = \sum_{i=1}^{N} \alpha_i y_i x_i, \tag{4}$$

where α_i are non-negative coefficients obtained by solving a quadratic optimization problem with linear inequality constraints [9]. Those training observations x_i with non-zero α_i lie on the boundary of the margin and are called support vectors.

Both LDA and SVM linear discriminant methods seek to find a decision boundary that separates data into different classes as well as possible. The LDA solution is a spectral matrix analysis of the data and is based on the assumption that each class can be represented by its distribution of data, that is, the corresponding mean vector and covariance matrix. In other words, LDA depends on all of the data, even points far away from the separating hyperplane and consequently is less robust to gross outliers [11]. The description of the SVM solution, on the other hand, does not make any assumption on the distribution of the data, focusing on the observations that lie close to the opposite class, that is, on the observations that most count for classification. As a consequence, SVM is more robust to outliers, zooming into the subtleties of group differences [12].

4 Discriminant Principal Components

We approach the problem of selecting the discriminant principal components as a problem of estimating a linear classifier. For this work, we assume that there are only two classes to separate and that a large number of features n is available.

To compose the PCA transformation matrix, that is $P_{pca} = [p_1, p_2, ..., p_m]$, we have retained all principal components with non-zero eigenvalues, that is, $m = N - 1$. The zero mean data vectors are projected on the principal components and reduced to m-dimensional vectors representing the most expressive features of each one of the n-dimensional data vector. Afterwards, the $N \times m$ data matrix and their corresponding labels are used as input to calculate the LDA and SVM separating hyperplanes. The most discriminant feature of each one of the m-dimensional vectors is obtained by multiplying the $N \times m$ most expressive features matrix by the $m \times 1$ discriminant vectors. Thus, the initial training set consisting of N measurements on n variables is reduced to a data set consisting of N measurements on only 1 most discriminant feature given by:

$$
\begin{aligned}
\tilde{y}_1 &= x_{11}w_1 + x_{12}w_2 + ... + x_{1m}w_m, \\
\tilde{y}_2 &= x_{21}w_1 + x_{22}w_2 + ... + x_{2m}w_m, \\
&... \\
\tilde{y}_N &= x_{N1}w_1 + x_{N2}w_2 + ... + x_{Nm}w_m,
\end{aligned}
\tag{5}
$$

where $[w_1, w_2, ..., w_m]$ are the weights corresponding to the principal component features calculated by the LDA and SVM separating hyperplanes, and $[x_{i1}, x_{i2}, ..., x_{im}]$ are the attributes of each data vector i, where $i = 1, ..., N$, projected on the full rank PCA space.

We can determine the discriminant contribution of each feature by investigating the weights $w^T = [w_1, w_2, ..., w_m]$ of the respective directions. Weights that are estimated to be 0 or approximately 0 have negligible contribution on the discriminant scores $\tilde{y}_i$ described in equation (5), indicating that the corresponding features are not significant to separate the sample groups. In contrast, largest weights (in absolute values) indicate that the corresponding features contribute more to the discriminant score and consequently are important to characterize the differences between the groups.

Thus, instead of selecting the principal components in decreasing order of their corresponding eigenvalues, as commonly done, we select as the first principal components the ones with the highest discriminant weights, that is,

$$P_{lda/svm} = [p_1, p_2, ..., p_m] = \arg\max_P \left|P^T S P\right| \tag{6}$$

where $\{p_i | i = 1, 2, \ldots m\}$ is the set of eigenvectors of S corresponding to the largest discriminant weights $|w_1| \geq |w_2| \geq \ldots \geq |w_m|$ described by either the LDA separating hyperplane

$$w_{lda} = \arg\max_w \frac{\left|w^T P_{pca}^T S_b P_{pca} w\right|}{\left|w^T P_{pca}^T S_w^* P_{pca} w\right|} \tag{7}$$

or the SVM separating hyperplane

$$w_{svm} = \sum_{i=1}^{N} \alpha_i y_i (x_i P_{pca}). \tag{8}$$

In short, we are selecting among the principal components the directions that are efficient for discriminating the sample groups rather than representing all the samples.

5 Experimental Results

In our experiments, we have used frontal and pre-aligned images of a face database maintained by the Department of Electrical Engineering of FEI [1], São Paulo, Brazil. Since the number of subjects is equal to 200 and each subject has two frontal images (one with a neutral or non-smiling expression and the other with a smiling facial expression), there are 400 images to perform the experiments. All faces are mainly represented by subjects between 19 and 40 years old with distinct appearance, hairstyle, and adorns.

The goal of the gender experiment is to evaluate the selection method on a discriminant task where the differences between the groups are evident. The facial expression experiment poses an alternative analysis where there are subtle differences between the groups. In these experiments, the total number of training examples N is limited and significantly less than the dimension of the feature space, that is, $N \ll n$ (*small sample size problem*) [3]. To address this problem for the Fisher's criterion, we use a regularized version of the LDA approach called MLDA [10].

[1] This database is publicly available on http://www.fei.edu.br/~cet/facedatabase.html

5.1 Weights of the Separating Hyperplanes

Table 1 lists the 20 principal components with the highest weights in absolute values for discriminating the gender and expression samples. It can be seen that SVM has selected some of the last principal components to compose its corresponding sets of top 20 most discriminant principal components, such as the 334th principal component for the gender experiment and the 385th, 382th, and 388th for the expression experiment. Since the last principal components describe particular information related to few samples, these results confirm the SVM approach of zooming into the subtleties of group differences. In contrast, the MLDA separating hyperplane has selected as discriminant principal components the ones based on information shared by more samples and, consequently, most common and representative to characterize group differences.

Table 1. Top 20 principal components ranked by MLDA and SVM hyperplanes

Gender		Expression	
MLDA	SVM	MLDA	SVM
20	83	18	18
1	20	22	22
25	64	16	14
37	78	20	17
31	31	14	16
19	25	17	20
35	87	19	119
64	112	30	28
8	108	28	30
39	36	23	56
46	125	25	385
83	148	36	19
36	334	31	70
44	56	13	382
41	92	5	388
22	1	44	275
12	19	56	202
29	62	37	151
7	39	24	245
62	46	11	189

5.2 Total Variance Explained

Figure 2 illustrates that as the dimension of the PCA most expressive subspace increases, there is an exponential decrease in the amount of total variance explained by the first principal components with the largest eigenvalues. This is a well-known behavior of the dimensionality reduction provided by the standard PCA [13].

However, the corresponding variances explained by the discriminant principal components do not follow the same behaviour. Figure 2 illustrates that in the gender experiment the first 20 MLDA and SVM most discriminant principal components represent a considerable amount of the total variance, that is, 42% and 36% respectively, but not as high as the 80% of the first 20 most expressive

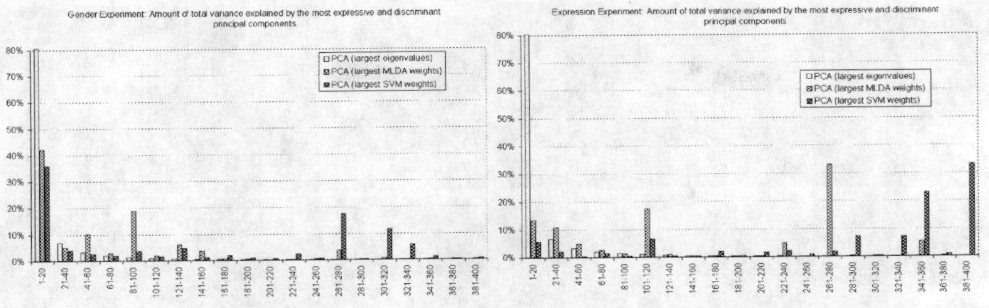

Fig. 2. Gender (left) and expression (right) experiments. Amount of total variance explained by PCA most expressive features components, and MLDA and SVM most discriminant principal components.

components. As the number of principal components increases, we can see that the next 20 most discriminant principal components (in the range of 21 - 40) for both MLDA and SVM approaches are not the ones with the second largest total variance explained. These results indicate that although the differences between male and female are substantial in the training set considered, there are some artifacts not related to gender characteristics that vary on several images and should not be considered as discriminant information.

In the expression experiment, this distinction between principal components that represent features that vary significantly and the ones that capture discriminant characteristics of sample groups are evident. Figure 2 shows that the first 20 MLDA and SVM most discriminant principal components represent a reduced amount of the total variance (14% and 6%, respectively) compared to the top 20 PCA most expressive components (80%). In fact, the principal components with the largest eigenvalues have been sorted as one of the last most discriminant components for both MLDA (range 261 - 280) and SVM (range 381 - 400) methods. Therefore, although some of the gender information can be captured by the most expressive components because differences between male and female are substantial, the expression differences that are subtle cannot be characterized by the first PCA most expressive components.

5.3 Visualisation

We have reconstructed the PCA most expressive features by changing each principal component separately using limits of $\pm 3\sqrt{\lambda_i}$, where λ_i is the corresponding eigenvalue. For the principal components selected by the MLDA and SVM separating hyperplanes, we have used the limits of $\pm 3\sqrt{max(\lambda_i, \bar{\lambda})}$, where $\bar{\lambda}$ is the average eigenvalue of the total covariance matrix S, because some λ_i can be very small in this case, showing no changes between the samples when we move along the corresponding principal components.

Figure 3 illustrates the transformations on the first two PCA most expressive components contrasted with the first two discriminant principal components

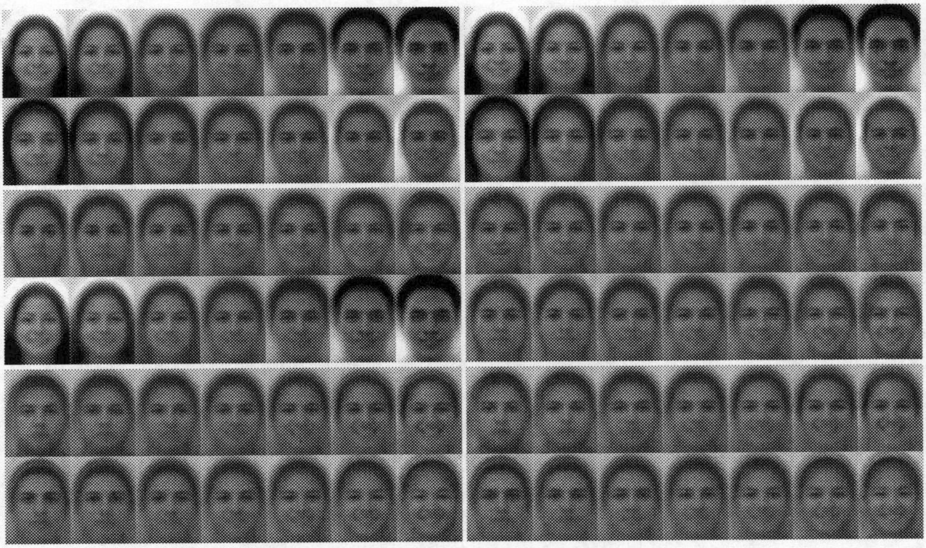

Fig. 3. From top to bottom: the first two PCA most expressive components, the first two PCA most discriminant principal components selected by MLDA (left) and SVM (right) to separate females from males, and the first two PCA most discriminant principal components selected by MLDA (left) and SVM (right) to separate non-smiling from smiling samples

ranked by the MLDA and SVM hyperplanes to separate males from females and smiling from non-smiling faces. The first two PCA most expressive directions capture essentially the changes in illumination and gender, which are the major variations of all the training samples. Since we have used the same training set for both gender and expression experiments, the PCA most expressive components are equal in both experiments. Owing to the fact that changes in facial expression are much less significant than the illumination and gender ones, the standard PCA is unable to capture such minor variations in its first most expressive components. However, when we compare these results with the ones reconstructed by the MLDA and SVM most discriminant principal components, illustrated in the last two rows of Figure 3, we can see that in fact other principal components do carry information about expression variations. In situations where the classes are not well separated, the discriminant principal components ranked by either MLDA or SVM have shown to be more effective with respect to extracting group-differences information than the most expressive ones without enhancing the image artifacts.

5.4 Recognition Rates

We have assigned each sample to the class mean with minimum Mahalanobis distance and adopted the leave-one-out method to evaluate the classification performance of the most expressive and discriminant principal components. Figure 4 shows that the MLDA and SVM discriminant principal components provide higher recognition rates than the standard PCA when less linear features are used to classifying gender and facial expression sample groups. For instance, when using only 20 principal components, the MLDA and SVM discriminant principal components have achieved gender recognition rates of 93.5% and 92.5% respectively compared with 90.3% of the standard PCA, and expression recognition rates of 89.8% compared with only 83.3%. These results confirm the classification power of the discriminant principal components especially in situations where the sample groups are not well separated.

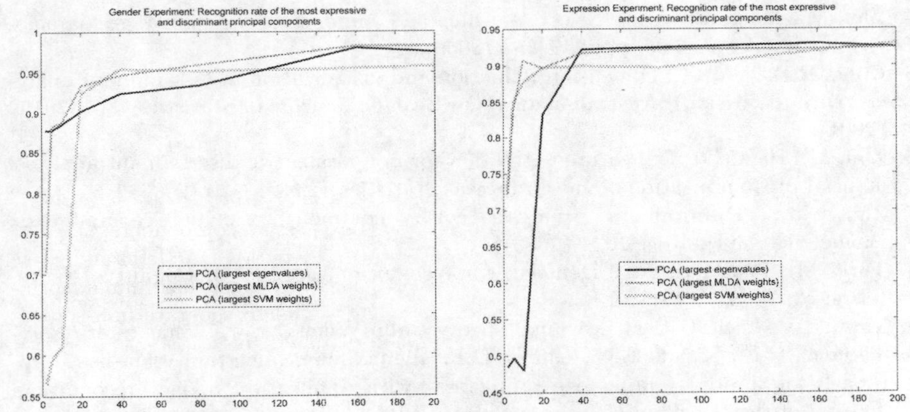

Fig. 4. Gender (left) and expression (right) experiments. Recognition rate of PCA most expressive components, MLDA and SVM most discriminant principal components.

6 Conclusion

This paper proposed a new selection method for the principal components given by the group-differences extracted by separating hyperplanes. The discriminant principal components were, among the principal components, the directions most efficient for separating the sample groups rather than representing all the samples. Moreover, the number of meaningful discriminant principal components was not limited to the number of groups, providing additional information to understand the group differences extracted from the small sample size problem of face image analysis.

Acknowledgements

We would like to thank Dr. Paulo S. S. Rodrigues for providing the SVM code based on the quadratic programming solution created by Alex J. Smola. Also we would like to thank Leo L. de Oliveira Jr. for acquiring and normalizing the FEI database (FEI-PBIC 32-05). In addition, we would like to thank the support provided by PCI-LNCC, FAPESP (2005/02899-4) and CNPq (472386/2007-7).

References

1. Swets, D., Weng, J.: Using discriminants eigenfeatures for image retrieval. IEEE Trans. Patterns Anal. Mach Intell. 18(8), 831–836 (1996)
2. Devijver, P., Kittler, J.: Pattern Classification: A Statistical Approach. Prentice-Hall, Englewood Cliffs (1982)
3. Fukunaga, K.: Introduction to Statistical Pattern Recognition. Academic Press, New York (1990)
4. Zhu, M., Martinez, A.M.: Selecting principal components in a two-stage lda algorithm. In: CVPR 2006, pp. 247–254 (2006)
5. Cook, R.D., Yin, X.: Dimension reduction and visualization in discriminant analysis (with discussion). Australian and New Zealand Journal of Statistics 43, 147–199 (2001)
6. Zhu, M., Hastie, T.J.: Feature extraction for nonparametric discriminant analysis. Journal of Computational and Graphical Statistics 12, 101–120 (2003)
7. Zhu, M.: Discriminant analysis with common principal components. Biometrika 93(4), 1018–1024 (2006)
8. Turk, M., Pentland, A.: Eigenfaces for recognition. Journal of Cognitive Neuroscience 3, 71–86 (1991)
9. Vapnik, V.N.: Statistical Learning Theory. John Wiley & Sons, Chichester (1998)
10. Thomaz, C.E., Kitani, E.C., Gillies, D.F.: A maximum uncertainty lda-based approach for limited sample size problems - with application to face recognition. Journal of the Brazilian Computer Society 12(2), 7–18 (2006)
11. Hastie, T., Tibshirani, R., Friedman, J.: The Elements of Statistical Learning. Springer, Heidelberg (2001)
12. Davatzikos, C.: Why voxel-based morphometric analysis should be used with great caution when characterizing group differences. NeuroImage 23, 17–20 (2004)
13. Johnson, R., Wichern, D.: Applied Multivariate Statistical Analysis. Prentice Hall, New Jersey (1998)

Enhancing the Interaction between Agents and Users*

Marcelo Armentano, Silvia Schiaffino, and Analía Amandi

ISISTAN Research Institute, Fac. Cs. Exactas, UNCPBA
Campus Universitario, Paraje Arroyo Seco, Tandil, 7000, Argentina
CONICET, Consejo Nacional de Investigaciones Científicas y Técnicas, Argentina
{marmenta,sschia,amandi}@exa.unicen.edu.ar

Abstract. A key aspect when interface agents provide personalized assistance to users, is knowing not only a user's preferences and interests with respect to a software application but also when and how the user prefers to be assisted. To achieve this goal, interface agents have to detect the user's intention to determine when to assist the user, and the user's interaction and interruption preferences to provide the right type of assistance at the right time. In this work we describe a user profiling approach that considers these issues within a user profile, which enables the agent to choose the best type of assistance for a given user in a given situation. We also describe the results obtained when evaluating our proposal in a calendar application.

Keywords: intelligent agents, user profiling, human-computer etiquette.

1 Introduction

In the last years, there has been an increasing interest in the area of human-computer etiquette [9]. Particularly, when talking about interface agents, researchers and developers are directing efforts towards learning how to best assist the user, that is providing the right help, at the right time and in the right way, without hindering the user's work. Learning users' habits, preferences and interests to provide them personalized assistance with a software application is not the only goal.

When working with a software application, a user can perform different tasks. Consequently, it is very important for an interface agent to know exactly the task the user is carrying out because it gives the context in which the user is working. By taking this context into account, the agent may infer the user's intention and try to collaborate with him. In addition, if the agent knows the user's intention, it will avoid interrupting him at an improper time. Users generally do not want to be interrupted while working on a specific task, unless this interruption is strongly related to the task they performing, or it has a high priority. Also, users differ in their preferences about how and when they want to be assisted, and even a single user may differ in the type of assistance he prefers for different contexts [11,12]. For example, if the agent observes that the user is scheduling a work meeting for the following day, the agent can offer to automatically complete the information required and send an email on the

* This work was also supported by ANPCyT, Argentina, through PICT Project 20178.

G. Zaverucha and A. Loureiro da Costa (Eds.): SBIA 2008, LNAI 5249, pp. 53–62, 2008.

user's behalf to each participant of the meeting, provided that it knows the user's preferences about the kind of meeting he is scheduling. The user might accept this type of assistance for a certain type of meeting, but he might prefer only a suggestion or no assistance at all in a different context.

In this work, we propose a new definition for a user profile considering these issues. We also propose a profiling approach to acquire the different components of the proposed profile. This enhanced profile enables interface agents to decide how to best assist a user. Our profiling approach uses, first, plan recognition to detect a user's intentions. Plan recognition aims at identifying the goal of a subject based on the actions he performs [2]. Then, we use two user profiling algorithms we developed, namely *WATSON* and *IONWI*, to learn a user's interaction and interruption preferences [12]. Finally, we combine the different components of the user profile in a decision making algorithm that enables an interface agent to decide how to best assist a user in a given situation.

The rest of the work is organized as follows. Section 2 presents an overview of our proposed approach. Section 3 describes how to detect a user's intention using plan recognition. Section 4 describes how to learn a user's interaction and interruption preferences. Section 5 presents the results obtained when evaluating our approach. Then, Section 6 analyzes some related works. Finally, we present our conclusions.

2 Overview of Our Proposed Approach

A user profile typically contains information about a user's interests, preferences, behavioral patterns, knowledge, and priorities, regarding a particular domain. However, such information is not enough to personalize the interaction with a user. The user's intentions with a software application and his interaction preferences play a relevant role in user-agent interactions.

Consider for example the following situation. A user of a calendar application has the intention of arranging a work meeting with John Smith, his project manager. To achieve this, he has to perform a set of tasks, such as selecting John Smith from his contact list, creating a new event, entering the subject, date, place and all the information required about the meeting, and sending an email to John Smith. The sooner the agent detects the user's intention, the better it will assist him in accomplishing his intention. We propose to use Plan Recognition to detect the user's intention. Plan recognition aims at identifying the goal of a subject based on the actions he performs [2]. Once the agent has detected that the user wants to arrange a work meeting with John Smith, it can use the information contained in the classic user profile to assist him. However, different users may have different preferences about the type of assistance they welcome from an interface agent. For example, some users may prefer the agent to automatically complete all the tasks it can, while others just prefer to receive suggestions. Moreover, this information is strongly dependent on the situation in which the agent is about to assist the user. In our approach, the information needed to determine what type of assistance a user wants to receive in a given situation is contained in the user interaction profile. This profile also comprises the expected modality of the assistance. In a certain context the user might want just a notification containing the suggested meeting date or place, while in a different context the user might prefer an interruption.

To obtain the proposed components of a user profile, we developed two profiling algorithms: *WATSON* and *IONWI*. *WATSON* learns a user's assistance preferences, that is, when a user wants a suggestion, a warning, an automated action or no assistance. *IONWI* learns a user's interruption preferences, that is, when a user prefers an interruption and when a notification. To achieve their goals, these user profiling algorithms analyze the user's interactions with the agent recorded when observing the user's behavior, and they consider the feedback provided by the user after the agent assisted him. An overview of our proposal is shown in Figure 1.

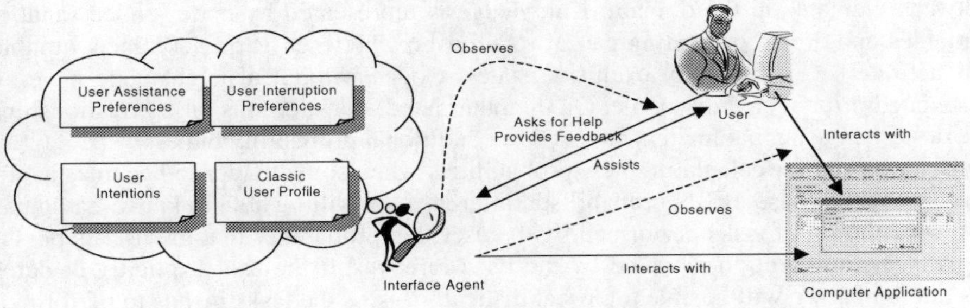

Fig. 1. Proposed user profiling approach

We define a user profile as: *User Profile = Classic User Profile + User Intentions + User Interaction Profile*, and *User Interaction Profile = User Assistance Preferences + User Interruption Preferences.* The user intentions are the set of all the possible intentions the user can be trying to achieve, each of them with a degree of certainty: *User Intentions = {<User intention, Certainty>}*. The assistance preferences are a set of problem situations or contexts with the required assistance action and a parameter (certainty) indicating how sure the agent is about the user wanting that assistance action in that particular situation: *User Assistance Preferences = {<Situation, Assistance Action, Certainty>}*.

We define the interruption preferences as a set of situations with the preferred assistance modality (interruption or no interruption). They might also contain the type of assistance action to execute. A parameter indicates how certain the agent is about this user preference: *User Interruption Preferences ={<Situation, [Assistance Action], Assistance Modality, Certainty>}*. The following sections explain how we obtain these components.

3 Detecting a User's Intentions

Plan recognition can be used to infer the user's intentions based on the observation of the tasks the user performs in the application. Plan recognizers take as inputs a set of goals the agent expects the user to carry out in the domain, a plan library describing the way in which the user can reach each goal, and an action observed by the agent.

The plan recognition process itself, consists in foretelling the user's goal, and determining how the observed action contributes to reach it. There are two main aspects that make classical approaches [2,10] to plan recognition unsuitable for being used by interface agents. First, the agent should deal with transitions and changes in the user intentions. Second, we have to take into account the influence of a user's preferences in the plan recognition process.

We propose Bayesian Networks (BN) [6] as a knowledge representation capable of capturing and modeling dynamically the uncertainty of user-agent interactions. We represent the set of intentions the user can pursue in the application domain as an Intention Graph (IG). An IG is materialized as a BN and represents a context of execution of tasks. BN are directed acyclic graphs representing probabilistic relationships between elements in the domain. Knowledge is represented by nodes called random variables and arcs representing causal relationships between variables. Each variable has a finite set of mutually exclusive states. Nodes without a parent node have an associated prior probability table. On the other hand, the strengths of the relationships are described using parameters encoded in conditional probability tables.

BN are used for calculating new probabilities when some evidence becomes available. By making use of BN probabilistic inference we will be able to know, having as evidence the set of tasks performed by the user, the probability that the user is pursuing any given intention modeled by the IG. Moreover, if the user explicitly declares his intentions, we will be able to probabilistically assess the tasks he has to perform to achieve his goal.

In our IG variables correspond to goals that the user can pursue and to tasks the user can perform in the application to achieve those goals. The two possible states of these variables are true, indicating that the user is pursuing that goal or that the user performed that task, and false. We call certainty of an intention to the probability of a variable being in a true state. Evidence on a task node will be set when the user interacts with a widget in the application GUI that is associated to the execution of that task. Our IG includes a third kind of variable: context variables. This kind of variables will be used to personalize the intention detection process by learning new relations that may arise between the attributes of the tasks performed by the user and the intention nodes in the IG.

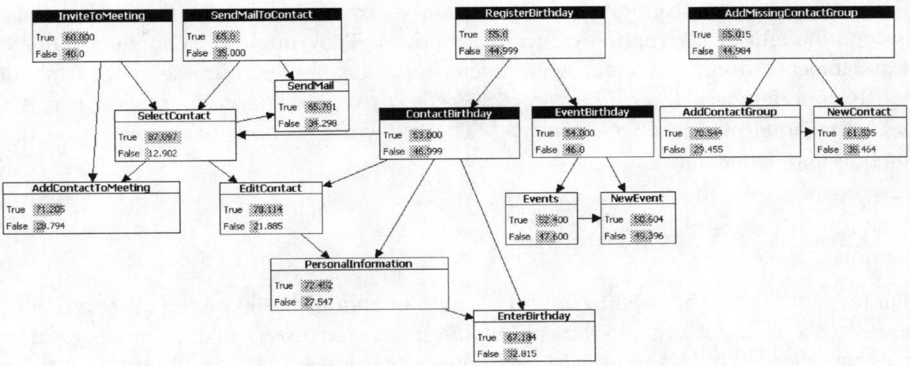

Fig. 2. Example of an intention graph

For example, in a calendar application, the user can select a contact from the address book with the objective of sending this contact an email or with the objective of scheduling a meeting with this contact, as shown in the IG presented in Figure 2. The IG constructed manually by a domain expert will allow the agent to rank which of the two goals is more probable, given that the user selected a contact in his address book. However, the information of the selected contact can be relevant in discerning which goal the user is pursuing. To consider this information, we introduce into the definition of our IG, the concept of traceable nodes. A traceable node is a node in which we want to register the values taken by some attributes of the corresponding task performed by the user, with the aim of adding new variables that represent the context in which the user performs that task and to find new relations between these variables and the nodes in the IG. In the example above, the task corresponding to the selection of a contact in the address book is a traceable node. The designer of the IG should decide which attributes of this task are of interest (for example, the city in which the contact lives or the group the contact belongs to) for which set of intentions (sending a mail to the selected contact or scheduling a meeting with him or her).

Each time the user performs a task corresponding to a traceable node, the agent will observe the values taken by the attributes of the task (for example, the selected contact is from New York and belongs to the group of friends). Then, the agent will continue observing the user until it can infer which his intention(s) are and will record the experience in an interaction history. Each experience will be of the form: $<attribute_1,\ldots,attribute_n, intention_1, \ldots, intention_k>$, where $attribute_i$ is the value taken by the $attribute_i$ and $intention_j$ is true if the agent infers that the user was pursuing $intention_j$, or false otherwise. This database of experiences is then used by the agent to run a batch learning algorithm and a parametric learning algorithm to find relations between the attributes and between the attributes and the intentions [6]. To adapt the probabilities (set by the domain expert who constructed the network) to a particular user's behavior, we take an statistical on-line learning approach [6].

Finally, as stated in Section 2, most of previous plan recognition approaches do not consider the uncertainty related to the moment in which the user starts a new plan to achieve a new goal. Those which consider this issue limit the memory of the plan recognizer by making evidence to be present in a fixed interval of time and then completely disregarding it. We take a different approach in which evidence is gradually forgotten. We adopt the concept of soft evidence to fade the evidence we entered to the BN as the user performs further tasks [6]. To do this, we use a fading function to gradually forget the tasks performed by the user. Evidence is faded according to this function until it reaches its original value, that is until the probability of a given node becomes less than the value that it would have if we would not have observed the execution of the corresponding task in the application. Fading functions can be any function that, given the IG and the evidence on tasks performed so far, decrements the certainty of the evidence gradually, according to some heuristic. For example, we can decrement current evidence by a fixed factor $0 \leq \nabla \leq 1$ every time the user performs a task in the application.

4 Learning a User's Interaction and Interruption Preferences

To learn a user's interaction and interruption preferences, the information obtained by observing a user's behavior is recorded as a set of user-agent interaction experiences. An interaction experience *Ex=<Sit, Act, Mod, UF, E, date>* is described by six arguments: a situation *Sit* that originates an interaction; the assistance action *Act* the agent executed to deal with the situation (warning, suggestion, action on the user's behalf); the modality *Mod* that indicates whether the agent interrupted the user or not to provide him/her assistance; the user feedback *UF* obtained after assisting the user; an evaluation *E* of the assistance experience (success, failure or undefined); and the *date* when the interaction took place. For example, if we consider an agent assisting a user of a calendar management system, an assistance experience could be the following. John Smith is scheduling a new event: a meeting to discuss the evolution of project A with his employees Johnson, Taylor and Dean. The event is being scheduled for Friday at 5 p.m. at the user's office. The agent has learned by observing the user's actions and schedules that Mr. Dean will probably disagree about the meeting date and time because he never schedules meetings on Friday evenings. Thus, it decides to warn the user about this problem. In reply to this warning, the user asks the agent to suggest him another date for the event.

To obtain a user's interruption and assistance preferences, we developed two algorithms: *WATSON* and *IONWI*. These algorithms use association rules to discover the existing relationships between situations or contexts and the assistance actions a user requires to deal with them, as well as the relationships between a situation, a user task, and the assistance modality required.

Association rules imply a relationship among a set of items in a given domain. As defined by [1], association rule mining is commonly stated as: Let $I = i_1, \ldots, i_n$ be a set of items and D be a set of transactions, each consisting of a subset X of items in I. An association rule is an implication of the form $X \rightarrow Y$, where $X \subseteq I$, $Y \subseteq I$, and $X \cap Y = \emptyset$. X is the rule's antecedent and Y is the consequent. The rule has support s in D if s percent of D's transactions contains $X \cup Y$. The rule $X \rightarrow Y$ holds in D with confidence c if c percent of D's transactions that contain X also contain Y. Given a transaction database D, the problem of mining association rules is to find all association rules that satisfy minimum support and minimum confidence.

We use the Apriori algorithm [1] to generate association rules from a set of user-agent interaction experiences. Then, we automatically post-process the rules Apriori generates so that we can derive useful information about the user's preferences from them. Post-processing includes detecting the most interesting rules according to our goals, eliminating redundant rules, eliminating contradictory rules, and summarizing the information obtained. To filter rules, we use templates or constraints [7] that select those rules that are relevant to our goals. For example, we are interested in those association rules of the forms: *situation, assistance action* → *user feedback, evaluation,* in the *WATSON* algorithm, and s*ituation, modality, [assistance action]* → *user feedback, evaluation*, in the *IONWI* algorithm, where brackets mean that the attributes are optional. Rules containing other combinations of attributes are not considered. To eliminate redundant rules, we use a subset of the pruning rules proposed in [13]. Basically, these pruning rules state that given the rules A,B→C and A→C, the first rule is redundant because it gives little extra information. Thus, it can be deleted if the two rules have similar confidence values. Similarly, given the rules A→B and A→B,C, the first rule is redundant since the second consequent is more specific. Thus, the redundant can be deleted provided that both rules have similar confidence values. Then, we eliminate contradictory rules. We define a contradictory rule in *WATSON* as one indicating a different assistance action for the same situation and having a small confidence value with respect to the rule being compared. Similarly, in *IONWI*, a contradictory rule is one that indicates a different assistance modality for the same context. After pruning, we group rules by similarity and generate a hypothesis that considers a main rule, positive evidence (redundant rules that could not be eliminated), and negative evidence (contradictory rules that could not be eliminated). The main rule is the rule in the group with the greatest support value. Once we have a hypothesis, the algorithm computes its certainty degree by taking into account the main rule's support values and the positive and negative evidence. To compute certainty degrees, we use equation 1:

$$Cer(H) = \alpha Sup(AR) + \beta \frac{\sum_{k=1}^{r} Sup(E+)}{\sum_{k=1}^{r+t} Sup(E)} - \gamma \frac{\sum_{k=1}^{t} Sup(E-)}{\sum_{k=1}^{r+t} Sup(E)} \quad (1)$$

where α, β, and γ are the weights of the terms in the equation (we use α=0.7, β=0.15 and γ=0.15), *Sup*(*AR*) is the main rule support, $Sup(E^+)$ is the positive evidence support, $Sup(E^-)$ is the negative evidence support, *Sup*(*E*) is the support of a rule taken as evidence (positive or negative), *r* is the amount of positive evidence, and *t* is the amount of negative evidence. If the certainty degree of the hypothesis is greater than a given threshold value δ, it becomes part of the user profile. Otherwise, it is discarded.

5 Experimental Results

We carried out two types of experiments. First, we studied the influence of users' preferences on the detection of users' intentions with plan recognition. Then, we analyzed the precision of our profiling approach at assisting users.

5.1 Evaluation of Our Plan Recognition Approach

To analyze the influence that the user's preferences have on the detection of the user's intention, we selected an scenario in which a user used a calendar application to organize a meeting with some contact in his address book, and then selected another contact to register his or her birthday. To achieve these goals, the user performed the following sequence of tasks: Select Contact, Add Contact To Meeting, Select Contact, Edit Contact, Personal Information, Enter Birthday. We recorded the certainty values for all the intentions both in the IG containing the user's preferences information and in the same IG without context nodes (the one owned originally by the agent). Figure 3 shows the evolution of the certainty values of related intentions. Intentions in the original IG are indicated with completed lines. In the first time slice, we show the a priori probabilities of each intention when the user has not perform any tasks in the application yet. "Send Mail To Contact" is the most probable intention, while "Contact Birthday" is the least probable one. When the user performed the first task,

“Select Contact”, the ranking remained unchanged, although there was a small increment in those intentions that contained this task. Then the user performed "Add Contact To Meeting", and "Invite Contact To Meeting" became the most probable intention. With the next set of tasks performed by the user, "Contact Birthday" increased its certainty while the other intentions decreased them. The agent considered a threshold value of 0.7 to believe in the intention pursued by the user. It needed both tasks to be performed to detect "Invite To Meeting", and three tasks out of four to detect "Contact Birthday".

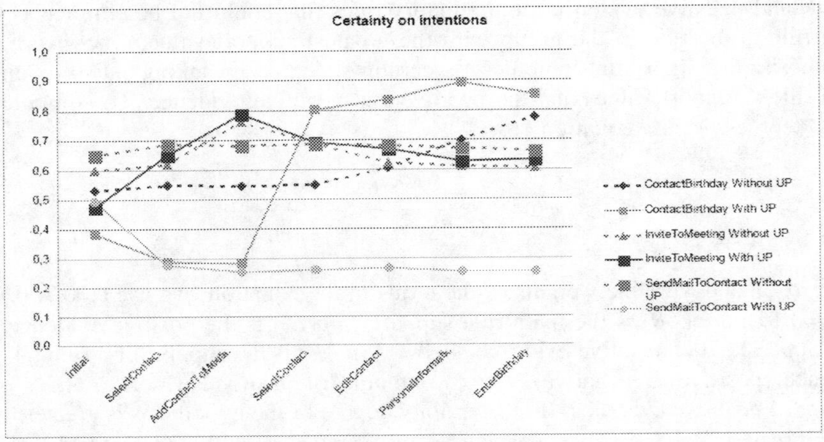

Fig. 3. Experimental results obtained with plan recognition

Dotted lines in Figure 3 show the same scenario but using the IG with context nodes merged. The first "Select Contact Task" was performed when the user selected a contact from the "Friends" group, living in "New York", who already had the birthday registered. We can see that the certainty for "Invite To Meeting" is higher only with the first task performed. We can also see that the other intentions dramatically decreased their certainty values. Obviously the user would not register a birthday in the contact information because the selected contact already had a birthday date.

The second contact selected by the user was also from the "Friends" group and its city was "New York", but the birthday was not registered yet in this case. So, we can see that with the mere selection of the contact, "Contact Birthday" intention could be predicted.

It is worth noting that the curves corresponding to the possible user’s intentions are closer when we do not consider the user’s preferences than when we do. Another interesting fact that can be observed in Figure 3 is that the certainty of finished intentions gradually decrements to its original value, as happens with "Invite To Meeting" intention. This is due to the fading function used by the IG that gradually decrements by a fixed constant to the strength of the evidence on the performed tasks. Similar experiments were carried out with different scenarios, with similar results.

5.2 Evaluation of Our Profiling Algorithms

To evaluate the precision of our interaction profiling approach at assisting users, we studied the number of correct assistance actions, where the "correctness" is determined by the user through explicit and implicit feedback. To do this, we used a precision metric that measures an agent's ability to accurately assist a user. We used this metric to evaluate the agent's performance in deciding between a warning, a suggestion, or an action on the user's behalf; and between an interruption or a notification. For each problem situation, we compared the number of correct assistance actions against the total number of assistance actions the agent executed. The resulting ratio is the agent's precision. To carry out the experiments, we used 33 data sets containing user-agent interaction experiences in the calendar management domain. Each database record contains attributes that describe the problem situation (or the situation originating the interaction), the assistance action the agent executed, the user feedback, and the user's evaluation of the interaction experience. The data sets contained anywhere from 30 to 125 user-agent interactions[1]. We studied four calendar-management situations: new event, the user is scheduling a new event, and the agent has information about the event's potential time, place, or duration; overlapping, the user is scheduling an event that overlaps with a previously scheduled event; time, not enough time exists to travel between the proposed event and the event scheduled to follow it; holiday, the user is scheduling a business event for a holiday.

We compared the precision values for three approaches: confidence-based (this algorithm is classical in agents [8]), *WATSON*, and *WATSON+IONWI*. The values were obtained by averaging the precision for the different datasets belonging to the different users. We used percentage values because the number of user-agent interactions varied from one user to another. Our approach had a higher overall percentage of correct assistance actions or interactions, 86% for *WATSON* and 91% for *WATSON+IONWI*, than the confidence-based algorithm, which got a 67% of precision. In some situations, although the etiquette-aware agent knew how to automate an action on the user's behalf it only made him a suggestion because it had learnt that the user wanted to control the situation by himself. The agent using the standard algorithm made an autonomous action in that case because it knew what the user would do. However, it did not consider that the user wanted to do the task by himself. When the problem is infrequent and warnings are required, the three algorithms behave similarly.

6 Related Work

Our work is related to those works that study the etiquette of human-computer relationships [9], since learning when to interrupt a user, detecting a user's intentions, and deciding how to best assist him can be considered as part of this etiquette. Considering these issues within interface agent development, and particularly within user profiles, is novel.

Regarding interruptions, they have been widely studied in the Human-Computer Interaction area, but they have not been considered in interface agent development.

---

[1] The datasets are available at http://www.exa.unicen.edu.ar/~sschia

With respect to agents using plan recognition to detect a user's plans, some works have been done in this direction [4,10]. However, they do not consider the user's preferences within the process. Some algorithms have been proposed to decide which action an agent should execute next. These algorithms adopt mainly one of two approaches: some use decision and utility theory [3,5], and others use confidence values attached to different actions [8]. However, these works do not consider a user's interaction preferences, the possibility of providing different types of assistance, or the particularities of the situation at hand.

7 Conclusions

In this article, we presented an approach to enhance the interaction with users that considers the user's intentions and the user's interaction preferences. To detect a user's intentions we propose a plan recognition approach, which considers the user's preferences to allow an earlier detection. To learn a user's interaction preferences, we proposed two profiling algorithms. We evaluated our proposal with promising results. As a future work, we will evaluate our approach in a different application domain.

References

1. Agrawal, R., Srikant, R.: Fast Algorithms for Mining Association Rules. In: Proc. 20th Int. Conf. on Very Large Data Bases (VLDB 1994), pp. 487–499 (1994)
2. Charniak, E., Goldman, R.P.: A bayesian model of plan recognition. Artificial Intelligence 64(1), 53–79 (1993)
3. Fleming, M., Cohen, R.: A user modeling approach to determining system initiative in mixed-initiative AI systems. In: Proc. 18th Int. Conf. On User Modeling, pp. 54–63 (2001)
4. Horvitz, E., Heckerman, D., Hovel, D., Rommelse, K.: The Lumière Project: Bayesian User Modeling For Inferring The Goals And Needs Of Software Users. In: Proc. 14th Conf. on Uncertainty in Artificial Intelligence, pp. 256–265 (1998)
5. Horvitz, E.: Principles of Mixed-Initiative User Interfaces. In: Proc. ACM Conf. Human Factors in Computing Systems (CHI 1999), pp. 159–166 (1999)
6. Jensen, F.: Bayesian Networks and Decision Graphs. Springer, New York (2001)
7. Klementinen, M., Mannila, H., Ronkainen, P., Toivonen, H., Verkamo, A.I.: Finding interesting rules from large sets of discovered association rules. In: 3rd Int. Conf. on Information and Knowledge Management, pp. 401–407 (1994)
8. Maes, P.: Agents That Reduce Work And Information Overload. Communications of the ACM 37(7), 30–40 (1994)
9. Miller, C.: Human-computer etiquette: Managing expectations with intentional agents. Communications of the ACM 47(4), 31–34 (2004)
10. Rich, C., Sidner, C., Leash, N.: Collagen: Applying Collaborative Discourse Theory To Human-Computer Interaction. Artificial Intelligence 22(4), 15–25 (2001)
11. Schiaffino, S., Amandi, A.: User – interface agent interactions: personalization issues. International Journal of Human Computer Studies 60(1), 129–148 (2004)
12. Schiaffino, S., Amandi, A.: Polite Personal Agents. IEEE Intelligent Systems 21(1), 12–19 (2006)
13. Shah, D., Lakshmanan, L., Ramamrithnanm, K., Sudarshan, S.: Interestingness and Pruning of Mined Patterns. In: Proc. Workshop Research Issues in Data Mining and Knowledge Discovery. ACM Press, New York (1999)

Re-routing Agents in an Abstract Traffic Scenario

Ana L.C. Bazzan and Franziska Klügl

[1] Instituto do Informatica, UFRGS
Porto Alegre, Brazil
bazzan@inf.ufrgs.br
[2] Dep. of Artificial Intelligence, University of Würzburg
Würzburg, Germany
kluegl@informatik.uni-wuerzburg.de

Abstract. Human drivers may perform replanning when facing traffic jams or when informed that there are expected delays on their planned routes. In this paper, we address the effects of drivers re-routing, an issue that has been ignored so far. We tackle re-routing scenarios, also considering traffic lights that are adaptive, in order to test whether such a form of co-adaptation may result in interferences or positive cumulative effects. An abstract route choice scenario is used which resembles many features of real world networks. Results of our experiments show that re-routing indeed pays off from a global perspective as the overall load of the network is balanced. Besides, re-routing is useful to compensate an eventual lack of adaptivity regarding traffic management.

1 Introduction

In traffic, human drivers – especially when equipped with a navigation system – do re-route when they are informed that a jam is on their planned route. In simple two-route scenarios it makes little sense to study re-routing. These scenarios are only adequate to study chaotic traffic patterns with drivers changing their routes in periodic ways, like in [3,12].

Scenarios in which agents can change their routes on the fly are just beginning to be investigated. It is unclear what happens when drivers can adapt to traffic patterns in complex traffic networks and, at the same time, there is a traffic authority controlling the traffic lights, trying to cope with congestions at local levels or at the level of the network as a whole. New questions are just arising such as: What happens if impatient drivers do change route when they are sitting too long in red lights? How can a traffic authority react to this behavior?

In this paper we discuss some issues related to on the fly re-routing, aiming at providing some answers to those questions. To this aim we use a simulation scenario already addressed [2,4]. The goal in these previous works was to investigate what happens when different actors adapt, each having its own goal. For instance, the objective of local traffic control is obviously to find a control

G. Zaverucha and A. Loureiro da Costa (Eds.): SBIA 2008, LNAI 5249, pp. 63–72, 2008.

scheme that minimizes queues. On the other hand, drivers normally try to minimize their individual travel time. Finally, from the point of view of the whole system, the goal is to ensure reasonable travel times for all users, which can be highly conflicting with some individual utilities. This is a well-known issue: Tumer and Wolpert [11] have shown that there is no general approach to deal with this complex question of collectives.

As mentioned, we now return to that scenario adding a new feature regarding drivers, namely that they can replan their routes on the fly. In the next section we review these and related issues. In Section 3 we describe the approach and the scenario. Results are shown and discussed in Section 4, while Section 5 presents the concluding remarks.

2 Agent-Based Traffic Simulation and Route Choice

Route choice simulation is an essential step in traffic simulation as it aims at assigning travel demand to particular links. The input consists of a matrix that expresses the estimated demand (how many persons/vehicles want to travel from an origin to a destination); of a network for which relevant information of nodes and links are given such as maximum speed and capacity; and of a function that relates density and speed/travel time. The output is an assignment of each traveler to a route. Thus, route choice is a traveler-focused view to the assignment step of the traditional four-step process of traffic simulation [8]. Traditionally, route choice simulations have been done as discrete choice models based on stochastic user equilibrium concepts. On the other hand, in agent-based traffic simulation there are basically two goals behind route choice simulation models: integration of simple route choice heuristics into some larger context, and the analysis of self-organization effects of route choice with traffic information. In large scale traffic simulations agents are capable of more than just solving one phase of the overall travel simulation problem: they plan their activities, select the locations and departure times, and connect activities by selecting routes to the respective destinations. A prominent agent-based example is MATSim [1]. Here route choice is tackled using some shortest path algorithm. In [6] a software architecture that may support driver adaptivity is described.

Agent-based route choice simulation has been intensively applied to research concerning the effects of intelligent traveler information systems. What happens to the overall load distribution, if a certain share of informed drivers adapt? What kind of information is best? These are frequent questions addressed by agent-based approaches. In these approaches, route choice simulation has been based on game-theory, e.g. minority games. Examples for such research line, which includes adaptive and learning agents, can be found in [3] for the Braess paradox, in [7] for an abstract two-route scenario, or in [5] where a complex neural net-based agent model for route choice is presented regarding a three route scenario. In [9], a simple network for fuzzy-rule based routing (including qualitative decisions) is used.

One problem with these approaches is that their application in networks with more than two routes between two locations is not trivial. The first problem is that, to model the route choice simulation as an agent choice problem, an option set consisting of reasonable alternatives has to be generated. The problem of generation of routes for route choice models is well known in traditional discrete choice approaches. Different solutions have been suggested. Using the n shortest paths is often a pragmatic solution, yet it yields routes that differ only marginally. Additionally, all approaches, including all agent-based ones, consider one route as one complete option to choose. On-the-fly re-routing during the simulation has hardly been a topic for research. We suspect this happens for two reasons. The first is that the set of routes corresponds to the set of strategies or actions that an agent may select for e.g. game-theoretic analysis. Agents learn to optimize the selection of an action or a route but are not able to modify the set of known routes as this would mean creating a new strategy on the fly. The second reason is of practical nature. Travel demand generation, route choice, and driving simulation have been tackled as subsequent steps for which different software packages are used. Often, there is no technical possibility to allow drivers to do re-routing from arbitrary positions in the network using different software packages. Even sophisticated agent architectures like that proposed in [10] lack the technical basis for re-routing during the actual traffic simulation of driving. Therefore the study of the effects of drivers re-routing is still an open question.

3 Approach and Scenario

In order to address a non trivial network, we use a grid with 36 nodes, as depicted in Figure 1. All links are one-way and drivers can turn to two directions in each crossing. Although it is apparently simple, from the point of view of route choice and equilibrium computation, it is a complex one as the number of possible routes between two locations is high.

In contrast to simple two-route scenarios, it is possible to set arbitrary origins (O) and destinations (D) in this grid. For every driver agent, its origin and destination are randomly selected according to probabilities given for the links: to

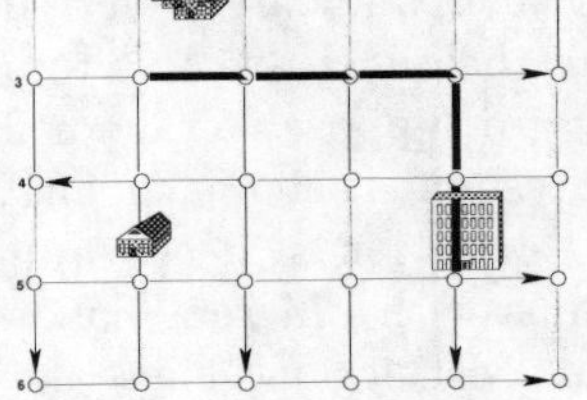

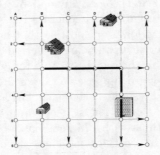

Fig. 1. 6x6 grid showing the main destination (E4E5), the three main origins (B5B4, E1D1, C2B2), and the "main street" (darker line)

render the scenario more realistic, neither the distribution of O-D combinations, nor the capacity of links is homogeneous. On average, 60% of the road users have the same destination, namely the link labeled as E4E5 which can be thought as something like a main business area. Other links have, each, 0.7% probability of being a destination. Origins are nearly equally distributed in the grid, with three exceptions (three "main residential areas"): links B5B4, E1D1, and C2B2 have, approximately, probabilities 3, 4, and 5% of being an origin respectively. The remaining links have each a probability of 1.5%. Regarding capacity, all links can hold up to 15 vehicles, except those located in the so called "main street". These can hold up to 45 (one can think it has more lanes). This main street is formed by the links between nodes B3 to E3, E4, and E5.

The control is performed via decentralized traffic lights. These are located in each node. Each of the traffic lights has a signal plan which, by default, divides the overall cycle time – in the experiments 40 time steps – 50-50% between the two phases. One phase corresponds to assigning green to one direction, either north/south or east/west. The actions of the traffic lights are to run the default plan or to prioritize one phase. Hence these particular strategies are fixed (always run the default signal plan) and greedy (to allow more green time for the direction with higher current occupancy).

Regarding the demand, the main actor is the simulated driver. In the experiments we have used 700 driver agents because this corresponds to a high occupancy of the network (72%) and proved a difficult case in a previous study [2].

This simple scenario goes far beyond simple two-route or binary choice scenario; we deal with route choice in a network with a dozens of possible routes. Additionally, the scenario captures properties of real-world scenarios, like interdependence of routes with shared links and heterogeneous capacities and demand throughout the complete network.

Every driver is assigned to a randomly selected origin-destination pair. The initial route is generated using a shortest path algorithm (from his origin to his destination), considering free flow. This means that the only route it knows in the beginning is the shortest path assuming no congestion. After generating this initial route, the driver moves through the network. Every link is modeled as a FIFO queue.

The occupancy of a link is the load divided by the capacity. It is used as cost or weight for some alternatives to generate the shortest path (see below). The lower the occupancy the lower the cost so that in a weighted shortest path generation, paths with lower occupancy are preferred. In every time step drivers behind the two first agents in the queue are able to perceive the traffic situation regarding the link that they are suppose to take next. If the occupation of this next link is higher than a threshold τ, the agent may trigger a replanning step.

Replanning involves the generation of a new route on the fly. This is done using shortest path algorithms with different flavors, from the link where the drivers wishes to replan to his destination. These flavors are summarized below:

UC ("uniform cost"): Every link between the current position and the destination has the same cost, except that the next link in the agent's plan is weighted badly to force it to avoid this link (after all, they should replan, not use the same old plan); however, if no other path is possible, the next link will be taken.

OCENL (occupancy-based, except next link): This second alternative differs from the first only by the costs that are used for computing the shortest path. Here, the current occupancy is used as cost for all links except for the next one which has a very high value so that it will be avoided if possible. This corresponds to a situation where agents have complete information about the current network status.

OCAL (occupancy-based for all links): In this third alternative the current occupancy of the next link is used as cost.

4 Experiments and Results

We have conducted several experiments using the alternative ways described in the previous section to re-generate routes on the fly. With these experiments the following questions can be addressed: In which context adaptation should be triggered? When should the driver re-evaluate its decision? What information is useful for finding the new route? How this form of driver agent adaption interferes with other agents adaptation (e.g. with intelligent traffic lights)?

In the following, we will discuss the cases in which traffic lights do not adapt (Sections 4.1 and 4.2), as well as cases where lights do adapt (Section 4.3). We also notice that these simulations run in a time frame of minutes.

4.1 Re-routing with Uniform Costs

In this experiment we have varied the re-planning threshold τ. If the perceived occupancy on the next link is higher than this threshold, re-routing is triggered. The procedure for computing the remaining route is the same as for computing

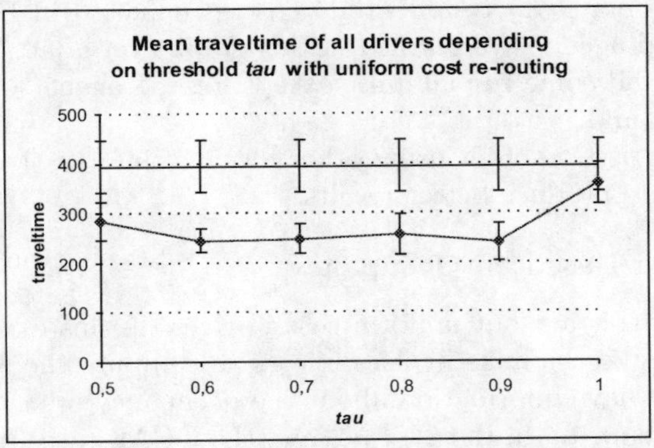

Fig. 2. Mean travel time over all drivers, varying the threshold for re-routing (τ); comparison of UC situation (circles) and no re-routing (dash)

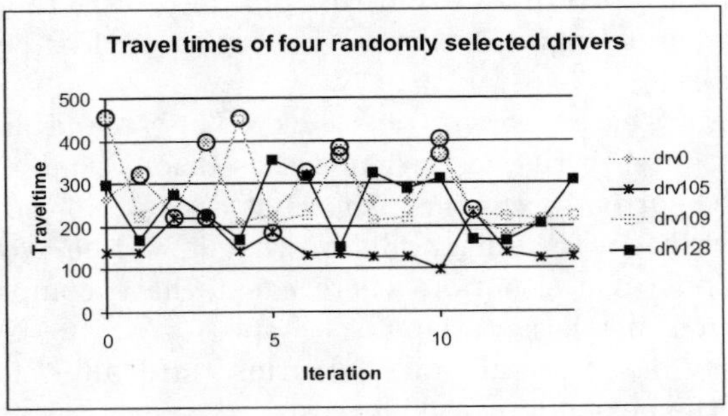

Fig. 3. Travel time for four randomly selected drivers. In rounds that are marked with small circles the driver re-planned (driver `drv128` re-plans in every round).

the initial route, namely shortest path assuming the same cost for every link. Figure 2 depicts the mean travel times over all drivers (mean of 20 runs with standard deviation between runs). In this figure, one can clearly identify that re-planning (even a quite basic one) is advantageous for the overall system. The distribution of load is improved when agents can avoid links with high occupancies. Interestingly, τ plays almost no role, except when $\tau = 1$ because in this case drivers re-plan only if the next link is completely occupied, which is of course an extreme situation. The fact that we observe no significant change in travel times with changing τ can be explained by the fact that occupancies are generally high due to the high number of agents in the network. A threshold lower or equal than 0.5 seems to be not as good as the higher ones as it triggers re-planning too often.

Having a deeper look into the dynamics of this basic simulation, it turns out that this overall improvement does not come at expense of only a few individuals. In an example simulation run with $\tau = 0.8$, in every iteration around half of the agents (371 ± 12) re-plan at least once. For the individual agents that replan, this may cause a quite severe individual cost as depicted in Figure 3 for four randomly selected agents. We marked in this graph every data point where the particular agent did some re-planning, except for the agent "drv128" that did replan in every round.

The question arises whether using other forms of information for computing the deviation route produce better results.

4.2 Re-routing Based on Occupancy

Instead of using arbitrary but uniform costs, drivers in this experiment use the current occupancy of the links in the network to compute the remaining route. To this aim, they have information about the occupancy of each link. Figure 4 depicts the results for both, the OCENL and the OCAL cases (mean of 20 runs depicted with standard deviation between runs).

These results show that allowing the drivers to access the current occupancies to compute the shortest path to their destination produces overall better mean

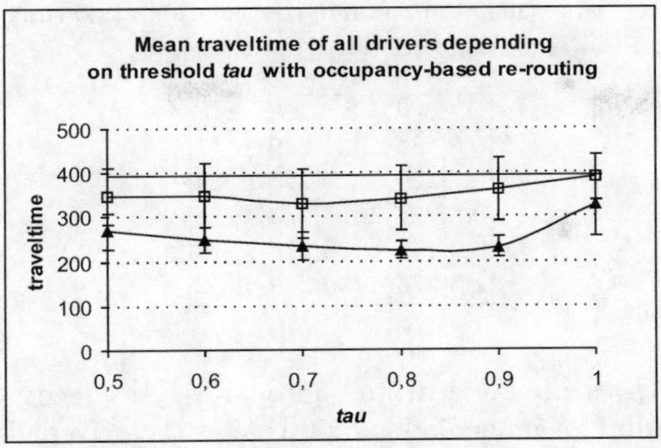

Fig. 4. Mean travel time when agents calculate the remaining of the route based on occupancy of links; comparison between three situations: no re-routing (dashes), OCENL (triangles), and OCAL (squares)

travel time than without online route adaptation. However, compared to the results observed in the UC situation, there is hardly any improvement, except for large values of τ.

In the OCAL the mean travel time is worse than OCENL for all values of τ. Taking standard deviations into consideration, there is even no significant difference from the corresponding situation where drivers cannot re-plan if they encounter a jam during their travel. This is due to the fact that the tradeoff between taking a deviation and the original route seems to be not high enough. Reasons for this might be that we use occupancy as link weight instead of some travel time forecast for computing the route from the current position.

Nevertheless, the next steps in our research agenda concern more intelligent forms of planning and re-planning, as discussed in Section 5. Because there seems to be no clear difference between UC and OCENL, we left re-routing techniques based on individually experienced travel times to future work as we expect no significant improvement in comparison to both techniques that we have examined here.

4.3 Re-routing Drivers in Adaptive Context

Adaptive and learning traffic light agents are able to improve the overall traffic flow. The main objective of this paper is to test whether adapting drivers can provide more improvement or if there are hindering interferences. To this aim, we have performed simulation experiments similar to the ones shown above, this time with adaptive traffic lights. As mentioned above, we selected a reactive form of adaptation called "greedy". The traffic light immediately increases the green time share for directions with higher occupancy. In our previous work, we showed in the same artificial scenario that we are using here, that this form of adaptivity produces at least similar results than other more elaborate forms

Table 1. Mean travel time (mean and standard deviation of 20 runs) for greedy traffic lights and drivers re-routing using UC and OCAL

τ	UC	OCAL
0.6	234 ± 21	239 ± 45
0.7	229 ± 27	223 ± 24
0.8	237 ± 27	224 ± 22
0.9	243 ± 38	218 ± 18
1.0	276 ± 39	289 ± 52

of learning (e.g. Q-learning or learning automata). As greedy traffic lights use the same information for immediate adaptation as the drivers described here for re-routing, we expect co-adaption effects to play a significant role.

The results (in terms of travel times) are given in Table 1 and in Figure 5. In Table 1, traffic lights use a greedy strategy and drivers use either the uniform costs (UC) or occupancy based re-routing (OCAL). In Figure 5, we compare the mean travel times for traffic lights with fixed and greedy strategies, while drivers re-route using the OCAL strategy. For sake of comparison, we notice that when drivers do not replan (i.e. always use their shortest path route computed at the beginning) and traffic lights act greedily, the overall mean travel time is 260 ± 45; thus it is higher than when drivers do re-plan indicating a positive co-adaption effect.

Analyzing Table 1, one can see that the lowest travel time is different for UC and for OCAL. Whereas the travel time under UC is best around $\tau = 0.7$, OCAL re-routing is best when $\tau = 0.9$. Both are very similar for situations when τ is low. For higher thresholds, OCAL becomes better (although not significantly). The reason might be that the OCAL re-routing uses the current information on occupancy. However when one link has high occupancy, it is very likely that nearby links also

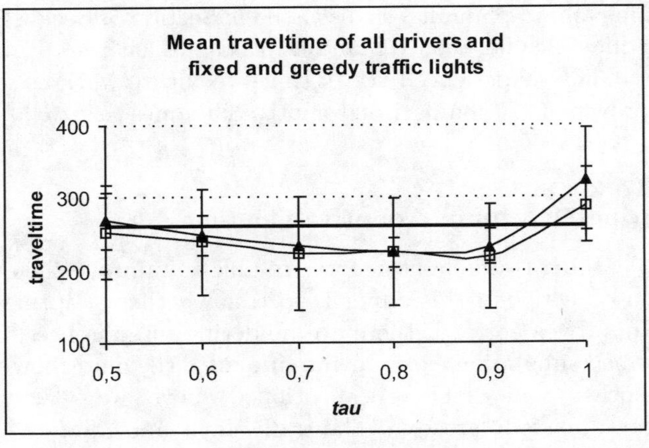

Fig. 5. Mean Travel time (mean of 20 runs with standard deviation) depending on the re-routing threshold τ, for fixed and greedy adaptive traffic lights

have high occupancy thus the driver has really not so much room to replan, but can use this room more intelligently when informed about it.

Finally, we can state that driver agents that are able to adapt their route on the fly as a reaction to traffic conditions may at least compensate an eventual lack of adaptivity by the traffic lights. If the drivers use more up-to-date information for generating their remaining routes, the situation improves slightly more. However, our experiments show that in the current driver re-routing model, the actual value of the threshold τ does not matter much – except for the extreme case of $\tau = 1$. Therefore, a further elaboration of the agent decision making model, turning it more realistic, may be helpful. This may show additional positive effects of adaptation in traffic scenarios thus permitting to really evaluate the effects of adaptive and learning traffic lights in realistic scenarios.

5 Conclusions and Future Work

In this paper we have analyzed the effects of drivers computing new routes on the fly. In most previous simulation scenarios route adaptation was only allowed before and after the actual driving. However, en-route modifications cannot be ignored in a realistic simulation of intelligent human decision making in traffic. In an artificial scenario that resembles all potentially problematic features of real-world scenarios, we have tested the effects of re-routing with results indicating that the overall system performance depends on different facets of the agents re-routing: on the criteria drivers use to decide whether to deviate from the originally planned route, and on the procedure they use to compute a new route. We have shown that re-routing may compensate less efficient traffic management and still increase the overall performance if traffic lights adapt to the traffic situation.

We are far from modeling intelligent human decision making about routing or even way-finding in traffic situations. However, this would be necessary for actually being able to evaluate re-routing effects in traffic networks. Our next steps will therefore comprise more intelligent ways regarding drivers perceiving that there is jam ahead and deciding whether to trigger a re-routing procedure. A simple extension is the integration of the status of other perceivable links into the decision.

Additionally we want to analyze the effects of an experience-based computation of the alternative route: a driver memorizes travel times that it has already experienced and uses this information to predict cost on the links that will potentially be contained in its deviation route. Also, we want to depart from the assumption that the complete network is known. Thus an important step in our research program is to use mental maps, thus moving from abstract game-theory based models to real-world application.

Acknowledgments

We thank CAPES and DAAD for their support to the bilateral project "Large Scale Agent-based Traffic Simulation for Predicting Traffic Conditions". Ana Bazzan is/was partially supported by CNPq and Alexander von Humboldt Stiftung.

References

1. Balmer, M., Cetin, N., Nagel, K., Raney, B.: Towards truly agent-based traffic and mobility simulations. In: Jennings, N., Sierra, C., Sonenberg, L., Tambe, M. (eds.) Proceedings of the 3rd International Joint Conference on Autonomous Agents and Multi Agent Systems, AAMAS, New York, USA, July 2004, vol. 1, pp. 60–67. IEEE Computer Society, New York (2004)
2. Bazzan, A.L.C., de Oliveira, D., Klügl, F., Nagel, K.: Adapt or not to adapt – consequences of adapting driver and traffic light agents. In: Tuyls, K., Nowe, A., Guessoum, Z., Kudenko, D. (eds.) ALAMAS 2005, ALAMAS 2006, and ALAMAS 2007. LNCS (LNAI), vol. 4865, pp. 1–14. Springer, Heidelberg (2008)
3. Bazzan, A.L.C., Klügl, F.: Case studies on the Braess paradox: simulating route recommendation and learning in abstract and microscopic models. Transportation Research C 13(4), 299–319 (2005)
4. Bazzan, A.L.C., Klügl, F., Nagel, K.: Adaptation in games with many co-evolving agents. In: Neves, J., Santos, M.F., Machado, J.M. (eds.) EPIA 2007. LNCS (LNAI), vol. 4874, pp. 195–206. Springer, Heidelberg (2007)
5. Dia, H., Panwai, S.: Modelling drivers' compliance and route choice behaviour in response to travel information. Special issue on Modelling and Control of Intelligent Transportation Systems, Journal of Nonlinear Dynamics 49(4), 493–509 (2007)
6. Illenberger, J., Flötteröd, G., Nagel, K.: Enhancing matsim with capabilities of within-day re-planning. Technical Report 07–09, VSP Working Paper, TU Berlin, Verkehrssystemplanung und Verkehrstelematik (2007)
7. Klügl, F., Bazzan, A.L.C.: Route decision behaviour in a commuting scenario. Journal of Artificial Societies and Social Simulation 7(1) (2004)
8. Ortúzar, J., Willumsen, L.G.: Modelling Transport, 3rd edn. John Wiley & Sons, Chichester (2001)
9. Peeta, S., Yu, J.W.: A hybrid model for driver route choice incorporating en-route attributes and real-time information effects. Networks and Spatial Economics 5, 21–40 (2005)
10. Rossetti, R., Liu, R.: A dynamic network simulation model based on multi-agent systems. In: Applications of Agent Technology in Traffic and Transportation, pp. 88–93. Birkhäser (2005)
11. Tumer, K., Wolpert, D.: A survey of collectives. In: Tumer, K., Wolpert, D. (eds.) Collectives and the Design of Complex Systems, pp. 1–42. Springer, Heidelberg (2004)
12. Wahle, J., Bazzan, A.L.C., Klügl, F., Schreckenberg, M.: Decision dynamics in a traffic scenario. Physica A 287(3–4), 669–681 (2000)

A Draughts Learning System Based on Neural Networks and Temporal Differences: The Impact of an Efficient Tree-Search Algorithm

Gutierrez Soares Caixeta and Rita Maria da Silva Julia

Department of Computer Science
Federal University of Uberlandia - UFU
Uberlandia, Brasil
gutierrez@umuarama.ufu.br, rita@ufu.br
http://www.ufu.br

Abstract. The NeuroDraughts is a good automatic draughts player which uses temporal difference learning to adjust the weights of an artificial neural network whose role is to estimate how much the board state represented in its input layer by NET-FEATUREMAP is favorable to the player agent. The set of features is manually defined. The search for the best action corresponding to a current state board is performed by minimax algorithm. This paper presents new and very successful results obtained by substituting an efficient tree-search module based on alpha-beta pruning, transposition table and iterative deepening for the minimax algorithm in NeuroDraughts. The runtime required for training the new player was drastically reduced and its performance was significantly improved.

Keywords: Draughts, Checkers, Temporal Difference Learning, Automatic Learning, Alpha-Beta Pruning, Transposition Table, Iterative Deepening, Table Hashing, Zobrist Key, Neural Network.

1 Introduction

The games are a great domain to study automatic learning techniques. The choice of draughts is due to the fact that it presents significant similarities with several practical problems, such as controlling the average number of vehicles over an urban network in order to minimize traffic jams and traveling time. To solve this kind of problem, an agent must be able to learn how to behave in an environment where the acquired knowledge is stored in an evaluation function, it must choose a concise set of possible attributes that best characterize the domain and, finally, it has to select the best action corresponding to a determined state [13].

In this context, Mark Lynch implemented the draughts playing program NeuroDraughts [12]. It uses an artificial neural network trained by the method of temporal difference (TD) learning and self-play with cloning, without any expert game analysis. In order to choose a good move from the current state, it employs the minimax algorithm. In addition, NeuroDraughts represents the

G. Zaverucha and A. Loureiro da Costa (Eds.): SBIA 2008, LNAI 5249, pp. 73–82, 2008.

game board states by means of a set of functions, called features, which captures relevant knowledge about the domain of draughts and uses it to map the game board (NET-FEATUREMAP mapping). In NeuroDraughts, the features set was manually defined.

In order to improve the NeuroDraughts abilities, Neto and Julia proposed the LS-Draughts [13]: an automatic draughts player which extends NeuroDraughts architecture with a genetic algorithm module that automatically generates a concise set of features, which are efficient and sufficient, for representing the game board states and for optimizing the training of the neural network by TD. The improvement obtained was confirmed by a tournament of seven games where LS-Draughts, with a more concise set of features, scored two wins and five draws playing against NeuroDraughts.

Still in order to improve the NeuroDraughts performance, the objective, here, is to exhibit the impact of replacing its search module (a minimax algorithm) by a very efficient tree-search routine which uses alpha-beta pruning, transposition table and iterative deepening.

The experimental results confirm that these improvements reduced more than 90% the execution time, which allowed to increase the depth search and to obtain a much better player in a reasonable training time.

Moreover, in a sequence of fourteen games against NeuroDraughts, the improved player scored five wins, eight draws and one loss, while against LS-Draughts, it scored four wins, eight draws and two losses.

2 Background on Computer Programs for Draughts

Chinook is the most famous and strongest draughts player in the world [2], [3], [4], [5]. It is the world man-machine draughts champion and uses a linear handcrafted evaluation function (whose role is to estimate how much a board state is favorable to it) that considers several features of the game, for example: piece count, kings count, trapped kings, turn and runaway checkers. In addition, it has access to a library of opening moves from games played by grand masters and to an endgame database, a computer-generated collection of positions with a proven game-theoretic value, actually, a collection of 39 trillion positions (all positions with ≤ 10 pieces on the board). This collection is compressed into 237 gigabytes, an average of 154 positions per byte. To choose the best action to be executed, Chinook uses a parallel iterative alpha-beta search with transposition tables and the history heuristic [3]. At the U.S. Open [2], the program averaged 20-ply searches (a ply is one move executed by one of the players).

The first great experiment in automatic learning for draughts is devoted to Samuel in 1959 [1]. He used a deterministic learning procedure where boards were evaluated based on a weighted polynomial comprising features considered important for generating good moves. This work remains a milestone in artificial intelligence research, since it was the pioneering automatic draughts player that succeeded competing against good human players.

Fogel [14] explored a co-evolutionary process to implement an automatic draughts player which learns without human expertise. He focused on the use of a population of neural networks, where each network counts on an evaluation function to estimate the quality of the current board state. In a tournament of 10 games against a novice level player of Chinook, its best neural network, called Anaconda, obtained 2 wins, 4 losses and 4 draws, which allowed it to be ranked as *expert*.

TD learning [6], [7] has emerged as a powerful reinforcement learning technique for incrementally tuning parameters. The basic idea is: a learning agent receives an input state that is continuously modified by means of the actions performed by the agent. Each current state is evaluated based on the previous one. At the end of the process, it outputs a signal and then receives a scalar *reward* from the environment indicating how good or bad the output is (reinforcement). That is, the learner is rewarded for performing well (positive reinforcement) and, otherwise, it is punished (negative reinforcement).

Tesauro [6] introduced the use of this techniques in games proposing the TD-Gammon, a very efficient automatic gammon player corresponding to a neural network that learned by TD. In chess, KnightCap [7] learned to play chess against humans on the internet also using TD. In the draughts domain, in spite of the undeniable efficacy that Chinook obtained by manual adjustment of its evaluation function during several years, Schaeffer [4] showed that TD learning is capable of competing with the best human effort in the task of adjusting the evaluation function. In fact, Schaeffer's automatic player based on a neural network trained by TD proved to be a very good draughts player [4].

NeuroDraughts [12] and LS-Draughts [13] also represent efficient draughts player programs based on a neural network trained by TD. As the present work is inspired in both of them, the next section summarizes their general architectures.

3 LS-Draughts and NeuroDraughts Architectures

Figure 1 shows the general architecture of LS-Draughts. The **Artificial Neural Network** module is the core of LS-Draughts. It corresponds to a three layer feed-forward network whose output layer is composed of a single neuron. The role of this network is to evaluate to what extent the board state, represented by NET-FEATUREMAP in the input layer, is favorable to the agent (it is quantified by a real number comprised between 0 and 1, called **prediction**, which is available in the output of the third layer neuron).

The **Tree-Search Routine** module selects, using minimax and the computed predictions, the best move corresponding to the current board state. The **Genetic Algorithm** module is the feature generator, that is, it generates individuals which represent subsets of every available features in the NET-FEATUREMAP mapping. These subsets will be used by the **Game Board** module to represent the board state in the network input. The **TD Learning** module is responsible for training the player neural network.

According to Neto and Julia [13], the LS-Draughts expands the NeuroDraughts as it automatically generates a concise set of features using genetic algorithms. The training strategy is also modified, since, in the former, several individuals (players whose boards are represented by different sets of features) are trained, whereas, in the latter, just one is trained. The purpose of LS-Draughts is to analyze how much the insertion of a module capable of automating the choice of the features by means of genetic algorithms could improve NeuroDraughts performance.

Then, the NeuroDraughts architecture can also be illustrated by Figure 1 by cutting off the **Genetic Algorithm** module.

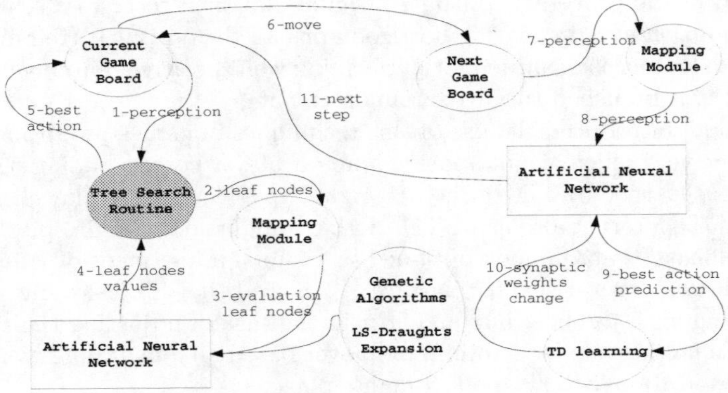

Fig. 1. The LS-Draughts Architecture

4 Improving the NeuroDraughts System by the Insertion of an Efficient Tree-Search Algorithm

At this point, it is possible to understand the objective of this work: the authors, taking advantage of search space properties with no explicit domain-specific knowledge, substitute a very efficient tree search module for the non-optimized search routine of NeuroDraughts. This modification will allow to analyze the impact of this new search strategy in the general performance of good automatic players which learn without counting on human expertise. The new search module is based on the following techniques: alpha-beta pruning, transposition table and iterative deepening.

The following facts motivated this work: first, there is too much expertise in Chinook [14] (that is why the authors preferred to use the non-supervised strategy of NeuroDraughts); second, TD learning opens up new opportunities for improving program abilities [4]; finally, taking into account the constraints to be respected in the implementation of an intelligent agent, it is not usually possible to spend several years handcrafting game functions, such as in the case of Chinook [3].

The architecture of the new player obtained is analogous to the one of NeuroDraughts summarized in the previous section, except for the improved **Tree-Search Routine** module, as described below.

4.1 Adapting Fail-Soft Alpha-Beta Pruning to the Player

In NeuroDraughts, the evolution of a game is represented by a tree of possible game states. Each node in the tree represents a board state and each branch represents one possible move [15]. To choose the best action corresponding to a current board state, NeuroDraughts uses the minimax algorithm with a max depth of search of 4-ply. This max depth of search is called look-ahead and 4-ply was chosen in order to satisfy the constraints imposed by the limitation of computational resources.

The minimax algorithm can be summarized as following: it is a recursive algorithm for choosing the best move performing a simple left-to-right depth-first traversal on a search tree [11] [16]. That is, given a node n, if n is a *minimizer* node, n receives the prediction value associated to the smallest child. Similarly, if n is a *maximizer* node, n receives the prediction value associated to the greatest child. For instance, see the left tree in the figure 2: at the deepest max level, the values 0.2 and 0.3 were bubbled up (black arrows) and, at min level, the value 0.4 was chosen.

The minimax algorithm examines more board states than necessary. So, it is common to use the alpha-beta algorithm to eliminate sections of the tree that can not contain the best move [15], such as discussed next.

The alpha-beta receives the following parameters: a board state, alpha and beta. Alpha and beta delimit the interval to which the prediction value of the best move corresponding to the input board state is supposed to belong to. The evaluation of the children of a *minimizer* node n can be interrupted as soon as the prediction computed for one of them (let's call it *besteval*) is less then alpha (**alpha prune**). For example, in the right tree of figure 2, the alpha-beta algorithm, differently from minimax, detects that is not necessary to compute the prediction associated to the highlighted nodes. Therefore, the best move

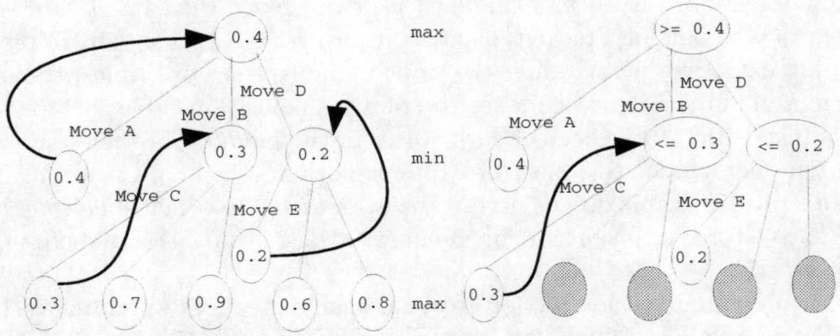

Fig. 2. Left: tree-search expanded by minimax algorithm. Right: tree-search expanded by fail-soft alpha-beta algorithm.

(Move A) is found much faster. Analogously, the evaluation of the children of a *maximizer* node n can be interrupted as soon as the alpha-beta algorithm detects that the prediction computed for one of them (*besteval*) is greater then beta (**beta pruning**). The meaning of the value returned by the alpha-beta algorithm as a prediction for the current board state depends on which variant of alpha-beta algorithm the system has been implemented: fail-soft or hard-soft version [10], [9].

In the hard-soft alpha-beta approach, if n is a *maximizer* node, whenever ***besteval*** $\geq$ **beta**, alpha-beta returns **beta**. Similarly, if n is a *minimizer* node, whenever ***besteval*** $\leq$ **alpha**, alpha-beta returns **alpha**. Then, the returned value represents either one limit on the minimax value or the limit imposed by the alpha-beta range. For instance, in the right tree of figure 2, when the algorithm looks at Move C, it would return the 0.4 (instead of 0.3) indicating that Move B would represent a value less than or equal to 0.4. The value 0.4 is imposed by the alpha-beta range.

In the fail-soft alpha-beta approach, if n is a *maximizer* node, whenever ***besteval*** $\geq$ **beta**, alpha-beta returns ***besteval***. Similarly, if n is a *minimizer* node, whenever ***besteval*** $\leq$ **alpha**, alpha-beta returns ***besteval***. This version returning a bound on true minimax value is called fail-soft alpha-beta because a fail high or fail low still returns useful information [10]. Then, the returned value always represents one limit on the minimax value. For instance, in the right tree of figure 2, when the algorithm looks at Move C, it returns the score 0.3 (despite of the fact that the 0.3 is out of alpha-beta range), meaning that move B represents a value at most equals to 0.3.

The both versions of alpha-beta algorithm returns, at the root, the same best action to be executed. In spite of this, the fail-soft version is essential in the proposed learning system in order to integrate with the transposition table, like described below.

4.2 Taking Advantage of Transposition Table and Iterative Deepening in the Search Module

The NeuroDraughts and the LS-Draughts reach the same board state several times in a game. For instance, a single checker can reach the same board square in 2 moves, by advancing first left, then right, or first right, then left. In order to avoid doing extra work expanding the same game tree several times, the search routine module in the figure 1 makes use of a transposition table.

The transposition table keeps board states in the memory in such a way as to indicate whether a game tree must be expanded or not (the board states available in the transposition table do not need to be evaluated). In order to check a board state against stored information in memory, a hash value is assigned to each board state [15].

There is a common technique for creating hash codes corresponding to board games that uses a set of fixed-length random bit (number) patterns stored for each possible state of each possible board square [8], [15]. The technique is called **zobrist hashing** and the bit (number) pattern is called **zobrist key**.

Entry	Piece	Random Sequence	Square
1	black man	14787540466645868636	Square 1
2	red man	2120251484556677534	
3	Black King	584882445155849028	
4	Red King	3760951787791404667	
...	...	...	...
125	black man	9487810991141940225	Square 32
126	red man	14639527447367762922	
127	Black King	8795549574004575914	
128	Red King	18030604617695974466	

Fig. 3. A Zobrist Key: a set of fixed-length (64 bits) random bit patterns

Draughts has 32 squares and each square can be empty or have 1 of 2 different pieces on it, each of 2 possible colors. Then, the **zobrist key** for draughts needs to have 32 x 2 x 2 = 128 entries. In this case, the **Tree-Search Routine** module, showed in the figure 1, makes use of the **zobrist key** in the figure 3.

In order to guarantee the randomness of the random sequence (third column in figure 3), it was used the Quantum Random Bit Generator [17], since the C language *rand* function does not perform well [15].

Using the **zobrist key**, the following process is used to create a hash code to be assigned to a board state: the value for each board square is determined based on its state (each line in the figure 3), then, a XOR operation is performed with these values to create the final hash code. This method has several advantages: it is reasonably collision resistant, it is easily implemented and the generated values are easily computed, once they can be incrementally updated [8].

The hash code assigned to a board state is the first field stored in a transposition table entry and is represented as following:

Example of a transposition table entry

```
struct TranspTable{
   int64    hash_value;
   int      score_type;
   int      score_value;
   int      depth;
   MOVE     best_move;
}
```

Moreover, when a tree is expanded by the alpha-beta routine, the algorithm output value, the search depth and the best move to be executed are stored in the following fields, respectively: ***score_value***, ***depth*** and ***best_move***. The ***depth*** field is responsible for assuring that the prediction for a board state is sufficiently accurate (for example, if the search has a look-ahead of ten moves, a table entry that holds a prediction computed from a look-ahead of three moves is not useful).

Furthermore, the alpha-beta routine rarely outputs the exact minimax value of a node (the board state represented by ***hash_value*** field), but the fail-soft variant always outputs the lower bound or the upper bound on the true minimax value (**prediction**). Then, it is important to store a flag that indicates what the ***score_value*** field means. For example, if the ***score_value*** field contains the value 0.3 and the ***score_type*** field contains the *exact* flag, this means that the value of the node is exactly 0.3 (that is, the **prediction** of the current board state is exactly 0.3). If the ***score_type*** field contains the *fail_low* flag, this means that the value of the node is at most 0.3. Similarly, if the ***score_type*** field contains the *fail_high* flag, this means that the value of the node is at least 0.3. So, the score type, or flag, is also recorded in the ***score_type*** field.

Iterative deepening is used in conjunction with *TranspTable*. The alpha-beta routine is called repeatedly with increasing depth until either the established time is over or the search reaches the maximum look-ahead previewed. In spite of the fact that this method apparently waste time performing shallow searches instead of just one deep search, actually it allows to improve the search efficiency, once the former iterations are used to obtain a higher quality of move ordering what allows more cutoffs [10].

Concluding, the tree-search module (figure 1), using *TranspTable*, avoids that a same board state is repeatedly and unnecessarily evaluated.

5 Experimental Results

This section presents the improvement obtained in NeuroDraughts by optimizing its search strategy. The left table in the figure 4 shows the results of comparing the training (learning) process of three distinct players: the original NeuroDraughts, a modified version of NeuroDraughts where the alpha-beta algorithm replaces the minimax module and the improved player proposed here (search module based on alpha-beta pruning and transposition table). The parameter taken into account for the comparison was the search execution time. Note that the improved player and the alpha-beta player spent, respectively, 2.27% and 5.83% of the time required by the original NeuroDraughts. Furthermore, compared to the alpha-beta player, the improved player only required 39% of its execution time, whereas NeuroDraughts required 1715.67% of the same execution time. Each player was trained using self-play with cloning strategy of training [12], [13], in 2 tournaments of 10 games each one, using depth of search of 8-ply and the following hardware configuration: Intel Pentium D CPU 2.66 GHz and 4 GB of RAM.

In the same way, the right table in the figure 4 also illustrates a competition between training process results, but now, considering an extended training composed of 10 tournaments of 200 games each one. Note that, in this case, the original NeuroDraughts player was not trained because of its impractical long runtime.

All these results confirm the significant contribution of the alpha-beta pruning and the impressive contribution of combining it with transposition tables during the training process of automatic players.

In order to check the performance of the new system proposed here, 2 tournaments of 14 games was executed with its best player: one, against the best player obtained by training NeuroDraughts and, other, against the best player

Algorithm	Execution Time (minutes)	(%) of Minimax Routine	(%) of Alpha-Beta Routine
MiniMax	441.27	100.00%	1715.67%
Alpha-Beta	25.72	5.83%	100.00%
Alpha-Beta + Transp Table	10.03	2.27%	39.00%

Algorithm	Execution Time (hours)	(%) of Alpha-Beta Routine
Alpha-Beta	46.52	100.00%
Alpha-Beta + Transp Table	16.79	36.09%

Fig. 4. Left: 2 matches of 10 games and depth equals to 8. Right: 10 matches of 200 games and depth equals to 8.

obtained by training LS-Draughts. Both results were very favorable to the proposed system: against NeuroDraughts, 5 wins, 8 draws and 1 loss; against LS-Draughts, 4 wins, 8 draws and 2 losses. During these tournaments, the trained player proposed here could, additionally, count on a iterative deepening routine for obtaining a better move ordering and for interrupting the search whenever the algorithm tended to run out of time.

6 Conclusions and Future Works

This paper presented how much an efficient search module based on alpha-beta pruning, transposition table and iterative deepening can improve the learning performance of intelligent automatic players.

This strategy was used to replace the minimax search algorithm of NeuroDraughts - a very good draughts player corresponding to a neural network whose learning process is conduced by temporal differences and whose board states are represented by a set of features manually defined.

A competition was executed, where the proposed draughts system played several games against the original NeuroDraughts and against LS-Draughts (another improved version of NeuroDraughts which automatically generates a concise set of features using genetic algorithms). The results of this competition confirm the high improvement obtained in the learning process as well as in the performance of the player agent. As future works, the authors intend to introduce the same efficient search module in the LS-Draughts, which corresponds to analyze the impact of inserting, in NeuroDraughts, both extensions: genetic algorithms and efficient tree-search algorithm based on alpha-beta pruning, transposition table and iterative deepening.

References

1. Samuel, A.L.: Some Studies in Machine Learning Using the Game of Checkers. IBM Journal on Research and Development, 210–229 (1959)
2. Schaeffer, J., Culberson, J., Treloar, N., Knight, B., Lu, P., Szafron, D.: A World Championship Caliber Checkers Program. Artificial Intelligence 53, 2–3 (1992)
3. Schaeffer, J., Lake, R., Lu, P., Bryant, M.: CHINOOK: The World Man-Machine Checkers Champion. AI Magazine 17(1), 21–29 (1996)

4. Schaeffer, J., Hlynka, M., Jussila, V.: Temporal Difference Learning Applied to a Highperformance Game-Playing Program. In: International Joint Conference on Artificial Intelligence, pp. 529–534 (2001)
5. Schaeffer, J., Burch, N., Bjornsson, Y., Kishimoto, A., Muller, M., Lake, R., Lu, P., Sutphen, S.: Checkers Is Solved (2007)
6. Tesauro, G.: Temporal Difference Learning and TD-Gammon. Communications of the ACM, 58–68 (1995)
7. Baxter, J., Tridgell, A., Weaver, L.: Knightcap: A Chess Program that Learns by Combining TD(λ) with Game-Tree Search. In: Machine Learning, Proceedings of the Fifteenth International Conference, Madison Wisconsin, pp. 28–36 (1998)
8. Zobrist, A.L.: A Hashing Method with Applications for Game Playing. Tech. Rep. 88, Computer Sciences Department, University of Wisconsin, Wisconsin (1969)
9. Shams, R., Kaindl, H., Horacek, H.: Using Aspiration Windows for Minimax Algorithms. In: Proceedings of the 8th International Joint Conference on Artificial Intelligence, Sydney, Australia, pp. 192–197 (1991)
10. Plaat, A.: Research Re: Search & Re-Search. PhD thesis, Erasmus University, Amsterdam (1996)
11. Schaeffer, J., Plaat, A.: New Advances in Alpha-Beta Searching. In: Proceedings of the 1996 ACM 24th Annual Conference on Computer Science, Philadelphia, Pennsylvania, United States, pp. 124–130 (1996)
12. Lynch, M.: An Application of Temporal Difference Learning to Draughts. Final Year Project Report, Department of Computer Science and Information Systems, University of Limerick, Ireland (1997)
13. Neto, H.C., Julia, R.M.S.: LS-DRAUGHTS - A Draughts Learning System based on Genetic Algorithms, Neural Network and Temporal Differences. In: 2007 IEEE Congress on Evolutionary Computation (CEC 2007), pp. 2523–2529. Research Publishing Services (RPS), Cingapura (2007)
14. Chellapilla, K., Fogel, D.: Anaconda Defeats Hoyle 6-0: A Case Study Competing an Evolved Checkers Program against Commercially Available Software. In: Proceedings of the 2000 Congress on Evolutionary Computation, pp. 857–863. IEEE Press, Piscataway (2000)
15. Millington, I.: Artificial Intelligence for Games. Morgan Kaufmann, San Francisco (2006)
16. Russell, S.R., Norvig, P.: Artificial Intelligence: A Modern Approach. Prentice Hall, New Jersey (1995)
17. Quantum Random Bit Generator Service, `http://random.irb.hr`

An Experimental Approach to Online Opponent Modeling in Texas Hold'em Poker

Dinis Felix[1] and Luís Paulo Reis[1,2]

[1] FEUP - Faculty of Engineering of the University of Porto
felixdinis@gmail.com
[2] LIACC – Artificial Intelligence and Computer Science Lab., University of Porto, Portugal
lpreis@fe.up.pt

Abstract. The game of Poker is an excellent test bed for studying opponent modeling methodologies applied to non-deterministic games with incomplete information. The most known Poker variant, Texas Hold'em Poker, combines simple rules with a huge amount of possible playing strategies. This paper is focused on developing algorithms for performing simple online opponent modeling in Texas Hold'em. The opponent modeling approach developed enables to select the best strategy to play against each given opponent. Several autonomous agents were developed in order to simulate typical Poker player's behavior and one other agent, was developed capable of using simple opponent modeling techniques in order to select the best playing strategy against each of the other opponents. Results achieved in realistic experiments using eight distinct poker playing agents showed the usefulness of the approach. The observer agent developed is clearly capable of outperforming all its counterparts in all the experiments performed.

Keywords: Opponent Modeling, Texas Hold'em, Poker, Autonomous Agents.

1 Introduction

Incomplete knowledge, risk management, opponent modeling and dealing with unreliable information are topics that identify Poker as an important research area in Artificial Intelligence (AI). Unlike in games of perfect information, in the game of Poker, players face hidden information resulting from the opponents' cards and future actions. In such a domain, to be successful, players face the need to use opponent modeling techniques in order to understand and adapt themselves to the opponents playing style [1] [2].

In a multi-player game with imperfect knowledge, where multiple competing agents must deal with risk management, unreliable information and deception, agent modeling is an essential element in successful agent play. In this kind of environment, agents act under uncertainty, and a crucial issue is to have a good opponent modeling (OM) system, learning and problem solving capabilities.

Opponent modeling allows determining a likely probability distribution for the opponent's hidden cards. However, the huge amount of possible playing strategies in Poker makes opponent modeling a very hard task in this domain.

G. Zaverucha and A. Loureiro da Costa (Eds.): SBIA 2008, LNAI 5249, pp. 83–92, 2008.

The main goal of this work was to prove that a poker agent that considers the opponent behaviour has better results, against players that use typical poker playing strategies, than an agent that doesn't, even when playing the same global betting strategy.

The rest of the paper is organized as follows. Section 2 describes games with incomplete information, Texas Hold'em Poker and some related work. Section 3 describes the opponent modeling strategies developed. Section 4 describes the Poker playing autonomous agents developed and section 5 the results achieved in controlled experiments. Section 6 contains the conclusions of the paper and pointers to future work.

2 Games with Incomplete Information

Games have proven to be both interesting and rewarding for research in Artificial Intelligence (AI). Many success stories like Chinook (checkers) [3], Logistello (Othello) [4], Deep Blue [5] and Hydra [6] (chess), TD Gammon (backgammon) [7], and Maven (Scrabble) [8] have demonstrated that computer programs can surpass all human players in skill. Games such as Poker are difficult because of the elements of imperfect information and partial observability [9].

Games with incomplete information are games where the player does not have complete knowledge of the entire game state. In Poker, the players only have access to the information of their own cards. Predicting opponent cards, probabilities for possible future card combinations and future opponent moves is a challenge in the Artificial Intelligence domain. Poker is also a stochastic game because the shuffling of the deck introduces the chance element into the game state.

Von Neumann introduced game theory [10] in 1940s and has since become one of the foundations of modern economics [11]. He used the game of poker as a basic model for 2-player zero-sum adversarial games, and proved the first fundamental result, the famous Minimax Theorem. However, all reasoning in poker must be probabilistic, as things are rarely ever certain. Also, the cumulative sum of a series of games matter more than any individual game [12][13]. Poker is also a non-cooperative multi-player game. Although multi-player games are inherently unstable, due in part to the possibility of coalitions (i.e., teams), those complexities are minimized in a non-cooperative game such as Poker [14].

Poker is a popular type of card game in which players bet on the value of the card combination ("hand") in their possession, by placing a bet into a central pot. The winner is the one who holds the hand with the highest value according to an established hand rankings hierarchy, or otherwise the player who remains "in the hand" after all others have folded. The game has many variations, all following a similar pattern of play. Depending on the variant, hands may be formed using cards, which are concealed from others, or from a combination of concealed cards and community cards. The hand ranking hierarchy starts whith Royal Flush, the highest of all poker hands (10, J, Q, K, A of the same suit), then Straight Flush (five cards in consecutive numerical order, all of the same suit), Four of a Kind (four cards of the same value and any other card), Full House (three cards of the same value and another two cards that form a pair), Flush (five non-consecutive cards of the same suit), Straight (five consecutive cards, but not of the same suit.), Three of a Kind (three cards of the same value, and two supporting cards that are not a pair.), Two Pair (two sets of pairs, and another random card.), One Pair (two cards of the same value and three random supporting cards.)

Texas Hold'em is the most popular Poker game in the world, and is thus the variant of Poker considered for this project. Hold'em is a community card game where each player may use any combination of the five community cards and the player's own two hole cards to make a poker hand, in contrast to Poker variants like Stud or Draw where each player holds a separate individual hand.

This project is based on previous work from the University of Alberta [15] and Billings [16, 17, 18] and a previously developed poker simulator multi-agent system [19].

There are 1326 possible hands prior to the flop. The value of one of these hands is called an income rate and is based on an off-line computation that consists of playing several million games where all players call the first bet [16, 17]. The basic betting strategy after the flop is based on computing the hand strength (HS), positive potential (PPot), negative potential (NPot), and effective hand strength (EHS) of agent's hand relative to the board. EHS is a measure of how well the agent's hand stands in relationship to the remaining active opponents in the game. The hand strength (HS) is the probability that a given hand is better than that of an active opponent. Suppose an opponent is equally likely to have any possible two hole card combination. Thus it is possible to calculate the hand strength as:

```
HandStrength(ourcards, boardcards) {
  ahead = tied = behind = 0
  ourrank = Rank(ourcards, boardcards)
  /*Consider all two-card combinations of remaining cards*/
  for each case(oppcards)
  {
       opprank = Rank(oppcards, boardcards)
       if (ourrank>opprank) ahead += 1
         else if (ourrank==opprank) tied += 1
              else behind += 1
  }
  handstrength = (ahead+tied/2) / (ahead+tied+behind)
  return(handstrength)
}
```

After the flop, there are still two more board cards to be revealed. On the turn, there is one and it's essential to determine the potential impact of these cards. The positive potential (PPot) is the chance that a hand that is not currently the best improves to win at the showdown. The negative potential (NPot) is the chance that a currently leading hand ends up losing.

PPot and NPot are calculated by enumerating over all possible hole cards for the opponent, like the hand strength calculation, and also over all possible board cards.

```
HandPotential(ourcards,boardcards,player_classification){
  int array HP[3][3], HPTotal[3]       /* initialize to 0 */
  ourrank = Rank(ourcards, boardcards)
  /*Consider all two-card combinations of remaining cards*/
  for each case(oppcards) {
      opprank = Rank(oppcards,boardcards)
      if(ourrank>opprank) index = ahead
```

```
            else if(ourrank=opprank) index = tied
                 else index = behind
        HPTotal[index] += 1
        /* All possible board cards to come. */
        for each case(turn)        {
               for each case(river) {
                  board = [boardcards,turn,river]
                  ourbest = Rank(ourcards,board)
                  oppbest = Rank(oppcards,board)
                  if(ourbest>oppbest) HP[index][ahead]+=1
                    else if(ourbest==oppbest)HP[index][tied]+=1
                         else HP[index][behind]+=1
               }
        }
    }
    /* PPot: were behind but moved ahead. */
    PPot = (HP[behind][ahead] + HP[behind][tied]/2
          + HP[tied][ahead]/2) / (HPTotal[behind]+HPTotal[tied]/2)
    /* NPot: were ahead but fell behind. */
    NPot = (HP[ahead][behind] + HP[tied][behind]/2
          + HP[ahead][tied]/2) / (HPTotal[ahead]+HPTotal[tied]/2)
     return(PPot,NPot)
}
```

The effective hand strength (EHS) combines hand strength and potential to give a single measure of the relative strength hand against an active opponent. One simple formula for computing the probability of winning at the showdown is:

$$Pr(win) = HS \times (1 - NPot) + (1 - HS) \times PPot \tag{1}$$

Since the interest is the probability of the hand is either currently the best, or will improve to become the best, one possible formula for EHS sets NPot = 0, giving:

$$EHS = HS + (1 - HS) \times PPot \tag{2}$$

These betting strategies, divided in betting strategy before and after the flop [16], were developed at University of Alberta [18] and are enough to develop a basic agent capable of playing poker.

3 Opponent Modelling

No poker strategy is complete without a good opponent modeling system [20]. A strong Poker player must develop a dynamically changing (adaptive) model of each opponent, to identify potential weaknesses. In Poker, two opponents can make opposite kinds of errors and both can be exploited, but it requires a different response for each [16]. The Intelligent Agent developed in this project observes the moves of the other players at the table. There are many possible approaches to opponent modeling [2,13,21], but in this work the observation model is based on basic observations of the starting moves of the players, so it could be created a fast, online estimated guess of their starting hands in future rounds.

3.1 Loose/Tight and Passive/Aggressive

Players could be classified generally in four models that depend of two variables: loose/tight and passive/aggressive. Knowing the types of hole cards various players tend to play, and in what position, is probably the start point of opponent modeling. Players are classified as loose or tight according to the percentage of hands they play. These two concepts are obtained analyzing the percentage of the time a player puts money into a pot to see a flop in Hold'em - VP$IP. The players are also classified as passive or aggressive. These concepts are obtained analyzing the Aggression Factor (AF) which describes the nature of a player. Figure 1 shows the target playing area for the agents developed as a factor of the number of starting hands played and the bet/raise size and frequency.

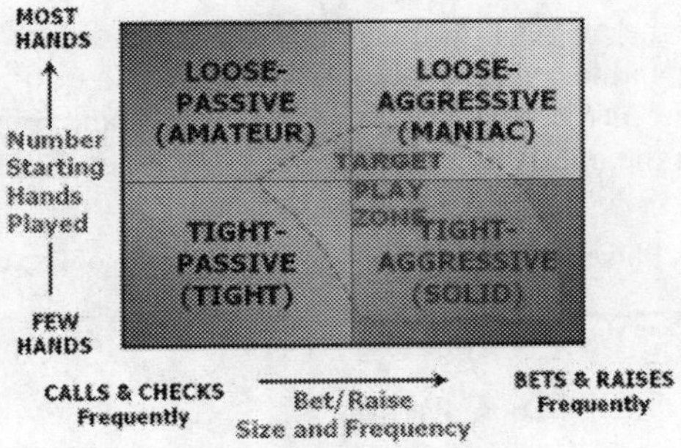

Fig. 1. Player Classification based on the number of starting hands played (VP$IP) and bet/raise size and frequency (AF)

3.2 Sklansky Groups

One of the most difficult and yet crucial decisions when playing Texas Hold'em is whether to even play or not the starting hand. David Sklansky and Mason Malmuth, co-authors of "Hold'em Poker and Advanced Hold'em Poker", were the first to apply rankings to the starting 2-card hands, and place them in groupings with advice on how to play those groups [22,23].

There are some computer simulations developed to test Sklansky's hand rankings that suggests some alterations. But in general, the classification is very similar. Considering a player loose/tight behavior and the Sklansky groups, it is easy to conclude what starting hands a tight player usually plays. If the VP$IP of the player is bellow 28%, he is probably playing most of the hands from higher groups and rarely from the other groups. On the other hand, if a player is a loose player, he's probably playing more hands from lower groups than a tight player. With this simple analysis, it is easy to exclude some of the hands that our opponents probably don't play. The percentage defined here, 28%, is an estimated approach that classifies players in loose or thight style. When using different player classification, with sub-levels for loose and tight (i.e. slightly loose or very loose), this percentage should be adapted.

4 Poker Playing Autonomous Agents

Based on the player classification developed, 8 intelligent agents were created, two for each player style:

- Two Loose Aggressive Agents (Maniac and Gambler);
- Two Loose Passive Agents (Fish and Calling Station);
- Two Tight Aggressive Agents (Fox and Ace);
- Two Tight Passive Agents (Rock and Weak Tight).

A general observer agent was also created capable of keeping the information of every move made from the opponents and calculating playing information like the VP$IP and AF of each opponent in every moment of the game. The opponents are classified into 4 types of players. So, an opponent with VP$IP above 28% is considered loose, otherwise, the player is considered tight. With an AF above 1, the player is considered aggressive and less than 1 is considered passive (table 1).

When the observer's turn comes, he knows which of the opponents are in the game and predicts, based on the available information, what kind of player they are.

Table 1. Player Classification considered by the observer agent

	AF<=1	AF>1
VP$IP>=28%	Loose Passive (classification1)	Loose Aggressive (classification2)
VP%IP<28%	Tight Passive (classification3)	Tight Aggressive (classification4)

After player classification the agent consider a different range of possible hands for different opponents. These considerations are based in the study of each kind of poker player. A general consideration is that tight players have a small range of possible hands than loose agents.

In order to pass this information to Hand Strength calculation, for each player is determined a parameter that was called "sklansky". This parameter is a float number that represents the lowest value of a hand that belongs to the most probable range of hands that the player plays with that specific movement (call or raise).

With conscience that many times the correct hand of the opponent is wrongly ignored, the better approach of Effective Hand Strength calculation given with this technique should give a better result that compensates this.

```
Sklansky(player_classification, player_move) {
  random = (rand() % 10) + 1;
  switch(player_classification) {
    case(1): /*loose passive*/
      if(player_move==raise) {
          if(random<=3){return 26.2;} /*last hand from group3*/
                  else {return 44.2;} /*last hand from group1*/
      }
```

```
      else if(player_move==call)
      {return -100.0;} /* all the possible hands */
   case(2):
      ...
```

The calculation of the "sklansky" parameter considers two variables for each opponent analyzed. The first is the "player classification", each one of the opponents are classified as one of the four kind of players described in table 1. The second parameter is "player move", that is the action made by the opponent in the pre-flop round. Based on those variables it's possible to exclude some hands that the opponent probably doesn't play. The random function is used in order to get a more flexible and correct result of the hands to exclude. For example, a loose passive player usually raises in the pre-flop hands from the group 1 of the Sklansky groups, meanwhile a small percentage of the times, these players also raise hands from other groups.

The Hand Strength and Potential Hand Strength could now be calculated with a better approach. They are calculated only for active players and only considering the hands with a rank better than the "sklansky" parameter. The Hand Strength Formula presented in chapter 2 is reformulated as follows:

```
HandStrength(ourcards, boardcards) {
  ahead = tied = behind = 0
  ourrank = Rank(ourcards, boardcards)
  /*Consider all two-card combinations of remaining cards*/
  for each case(oppcards) {
      if(oppcards belong to player_starting_hands_range) {
            opprank = Rank(oppcards, boardcards)
            if (ourrank>opprank) ahead += 1
              else if (ourrank==opprank) tied += 1
                   else behind += 1
      }
  }
  handstrength = (ahead+tied/2) / (ahead+tied+behind)
  return(handstrength)
}
```

A similar reformulation is performed for Hand Potential Strength. The Effective Hand Strength for each one of the opponents is given by the equation 3.

$$EHSi = HSi + (1 - HSi) \times PPoti \qquad (3)$$

The observer developed with the strategy presented is an agent capable of observe the opponents and take decisions based on this observation. This new strategy only considers the possible cards of the opponent to calculate the Effective Hand Strength.

5 Results

The methodology used to test the approach was based on performing game simulations with poker agents playing different strategies. This was similar to a simulation of a real game with the objective to analyze the differences between the performance of an observer agent and a non observer agent. Both agents were set to play at the same table using the same strategy of pre-flop hand selection.

In order to obtain some results, several simulations were made with the agents created. There are 8 normal agents and 1 observer, so the simulations were performed with 9 players at each table.

The intention is to give the Observer Agent the possibility to play in a table with different kind of players. The Observer had the chance to test his new strategy against different players several times along a complete simulation. The observer was programmed to act like a Loose Aggressive in the first round of simulations, Loose Passive in the second, Tight Aggressive in the fourth and Tight Passive in the final round of simulations.

Figures 2, 3, 4 and 5 show the results achieved. Each of the figures compares the evolution of the observer and non-observers agent's bankroll during the simulations using a distinct behavior: Loose Aggressive (Gambler), Loose Passive (Calling Station), Tight Aggressive (Fox) and Tight Passive (Rock).

In the 12 tests done (more than 10 000 hands played) with observer agents, the Observer has better results than the non observer agent that uses the same hand selection in pre-flop. Even with no significant advantage in some of the simulations, the global result of Opponent Modeling reveals to be positive.

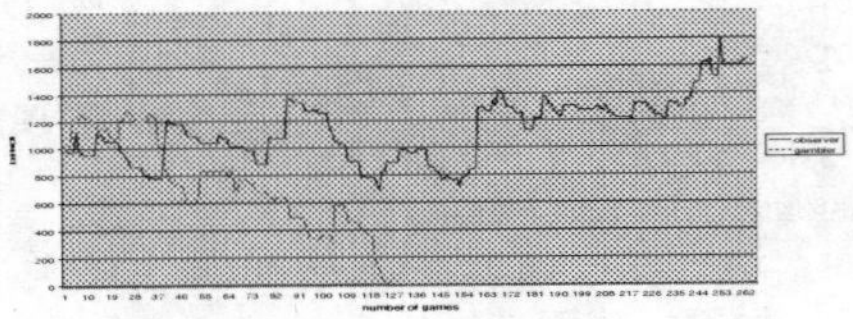

Fig. 2. Bankroll of the Loose Aggressive (Gambler) Observer and Non-Observer Agents

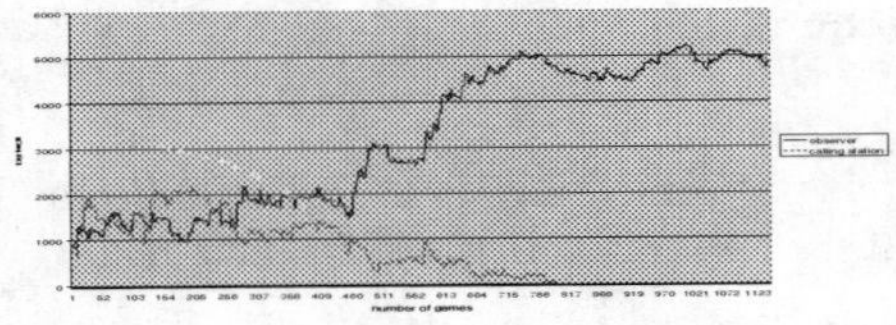

Fig. 3. Bankroll of the Loose Passive (Calling Station) Observer and Non-Observer Agents

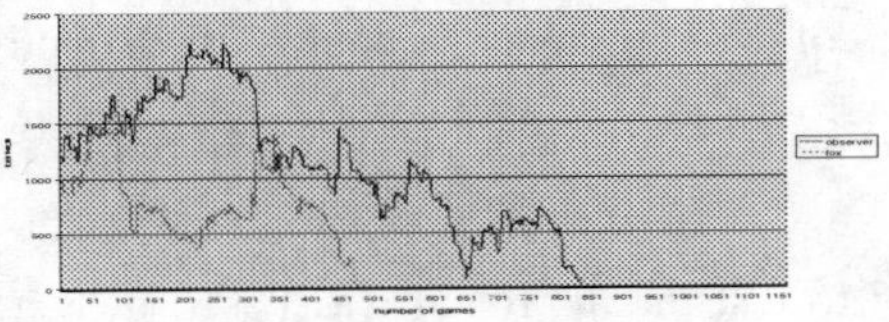

Fig. 4. Bankroll of the Tight Aggressive (Fox) Observer and Non-Observer Agents

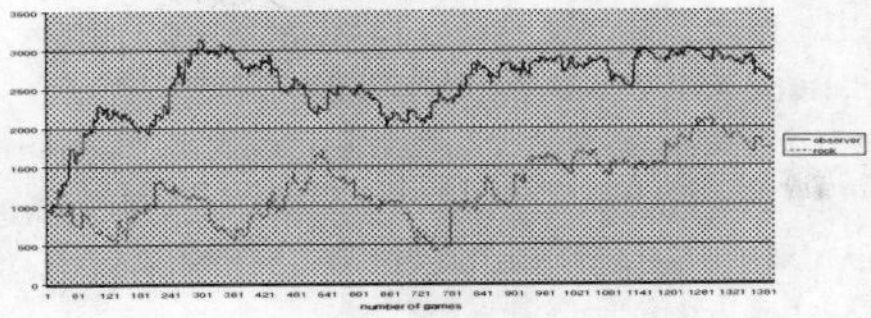

Fig. 5. Bankroll of the Tight Passive (Rock) Observer and Non-Observer agents

The most conclusive results are with passive agents, Observer besides having always a big advantage from non observer, the results are also very good, reaching a good level of bankroll. With aggressive agents, the simulations seem to be a bit inconclusive due to big variations of bankroll that sometimes causes the end of the game too soon for an agent. Although, we can conclude that Opponent Modeling could help these kinds of agents to keep in game for a long time.

6 Conclusions and Future Work

The results achieved with all the agents developed showed the usefulness of the opponent modeling techniques developed. In most of the tests it is possible to verify that the Observer agent has clearly better results than a non observer agent, even when the strategy of hand selection is not very good.

In this project several other techniques were not considered. So, the agent developed is not, globally, a great poker player if compared with good human poker players. However, the main objective was reached and the agent is capable of modeling opponents and effectively using the models to improve its playing style which is an added value to future work in this area.

Future work done in Artificial Intelligence applied to poker may use the work done in this project and the conclusions achieved. The agent developed till the moment must be explored in several other topics, like learning to play in function of the position at the table and bluffing. These topics could be better explored considering Opponent Modeling.

In the domain of player classification, future projects could tune the approach done in this work. The Opponent Modeling described intended to be very simple and basic like a first approach. Future work may include: to consider more than the 4 type of players; analyze other player style variables; and retrieve information from the cards shown at showdown.

References

[1] Billings, D., Papp, D., Schaeffer, J., Szafron, D.: Opponent modeling in poker. In: American Association of AI National Conference, AAAI 1998, pp. 493–499 (1998)

[2] Davidson, A.: Opponent modeling in poker. Master's thesis, Department of Computing Science, University of Alberta (2002)

[3] Schaeffer, J., Culberson, J.C., Treloar, N., Knight, B., Lu, P., Szafron, D.: A world championship caliber checkers program. Artificial Intelligence 53(2-3), 273–289 (1992)

[4] Buro, M.: The Othello match of the year: Takeshi Murakami vs. Logistello. International Computer Chess Association Journal 20(3), 189–193 (1997)
[5] Newborn, M.: Kasparov versus deep blue: computer chess comes of age. Springer, New York (1996)
[6] Donninger, C., Lorenz, U.: Hydra chess webpage (consulted) (March 2008), http://hydrachess.com
[7] Tesauro, G.: Temporal difference learning and TD-gammon. Communications of the ACM 38(3), 58–68 (1995)
[8] Sheppard, B.: World-championship-caliber scrabble. Artificial Intelligence 134(1-2), 241–275 (2002)
[9] Kan, M.: Post-game analysis of poker decisions. Master's thesis, Department of Computing Science, University of Alberta (2006)
[10] Wikipedia. Game theory. Wikipedia: The Free Online Encyclopedia (consulted) (February 2008), http://en.wikipedia.org/wiki/Game_theory
[11] von Neumann, J., Morgenstern, O.: The Theory of Games and Economic Behavior. Princeton University Press, Princeton (1944)
[12] Billings, D., Papp, D., Schaeffer, J., Szafron, D.: Poker as a testbed for machine intelligence research. In: Mercer, R., Neufeld, E. (eds.) Advances in Artificial Intelligence, AI 1998, pp. 228–238. Springer, Heidelberg (1998)
[13] Davidson, A., Billings, D., Schaeffer, J., Szafron, D.: Improved opponent modeling in poker. In: International Conference on Artificial Intelligence, ICAI 2000, pp. 1467–1473 (2000)
[14] Sturtevant, N.: A comparison of algorithms for multi-player games. In: Schaeffer, J., Müller, M., Björnsson, Y. (eds.) CG 2002. LNCS, vol. 2883, pp. 108–122. Springer, Heidelberg (2003)
[15] UA GAMES Group. The University of Alberta GAMES Group (consulted) (March 2008), http://www.cs.ualberta.ca/~games
[16] Billings, D., Davidson, A., Schaeffer, J., Szafron, D.: The challenge of poker. Artificial Intelligence 134(1-2), 201–240 (2002)
[17] Billings, D., Burch, N., Davidson, A., Holte, R.C., Schaeffer, J., Schauenberg, T., Szafron, D.: Approximating Game-Theoretic Optimal Strategies for Full-scale Poker. In: IJCAI 2003, pp. 661–668 (2003)
[18] Billings, D.: Ph.D. dissertation. Algorithms and Assessment in Computer Poker. Department of Computing Science, University of Alberta, Canada (2006)
[19] Afonso, D., Silva, H.: Aplicação para jogar Texas Hold'em Poker (2007)
[20] Southey, F., Bowling, M., Larson, B., Piccione, C., Burch, N., Billings, D., Rayner, C.: Bayes' bluff: Opponent modelling in poker. In: 21st Conference on Uncertainty in Artificial Intelligence, UAI 2005, pp. 550–558 (July 2005)
[21] Carmel, D., Markovitch, S.: Incorporating opponent models into adversary search. In: American Association of AI National Conference, AAAI 1996, pp. 120–125 (1996)
[22] Sklansky, D., Malmuth, M.: Hold'em Poker for Advanced Players, 2nd edn. Two Plus Two Publishing (1994)
[23] Sklansky, D.: The Theory of Poker. Two Plus Two Publishing (1992)

Individual and Social Behaviour in the IPA Market with RL

Eduardo Rodrigues Gomes and Ryszard Kowalczyk

Swinburne University of Technology,
Faculty of Information and Communication Technologies,
John Street, Hawthorn, VIC 3122, Australia
{egomes,rkowalczyk}@ict.swin.edu.au
http://www.swinburne.edu.au/ict

Abstract. Market-based mechanisms offer a promising approach for distributed resource allocation. Machine Learning has been proposed to influence and optimize market-based resource allocation. In particular, Reinforcement Learning (RL) has been used to improve the allocation in terms of utility received by resource requesting agents in the Iterative Price Adjustment (IPA) mechanism. This paper analyses the individual and social behaviour of agents in the IPA market-based resource allocation with RL. In particular, it presents results of experimental investigation on the influences of the amount of learning in the agents' behaviour aiming at determining how much learning is sufficient and the theoretical-experimental explanation of the agents' behaviours using game theory.

Keywords: Market-based Resource Allocation, Reinforcement Learning, Multiagent Systems.

1 Introduction

One of the main challenges faced by the Grid and other large-scale distributed systems is the efficient allocation of resources. Several features contribute to complicate the problem. Computational and geographical distribution, dynamic architecture, lack of coherent global knowledge and lack of centralized control are just some of them. Current research in this area points to the use of market-based mechanisms as the most promising approach [1,2,3,4].

In this paper we consider the Iterative Price Adjustment (IPA) mechanism [5]. The IPA is a *tâtonnement*-like pricing mechanism that can be used in commodity-market resource allocation systems. Pricing mechanisms are responsible for defining the price level at which resources will be traded. In the IPA, the price is adjusted iteratively. Under standard assumptions, the interests of resource requesting agents in the IPA are described by means of demand functions. Demand functions specify mappings from price levels to demand requests but do not account for possible preferences of the agents over attributes of the allocation, for example, the price and the resource levels themselves. It makes

G. Zaverucha and A. Loureiro da Costa (Eds.): SBIA 2008, LNAI 5249, pp. 93–102, 2008.

it difficult to influence and optimize the resource allocation in terms of utility received by the agents.

An approach to address that problem was proposed in Gomes & Kowalczyk [6,7]. In that research, agents use utility functions to describe preferences over resource attributes and learn demand functions optimized for the market by Reinforcement Learning (RL) [8]. The reward functions used during the learning process are based either on the individual utility of the agents or the Social Welfare (SW) resulting from the final allocation. An interesting outcome revealed by that investigation is that both reward functions delivered similar results when only learning agents were used in the market. This result is potentially important for a series of domains where social utility should be maximized but agents are unwilling to reveal private preferences. However, the approach has been evaluated only in two- and three-agent scenarios and the results obtained after a large amount of learning. Moreover, the authors have not investigated the reasons for such results.

In this paper we analyse the individual and social behaviour of the agents in the IPA market-based resource allocation with RL. Our contribution is threefold. First, we evaluate the quality of demand functions obtained throughout the learning process and show that it can be shorter than originally believed. Second, we consider more agents in the market and show that the results observed above hold for a higher number of agents. And third, we develop a theoretical-experimental explanation for the results using game-theory.

2 The IPA Market with RL

We address the scenario in which a limited amount of resources has to be allocated to a set of self-interested agents in a commodity-market resource allocation system using the IPA mechanism. Agents use utility functions to describe preferences in the allocation and learn their demand functions from interaction with the market using RL.

The IPA decomposes the resource allocation optimization problem into smaller and easier sub-problems. Its behaviour mimics the law of demand and supply. The price is increased if the demand exceeds the supply and decreased otherwise. The

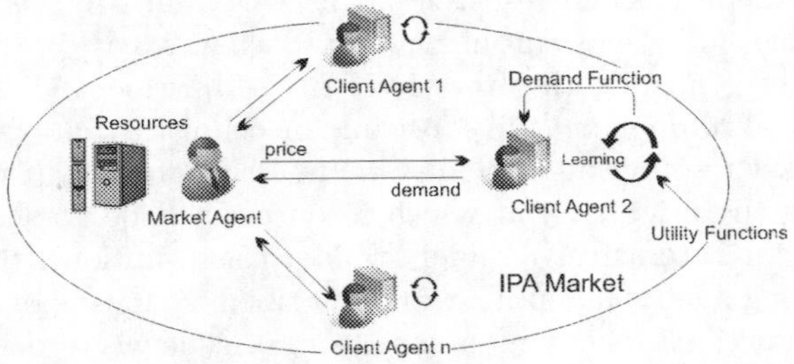

Fig. 1. The IPA mechanism with RL

process is a cycle that begins with a facilitator (the market) announcing the initial prices for the resources. Based on the initial price, agents decide on the amount of resources that maximize their private utilities (the sub-problems) and send these values to the facilitator. The facilitator adjusts the prices according to the total demand received and announces the new prices. New prices are given by $p_i(t+1) = \max\{0, p_i(t) + \alpha(\sum_{j=1}^{n} d_{i,j}(t) - C_i)\}$, where $p_i(t)$ is the price of the resource i at time t, $d_{i,j}(t)$ is the demand request of agent j for resource i, C_i is the total supply of resource i, and α is a constant. The process continues until an equilibrium price is reached, when we say that the market is *cleared*. In the equilibrium, the total demand equals the supply or the price of the excessive supply is zero. Under some circumstances, the equilibrium price may not exist [9], but that problem is out of the scope of this paper.

RL agents learn how to map states of the environment to actions so as to maximize a numerical reward signal. Q-learning is probably the most common algorithm for RL. It is simple and easy to implement. In this algorithm, the agent maintains a table of Q(s, a)-values that are updated as it gathers more experience in the environment. In our case the environment states s represent the resource prices p and the actions a are the demand requests d of the agents, thus Q(p, d). Q-values are estimations of Q^*(p, d)-values, which are the sum of the immediate reward r obtained by requesting demand d at price p and the total discounted expected future rewards obtained by following the optimal policy (demand function) thereafter. By updating Q(p, d), the agent eventually makes it converge to Q^*(p, d). The optimal demand function π^* is then followed by selecting the actions where the Q^*-values are maximum. Q-values are updated using $Q(p,d) = Q(p,d) + \alpha(r(p,d) + \gamma \max_{d'} Q(p',d') - Q(p,d))$, where $0 < \alpha < 1$ is the learning rate and $0 < \gamma < 1$ is the discount rate.

An important component of Q-learning is the action selection mechanism. It is used to harmonize the trade-off between exploration and exploitation. We use the ϵ-greedy method, which selects a random action with probability ϵ and the greedy, the one that is currently the best, with probability 1-ϵ.

3 Learning Experiments

Gomes & Kowalczyk [6] have investigated the IPA market with RL in two different cases: one using a mix of learning and static agents and the other using only learning agents. In this paper we are most interested in this later case.

3.1 Experiments Setup

As in [6], we consider a single IPA market with one type of resource, e.g. memory, but we increase the number of agents, running experiments with 2, 4, 6 and 8 agents. They have preferences over price and amount of resources and use a utility function for each attribute, $U_1(p)$ for price and $U_2(m)$ for amount of resource. The total utility of an agent is given by the product of these two utility functions, $U(p,m) = U_1(p) * U_2(m)$. The actual utility functions used by the agents in the experiments are shown in Figure 2.

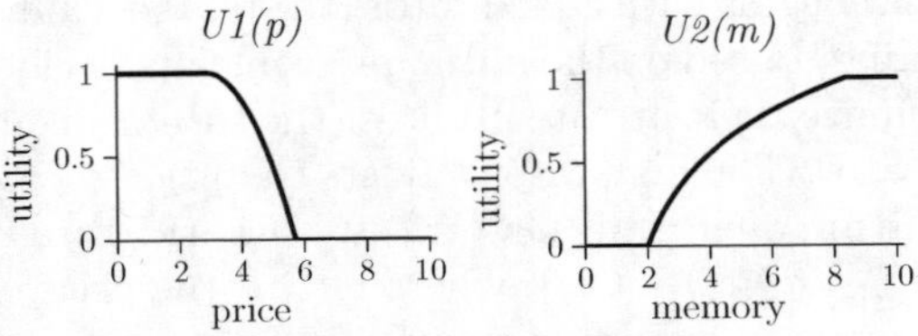

Fig. 2. Agents' utility functions

In the experiments, prices (states) and demand requests (actions) ranged from 0 to 10. Q-learning formally relies on discrete sets, so both prices and demand requests were rounded to 1 decimal place. Therefore, the market has 101 possible states and each agent has 101 actions to choose from. In the IPA market, the only information the agents have available is the current price of the resources. It means that they do not know the actions taken and the rewards received by the other agents.

Two different reward schemas are applied: a local reward function based on the individual utilities of the agents and a global reward function based on the SW of the allocation. In the individual reward function, agents receive a positive reward given by $U(p, m)$ when the market reaches an equilibrium state and is cleared, and zero for all the other states. In the social reward function, agents receive a reward equal to the SW of the allocation for the states where the market is cleared, and zero for the others. The SW is calculated using the Nash Product (NP) function, which is given by the product of the individual utility of the agents: $NP = \prod_{i=1}^{n} U_i$, where U_i is utility of agent i. The NP is suitable for the resource allocation domain because it encourages the improvement and the balance of the utility of the agents.

The market is set with 5 units of resources per agent. From the analysis of the utility functions, we can note that such an amount does not allow for all the agents to have a complete satisfaction in the allocation, but it permits the analysis of the market under a condition of limited supply, which is the most interesting case.

The agents' behaviour is evaluated using demand functions obtained at pre-defined intervals of the learning process. The quality of the learnt demand functions is measured under two aspects: the individual utility received by the agents and the SW of the market.

The evaluation is made with the trends of the actual demand functions learnt by the agents. One of the reasons for using the trends is that we transformed prices (states) and demands (actions) into discrete sets to implement the learning algorithm. However, in the IPA market, such a discretization may lead to small losses of economical efficiency. Other reason is that by using the trends we avoid local instabilities present in the learnt demand functions [6]. The trends were obtained by a process of curve-fitting.

We ran learning experiments with both reward functions using 2, 4, 6 and 8 agents. Each configuration was run 20 times of 10 000 learning episodes.

The demand functions were extracted and evaluated in intervals of 100 learning episodes. The learning parameters were set to $\alpha = 0.1$, $\gamma = 0.9$ and $\epsilon = 0.4$, and the market constant set to $\alpha = 0.05$.

3.2 Market Results

Figure 3 shows the evolution of the market's average NP using the individual and the social reward functions. The main point to note is the quick evolution to a level of relative stability, suggesting that the learning process can be shorter than the 450 000 learning episodes applied in [6]. It evolves to that level before 1000 learning episodes and fluctuates around it afterwards. The fluctuation is mostly influenced by characteristics of Q-learning inherent to multi-agent scenarios [10] and we expect it to be reduced with the application of different decay rules and some other RL algorithms.

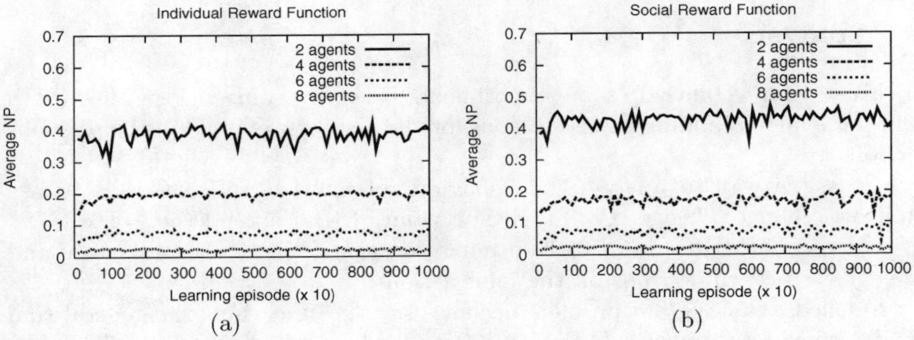

Fig. 3. Average NP obtained by the market using the individual and social reward functions over the learning episodes

Another point to note is that both reward functions presented similar results, approaching the optimal social welfare. This behaviour was previously identified in [6]. The results here show that it is maintained over the learning episodes and that it holds for a higher number of agents. The same type of result was found in the individual utilities received by the agents, shown in Figure 4.

The most important observation we can draw from these experiments is that, using both reward functions, there is a trend for the learning agents to divide the resources equally. While this strategy seems obvious for the case of the social reward function, since the optimal reward is received when the price is low enough and all the agents receive an equal share of the allocation, it is not so intuitive for the individual reward function. The next section further investigates this behaviour.

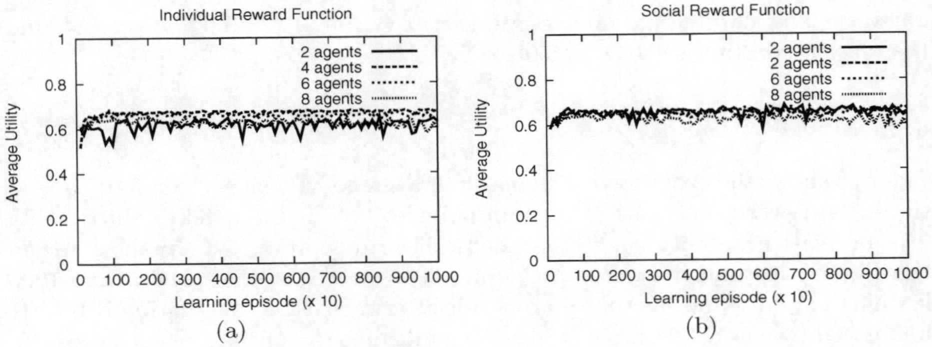

Fig. 4. Average utility obtained by the agents using the individual and social reward functions over the learning episodes

4 Analysis

In this section we develop some experiments based on game-theory in order to gain some understanding on the reasons for the results presented in the previous section.

A possible way to analyse the behaviour of multiagent learners is to use Stochastic Game Theory [11]. Stochastic games (SGs) extend the Markov Decision Process (MDP) framework to multiple agents, where the next state and one agent's reward depends on the joint actions.

Modelled as an SG, our problem becomes the one of learning the general-sum SG described by a tuple $(n, P, D_{1\cdots n}, T, R_{1\cdots n})$, where n is the number of learning agents (players), P is the set of possible resource prices (state space), D_i is the set of possible demand requests available to agent i (action space) with D being the joint demand requests space $D_1 \times \cdots \times D_n$, T is the IPA's price update rule $P \times D \times P \to [0, 1]$ (state transition function), and R_i is the reward function for the ith player $P \times D \to \mathbb{R}$.

For this game, the *best-response* function for agent i, $BR_i(\pi_{-i})$, is the set of all demand functions (policies) that are optimal given that the other agent(s) use the joint demand function π_{-i}. A demand function π_i is optimal given π_{-i} if and only if,

$$\forall \pi'_i \in \Pi_i \qquad V_i(\pi_i, \pi_{-i}) \geq V_i(\pi'_i, \pi_{-i})$$

where Π_i is the set of possible demand functions for agent i and V_i is the expected reward for using demand function π_i given π_{-i}.

A *Nash Equilibrium* (NE) occurs when all the players are using best-response policies. So, for us, an NE is a joint demand function, $\pi_{i=1,\cdots,n}$, with

$$\forall i = 1, \cdots, n \qquad \pi_i \in BR_i(\pi_{-i})$$

An important result shown by Fink [12] guarantees that *every n-player general-sum discounted stochastic game possesses at least one NE point in stationary strategies.* A stationary strategy is one that depends only on the current state.

According to Fulda & Ventura [13], from the system perspective, an *optimal solution* would be any one which is an NE and Pareto-Optimal (PO). A solution is Pareto-Optimal if there is no other solution that can improve one agents' outcome without deteriorating others'. As observed by Bowling & Veloso [11], Q-learning has been developed to learn best-response policies in MDPs. It is also well-known that Q-learning has no proof of convergence in multiagent learning scenarios. However, if Q-learning ever converges in those scenarios, it will do to an NE, since all the agents will be playing *best-response* policies. The remaining questions are to which NE the agents will converge to and how optimal this NE is.

In research addressing those questions, Claus & Boutilier [14] investigated the dynamics of RL in cooperative multiagent systems. Their analysis of *independent Q-learners* in cooperative games, where the agents have the same payoff table, show that the agents tend to converge to the most profitable equilibrium in simple games. Fulda & Ventura also studied the cooperative case, formulating conditions in which the optimal solution will be chosen. Those results are important to the analysis of our social reward function. However, it should be noted that the individual reward function generates a different game.

Analysing our actual games is quite complex, given the number of states and actions. For this reason, we use simplified versions of them. We defined a two-player five-action *stage game* for each reward function. A *stage game* is a special case of a *stochastic game* where the number of states is 1. Tables 1 and 2 show the new games. The actions represent the possible demand requests. Note that the payoffs simulate the reward functions, generating a competitive game with symmetric payoff table for the individual reward function and a cooperative game for the social reward function. In the tables, *pure* NEs are presented in **bold** and PO joint-actions in *italic*.

Table 1. Individual Reward Game

	0	*1*	*2*	*3*	*4*
0	0,0	0,0	0,0	0, 0	***0,4***
1	0,0	0,0	0,0	***1,3***	0,0
2	0,0	0,0	***2,2***	0,0	0,0
3	0,0	***3,1***	0,0	0,0	0,0
4	***4,0***	0,0	0,0	0,0	**0,0**

Table 2. Social Reward Game

	0	*1*	*2*	*3*	*4*
0	**0,0**	0,0	0,0	0,0	**0,0**
1	0,0	0,0	0,0	**3,3**	0,0
2	0,0	0,0	***4,4***	0,0	0,0
3	0,0	**3,3**	0,0	0, 0	0,0
4	**0,0**	0,0	0,0	0, 0	**0,0**

Considering a RL process, the agents do not know the payoff tables, they chose actions independently and play the games an infinite number of times. In this case, the most reasonable solution for the social reward game is learning to play $\langle 2, 2 \rangle$, which is the only Nash Pareto-Optimal equilibrium. For the individual reward game, however, all the joint-actions in the minor diagonal are NEs and PO. Suppose agent 1 plays action 4, whatever agent 2 plays, it will earn 0. Thus, agent 2 will select random actions, making agent 1 earn in average 4/5. Therefore, on this particular game, playing action 0 or action 4 seems to be the least reasonable strategy. Now suppose agent 1 plays action 3. In this case, agent 2 should play action 1 to earn 1, and agent 1 earns 3. Using Q-learning, however, each agent has a probability of choosing random actions in each turn. By exploring the game in this fashion, agent 2 will eventually discover that other actions are more attractive and may change its strategy, making agent 1 adapt to it. The same reasoning applies to agent 1.

To better understand how that dynamics proceeds, we performed a series of learning experiments. We set the Q-learning parameters with the same values used in the previous section, but modified the update rule to $Q(s, a) = Q(s, a) + \alpha(r(s, a) - (\gamma * Q(s, a)))$ since only one state is available. Figures 5 and 6 present the results, obtained out of 1000 experiments.

The main point to note is that *action 2* is the most chosen in both games. The agents develop the preference for this action very early. For the individual reward game, note that *action 4*, which is the one with best payoff, is the fourth action in preference. Not surprisingly, *action 0* is quickly ignored in both cases. In addition, figures 5-c and 6-c show equilibrium between the joint-actions $\langle 1, 3\rangle$ and $\langle 3, 1\rangle$. In the individual reward game, this equilibrium is maintained over the episodes, suggesting that the agents keep trying to achieve a higher payoff, even if the joint-action $\langle 2, 2\rangle$ is more attractive. In the social reward game, the likelihood of choosing $\langle 1, 3\rangle$ and $\langle 3, 1\rangle$ lose its strength over the episodes, which is quite obvious since both players can do better by choosing $\langle 2, 2\rangle$.

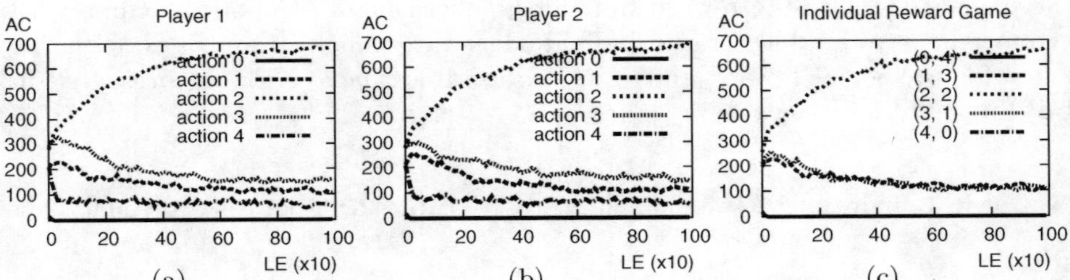

(a) (b) (c)

Fig. 5. Number of times each action was chosen (AC) by the players over the learning episodes (LE) in the individual reward game

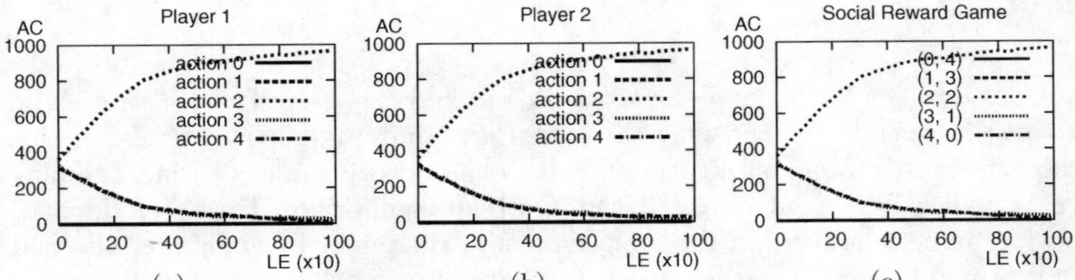

(a) (b) (c)

Fig. 6. Number of times each action was chosen (AC) by the players over the learning episodes (LE) in the social reward game

The results from these experiments are coherent with the behaviours found in the investigated IPA market. It is clearly seen that choosing to share the resources equally is Pareto-Optimal and also a Nash Equilibrium. Therefore, the convergence (or *pseudo*-convergence as no decay rules have been applied) to such a strategy via Q-learning is rational for both reward functions. As commented above, similar results have been found in cooperative games [14,13,15], which is the case of the social reward game. For the individual reward game, however, further investigations are necessary, especially to evaluate cases with non-symmetrical payoff tables.

5 Conclusions

In this paper we have investigated the individual and social behaviour of agents in the IPA market with RL. In particular, we presented results of experimental investigation on the influences of the amount of learning in the agents' behaviour aiming at to determine how much learning is sufficient and theoretical-experimental explanation of the agents' behaviour based on game theory.

The results of the experiments have shown that the learning process can be much shorter than the 450 000 learning episodes originally used in [6]. In addition, they have shown that the behaviour in which both the individual-based and the social-based reward functions deliver similar results when only learning agents are used in the market is maintained over the learning episodes and, most remarkably, holds for a higher number of agents. Investigating this behaviour, game-theoretical analysis of a simplified version of the scenario revealed the presence of Nash Equilibrium and Pareto-Optimality at the joint-actions in which the resources are shared equally, for both games. Experiments using Q-learning have shown the *pseudo*-convergence of the agents to those joint-actions, explaining the behaviour found in the actual IPA market.

Machine Learning has been used to optimize a series of different market-based resource allocation mechanisms, from bargaining [2] to auctions [16]. However, as far as we are aware, no work has addressed the problem of learning demand functions based on utility functions. Utility-based resource allocation has been proposed by Chunlin & Layuan [3]. Nonetheless, such an approach does not consider agents with preferences over price, as this work does.

There are several aspects in which this work could be extended. An area involves further reduction of the required learning episodes in the market, which in general is related to well-known scalability issues of RL. It is also necessary to evaluate the approach in extended scenarios, including agents with preferences described over multiple attributes, multiple markets, and the existence of resource provider agents. The later deals with a limitation of the IPA mechanism. In particular, the IPA does not model resource provider agents. Simply adding those agents to the model may change its theoretical implications as the resource supply becomes dynamic and, therefore, has to be carefully considered. An interesting alternative to the IPA is proposed by Wolski *et al.* [4].

Acknowledgments

This work has partially been supported by the Utility Grid Project (AU-DEST-CG1001511) funded by the Department of Education, Science and Training (DEST) from the *International Science Linkages programme* established under the Australian Government's innovation statement, *Backing Australia's Ability.*

References

1. Yeo, C.S., Buyya, R.: A taxonomy of market-based resource management systems for utility-driven cluster computing. Softw. Pract. Exper. 36(13), 1381–1419 (2006)
2. Schnizler, B., Neumann, D., Veit, D., Reinicke, M., Streitberger, W., Eymann, T., Freitag, F., Chao, I., Chacin, P.: Catnets - wp 1: Theoretical and computational basis (2005)
3. Chunlin, L., Layuan, L.: Pricing and resource allocation in computational grid with utility functions. In: Proceedings of the International Conference on Information Technology: Coding and Computing (ITCC 2005), Washington, DC, USA, April 2005, vol. II, pp. 175–180. IEEE Computer Society, Los Alamitos (2005)
4. Wolski, R., Plank, J.S., Brevik, J., Bryan, T.: Analyzing market-based resource allocation strategies for the computational grid. International Journal of High Performance Computing Applications 15(10), 258–281 (2001)
5. Everett, H.: Generalized lagrange multiplier method for solving problems of optimum allocation of resources. Operations Research 11(3), 399–417 (1963)
6. Gomes, E.R., Kowalczyk, R.: Learning the IPA market with individual and social rewards. In: Proceedings of the International Conference on Intelligent Agent Technology (IAT 2007), pp. 328–334. IEEE Computer Society, Los Alamitos (2007)
7. Gomes, E.R., Kowalczyk, R.: Reinforcement learning with utility-aware agents for market-based resource allocation. In: Proceedings of the 7th International Joint Conference on Autonomous Agents and Multiagent Systems (AAMAS 2007) (2007)
8. Sutton, R.S., Barto, A.G.: Reinforcement Learning: An Introduction. MIT Press, Cambridge (1998)
9. Jennergren, P.: A price schedules decomposition algorithm for linear programming problems. Econometrica 41(5), 965–980 (1973)
10. Sandholm, T.W., Crites, R.H.: On multiagent Q–learning in a semi–competitive domain. In: Weiß, G., Sen, S. (eds.) Adaptation and Learning in Multi–Agent Systems, pp. 191–205. Springer, Berlin (1996)
11. Bowling, M.H., Veloso, M.M.: Multiagent learning using a variable learning rate. Artificial Intelligence 136(2), 215–250 (2002)
12. Fink, A.M.: Equilibrium in a stochastic n-person game. Journal of Science in Hiroshima University Series A-I(28), 89–93 (1964)
13. Fulda, N., Ventura, D.: Predicting and preventing coordination problems in cooperative q-learning systems. In: Veloso, M.M. (ed.) IJCAI 2007, pp. 780–785 (2007)
14. Claus, C., Boutilier, C.: The dynamics of reinforcement learning in cooperative multiagent systems. In: Proceedings of the Fifteenth National Conference on Artificial Intelligence, pp. 746–752. AAAI, Menlo Park (1998)
15. Sen, S., Sekaran, M.: Individual learning of coordination knowledge. Journal of Experimental & Theoretical Artificial Intelligence 10(3), 333–356 (1998)
16. Preist, C., Byde, A., Bartolini, C.: Economic dynamics of agents in multiple auctions. In: AGENTS 2001: Proceedings of the fifth international conference on Autonomous agents, pp. 545–551. ACM Press, New York (2001)

Optimizing Preferences within Groups: A Case Study on Travel Recommendation

Fabiana Lorenzi[1,2], Fernando dos Santos[1], Paulo R. Ferreira Jr.[1,3], and Ana L.C. Bazzan[1]

[1] Instituto de Informática, UFRGS
Caixa Postal 15064, 91.501-970 Porto Alegre, RS, Brazil
{lorenzi,fsantos,prferreirajr,bazzan}@inf.ufrgs.br
[2] Universidade Luterana do Brasil
Av. Farroupilha, 8001 Canoas, RS, Brazil
[3] Instituto de Ciências Exatas e Tecnológicas, Centro Universitário Feevale
RS239, 2755, CEP 93352-000, Novo Hamburgo, RS, Brasil

Abstract. This work describes a multiagent recommender system where agents work on behalf of members of a group of customers, trying to reach the best recommendation for the whole group. The goal is to model the group recommendation as a distributed constraint optimization problem, taking customer preferences into account and searching for the best solution. Experimental results show that this approach can be sucessfully applied to propose recommendations to a group of users.

1 Introduction

The internet is a rich source of information where users search information about products and services related to their interests and preferences. However, locating the necessary information has become a hard task for the user [7]. Moreover, the information is usually distributed through several locations.

In order to deal with some of these issues, recommender systems have been developed [12]. These systems learn about user preferences over time and automatically suggest products that fit the user needs. Recommender systems are being applied in several domains in e-commerce to suggest products to their customers [2] such as book recommendation (amazon.com) or movie recommendation (netflix.com). The main advantage of recommender systems is the ability to aggregate information and to match the recommendations with the information people are looking for.

Group recommendation is a new challenge to the recommendation area because it is necessary to take into account all members preferences. Each group member elicits his preferences, which means that preferences within the group are not homogeneous. The recommender system needs to aggregate all preferences to formulate a recommendantion that suits the whole group.

Several group recommender systems have been developed in the past years, in different domains. *Let's Browse* [3], for instance, is a group recommender system which recommends web pages to a group of users who are browsing the web.

G. Zaverucha and A. Loureiro da Costa (Eds.): SBIA 2008, LNAI 5249, pp. 103–112, 2008.

Another example is *MusicFX* [9], a system used in a fitness center to adjust the selection of background music to best suit the preferences of people working out at any given time. A special feature found in this system is that a group is composed by people who happen to be in the place at the same time. *MusicFX* uses explicit preferences of all participants to make a music selection that will be listen by everyone who is present. In this case, the group is composed by strangers rather than family members or friends.

Intrigue [1] is another example of group recommender system. The system recommends attractions and itineraries by taking into account preferences of heterogeneous groups of tourists (such as families with children) and explains the recommendations by addressing the requirements of the group members. Attractions are separately ranked by first partitioning a user group into a number of homogeneous subgroups with the same characteristics. Then each subgroup may fit one or more stereotypes and the subgroups are combined to obtain the overall preference, in terms of which attractions to visit for the whole group.

The group recommendation task may become more difficult according to the complexity of the domain. Recommendation of travel packages, for instance, is composed by several information components such as flights, hotels, and attractions [13] [5]. Besides the specific knowledge necessary to assemble all the components, each user has different preferences that need to be considered during the recommendation process. A group member may prefer flying during the day and staying in a four-star hotel; while another member prefers flying at night and staying in a hostel to save money. Besides the fact that group members have different and hard criteria to make decisions, sometimes group members do not want to let other members know their preferences. In these cases, the privacy is very important and the system cannot allow users to see each other preferences. Thus, finding the best option for the entire group according to the preferences of each individual and keeping the privacy of each member is a tricky task. The system needs to find the global optimal recommendation in a distributed fashion.

The formalism of distributed constraint optimization problem (DCOP) has been proposed to deal with the problem of coordinating and optimizing agents' interactions. It has been used in multiagent systems (MAS) in several domains [8] [10] [11]. DCOP is related to the constraint satisfaction problems (CSP), a well known technique in AI, in the sense that it deals with assignment of values to variables under certain constraints. However, DCOP is more difficult. First, it deals with optimization (not only with satisfaction) meaning that the best solution must be found (not any one). Second, the assignment is computed in a distributed way.

CSP was also used in [6], where authors proposed a multiagent recommender system to arrange meetings for several participants taking into account constraints for personal agendas. Three different agents were proposed: the *personal assistant agent* is the interface between the user and the MAS; the *flight travel agent* is connected to a database of flights; and the *accommodation hotel agent* is responsible for finding an accommodation on the cities involved in the meeting. However, the system does not provide the best recommendation to the group, due to the limitations of the CSP formalism (any recommendation is acceptable instead of the best).

We propose a DCOP-based multiagent recommender system to perform the travel group recommendation task. Aspects of travel group recommendation task such as customers, their preferences, and the need of coming out with a recommendation for the whole group can be viewed as components of a DCOP instance. Each user is represented by an agent that is responsible for negotiating with a recommender agent who holds the information about the travel services, trying to get the best recommendation according to customers.

Experimental evaluations were performed to verify the feasibility of the DCOP-based multiagent recommender system. Test cases were generated and solved with a DCOP algorithm, namely the *distributed pseudotree optimization* (DPOP) [11]. The results obtained have shown that using DCOP to optimize the problem of finding a recommendation to a group is feasible both in terms of running time and communication load.

The rest of the paper is organized as follows: section 2 describes the DCOP framework. Section 3 shows the proposed multiagent recommender system, providing details about the use of DCOP to specify the travel groups recommendation process. Section 4 discusses the feasibility of the proposed multiagent recommender system using the DPOP algorithm. Section 5 summarizes the contributions and shows future direction for this work.

2 Distributed Constraint Optimization Problem

DCOP represents a generic framework for the resolution of ditributed problems with a significant application in MAS. The challenge is to find the best distribution and value attribution of a set of variables to a set of agents with interdependencies. In a DCOP, differently from a distributed constraint satisfaction problem (DisCSP) [14], the interest is to optimize the restrictions and not only to satisfy them. DCOP is associated with a global function and the objective is to maximize or minimize it. This function depends on a cost value associated to each restriction.

The approaches for dealing with DCOP in real life problems should consider that the agents must be able to optimize the global function in a distributed way, using only local communication. It is not acceptable to use a central agent responsible for all the processing. Also a DCOP algorithm should be capable of finding the solution with the agents working in an asynchronous way. Finally, the approach should provide quality guarantees.

DCOP is a formalism to model a range of agents coordination issues. A DCOP consists of n variables $V = \{x_1, x_2, ..., x_n\}$, where each can assume values in a finite, discrete domain $D_1, D_2, ..., D_n$. Each variable is assigned to one agent that has the control over the values of the variable. The goal of the agents is to choose values for the variables to optimize a global objective function. This function is described as an aggregation over a set of cost functions related to pairs of variables (in the case of binary constraints). Thus, for each pair of variables

x_i, x_j, there is a cost function defined as $f_{ij} : D_i \times D_j \rightarrow \mathbb{N}$ [10]. A DCOP can be represented by a constraint graph, where vertices are variables and edges are cost functions between variables.

Although some algorithms have been proposed to deal with DCOP, here we use the DPOP [11] algorithm. It must be said that we have also used other algorithms such as Adopt [10]. Due to lack of space we do not show the results here but notice that DPOP has performed better than the others in this setting. DPOP uses a message-passing scheme to allow the communication among the agents and to achieve the solution in a distributed fashion. DPOP is based on the dynamic programming technique. It provides solutions quickly with a low number of messages since it is not fully asynchronous. On the other hand, it requires an exponential memory space. It is a complete algorithm, i.e., it always finds the optimal solution.

3 The Multiagent Recommender System

In our DCOP-based multiagent recommender system, a community of agents share a common goal (the travel package recommendation) as well as individual goals (the individual preferences). The proposed multiagent recommender system is composed by two different kind of agents: the user agent (UA) and the recommender agent (RA). The UAs work on behalf of each user and represent their travel preferences. Each UA knows the user individual preferences for each travel service (hotel, flight companies, tour operators, etc). In the recommendation process, these agents interact with the recommender agent to reach the global optimal recommendation. The RAs work on behalf of suppliers of travel services and each one has information about a different kind of service. Each RA is responsible for collecting information from a supplier. This step is reported elsewhere [4].

Figure 1 illustrates the scheme of the proposed multiagent recommender system. First, it is required that each user set up his preferences and inform his UA. After, the UAs start an interaction with the RA to get a list of possibilities for the particular service. This list is ranked according to the user preferences for this service. It is important to mention that a travel package recommendation is composed by several components (flight, hotel, attractions and so on). However, we are considering in our example (shown in Figure 2) just one service (e.g. hotel) to recommend.

The major challenge in this approach is to guarantee the best recommendation for the whole group of users. The recommendation process is easy when it deals with an individual user. However, when we have a group of users, the system has to find the best recommendation for the whole group, i.e., the system has to take into account all preferences of the users .

Group recommender systems reported in the literature are based on different approaches in what regards how to reach to the final decision. These approaches were already discussed in section 1. Here we revisit them to draw comparisons with our approach. In *MusicFX*, for example, the system selects and plays music autonomously based on the preferences of the group members who are present in the gym. In *Let's Browse*, there is a group member responsible for making the final decision. This member decides the best option for the group. *Intrigue* uses a similar approach, but it is the tour guide who decides which tour should be taken by the group. Another existing approach is to assume that group members will arrive at the best decision through conventional discussion. In this case, group members could chat or exchange e-mails, trying to reach a consensus.

As we propose a multiagent recommeder system to recommend travel packages to groups, in our point of view, none of the approaches mentioned is adequate to our scenario. First, a travel package has different components and a group has several members. The group would take too long to make a final decision. Second, the approach based on a member who is responsible for the final decision is a centralized one.

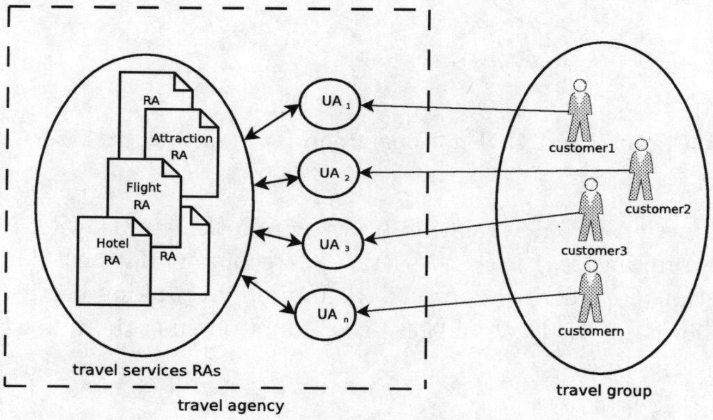

Fig. 1. DCOP-based multiagent recommender sytem

As mentioned, each UA represents one user that ranks all information obtained from the RA. $\mathcal{R}$ is the set of possible recommendations defined as $\{r_1, r_2, ..., r_{|\mathcal{R}|}\}$. The individual ranking $S_a(r)$ for the set $\mathcal{R}$ is defined as follows: we compute the preference $P_a(r)$ of UA a regarding a recommendation r. Assuming that there is a permutation of $\mathcal{R}$ such that $P_a(r'_1) > P_a(r'_2) > \ldots > P_a(r'_{|\mathcal{R}|})$, we defined an individual ranking $S_a(r)$ for the set $\mathcal{R}$ as $S_a(r'_1) = 0, S_a(r'_2) = 1, \ldots, S_a(r'_{|\mathcal{R}|}) = |\mathcal{R}| - 1$. The preference values $\{0, 1, \ldots, |\mathcal{R}| - 1\}$ are used to simplify the notation and save space in the description. However, any other form of preferences representation (e.g. a table) could be used. These values can be any positive integers and the most preferable recommendation must have the lowest value.

We map the UA rankings into cost functions and these cost functions are aggregated to describe the global objective function. The optimum of this objective function provides the best recommendation of a travel package to the group.

For each travel group, the DCOP is defined as follow: a set of variables $\{a_0, a_1, ..., a_n\}$, where the variable a_0 is assigned to the RA and a_i is assigned to the UA_i (where $0 < i \leq n$). These variables can take values from the domain $D = \mathcal{R}$. For sake of illustration, Figure 2 shows a simple example of a DCOP of a travel group recommendation, with four UAs $\{a_1, a_2, a_3, a_4\}$ and one RA $\{a_0\}$ that is representing hotels information. This RA has three option of hotels that can be recommended to the group, thus the domain D of each variable is $\{1, 2, 3\}$. Each UA has a ranking of these hotels according to its user preferences.

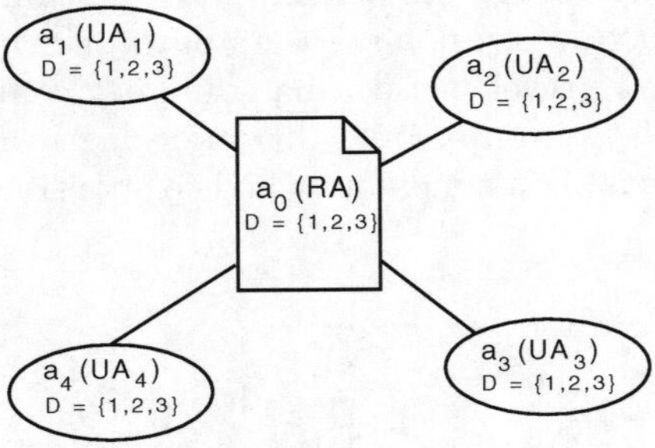

Fig. 2. The DCOP for one group recommendation case

The cost function for each relation between one UA and the RA is computed based on the ranking $S_a(r)$ for the set $\mathcal{R}$ of recommendations. Thus, there is one cost function $f(a_0, a_i)$ to represent the individual ranking between each pair (a_0, a_i) for all $0 < i \leq n$. Each of these cost functions has the following form

$$f(a_0, a_i) = \begin{cases} S_{a_i}(r) & \text{when } a_i = r \text{ and } a_0 = r \\ \infty & \text{otherwise} \end{cases} . \tag{1}$$

The first case of Equation 1 captures the situation where the agents a_0 and a_i agree on the recommendation. The second case captures the situation where some agent disagrees with the RA. Equation 2 shows the global objective function, which is the aggregation over all cost functions.

$$\mathcal{F} = \sum_{i=1}^{n} f(a_0, a_i) . \tag{2}$$

Since the individual ranking is defined in an ascendant way, the solution to the DCOP (and thus the best recommendation to the travel group) is the one that minimizes this global objective function. It is important to note that our DCOP-based multiagent recommender system can be extended to have more than one RA.

4 Experimental Evaluation

We have conducted a set of experiments to empirically evaluate the feasibility of the proposed multiagent recommender system. We use two measures to evaluate the performance:

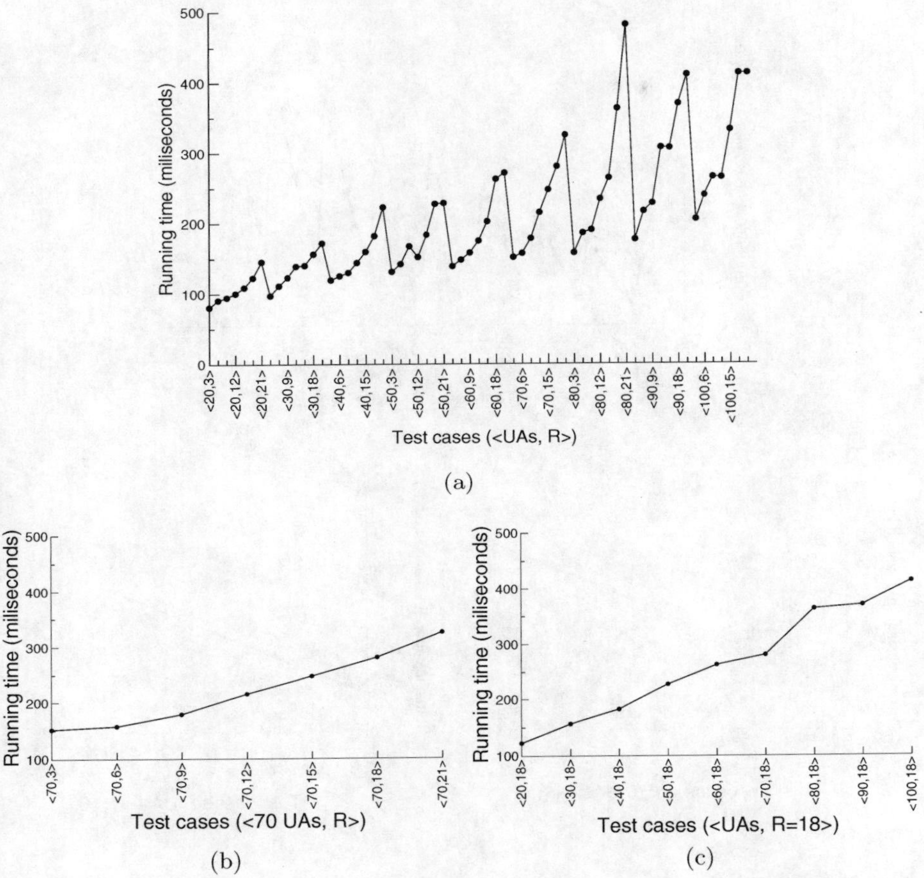

Fig. 3. (a) Running time, in milliseconds, for each test case. (b) Details of the nearly linear running time in the domains size for a fixed number of 70 UAs. (c) Details of the linear running time in the number of agents for a fixed $\mathcal{R}$ with 18 recommendations.

- *Running time*: time required by the algorithm to solve the recommendation for the group taking into account user preferences.
- *Communication (network usage)*: number of bytes transferred through message exchanges among the agents during the algorithm execution.

Test cases have 20, 30, 40, 50, 60, 70, 80, 90, and 100 agents, each representing one individual in the travel group (one UA). In addition, we have one RA. For each of these cases, $\mathcal{R}$ takes the following values: 3, 6, 9, 12, 15, 18, and 21 (representing possible recommendations).

In each experiment the data was averaged over 20 runs. We run the original implementations of the DPOP algorithm, which are available on the internet[1].

[1] DPOP is available at `http://liawww.epfl.ch/frodo/`

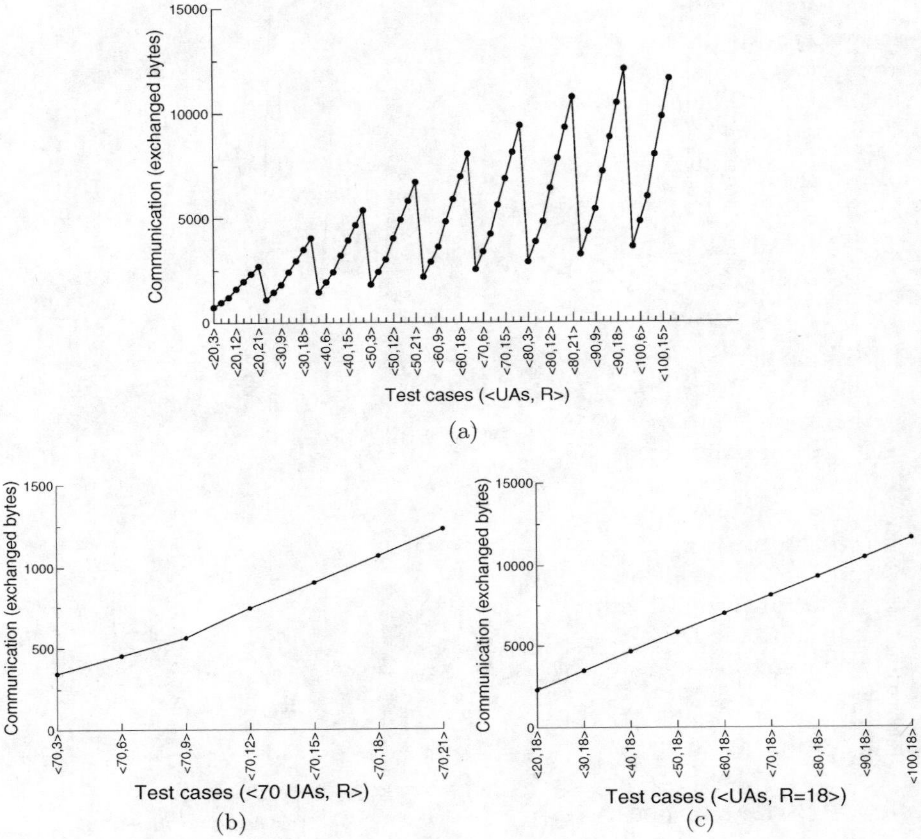

Fig. 4. (a) Communication (sum of exchanged bytes) for each test case. (b) Details of the linear communication in the domains size for a fixed number of 70 UAs. (c) Details of the linear communication in the number of agents for a fixed $\mathcal{R}$ with 18 recommendations.

Figure 3(a) shows the running time in milliseconds for each test case. We can see that the running time is nearly linear in the number of UAs and in the domain ($\mathcal{R}$) sizes. To show with more details these increasing rates, Figure 3(b) shows how the running time is nearly linear in the domains size for a fixed number of 70 UAs and Figure 3(c) shows the running time for an increasing number of UAs, for a fixed $\mathcal{R}$ with 18 recommendations. The algorithm provides very satisfactory results regarding these metrics and proves the feasiblility of using DCOP in the group recommendation task. In the hardest test case, with 100 UAs and 21 travel recommendation options the algorithm takes only about 413 miliseconds to solve the DCOP and provides the optimal recommendation to the group.

Figure 4(a) shows the total number of bytes transferred. The results are both nearly linear in the number of UAs and in the domain ($\mathcal{R}$) sizes. Figure 4(b) shows details of the increasing in the domains size for a fixed number of 70 UAs while figure 4(c) depicts the linear growing rate in the number of UAs for a fixed $\mathcal{R}$ with 18 recommendations. For example, in the case of 100 UAs and 18 domain values, the total number of bytes transferred is about just 11625 bytes.

5 Conclusion and Future Work

Recommender systems are been used in e-commerce to help users to find out better products according to their needs and preferences. Multiagent recommender systems use agents to help in the solution process, trying to improve the recommendation quality. Agents cooperate in order to reach the best recommendation for the whole group. The recommendation for group of users is considered a challenge to recommender systems due the fact that is not an easy task to find out the best solution that satisfies all the users in the group.

In this paper we have shown that the DCOP formalism can be successfully used to help the group recommendation process. Agents work on behalf of the users and try to get the best recommendation according to the user preferences. Experimental evaluations have shown the feasibility of using the DPOP algorithm to solve the DCOPs generated for each recommendation case.

As a future work we intend to adapt and test our system to recommendations where the preferences are interrelated. These interrelation appears for instance when the users define the preferences for a travel option A as a result of the available travel options B.

References

1. Ardissono, L., Goy, A., Petrone, G., Segnan, M., Torasso, P.: Intrigue: Personalized recommendation of tourist attractions for desktop and handset devices. Applied Artificial Intelligence: Special Issue on Artificial Intelligence for Cultural Heritage and Digital Libraries 17(8-9), 687–714 (2003)
2. Goy, A., Ardissono, L., Petrone, G.: Personalization in e-commerce applications. In: The Adaptive Web, pp. 485–520. Springer, Heidelberg (2007)
3. Lieberman, H., Dyke, N.W.V., Vivacqua, A.S.: Let's browse: A collaborative web browsing agent. In: Maybury, M.T. (ed.) International Conference on Intelligent User Interfaces - IUI 1999, New York, pp. 65–68 (1999)
4. Lorenzi, F., Bazzan, A.L.C., Abel, M.: Truth maintenance task negotiation in multiagent recommender system for tourism. In: Proceedings of the AAAI Workshop on Intelligent Techniques for Web Personalization and Recommender Systems in E-Commerce (AAAI 2007), pp. 122–125 (July 2007)
5. Lorenzi, F., Santos, D.S., de Oliveira, D., Bazzan, A.L.C.: Task allocation in case-based recommender systems: a swarm intelligence approach. Information Science Reference, pp. 268–279 (2007)
6. Macho, S., Torrens, M., Faltings, B.: A multi-agent recommender system for planning meetings. In: Workshop on Agent-based recommender systems (WARS 2000) (2000)

7. Maes, P.: Agents that reduce work and information overload. Commun. ACM 37(7), 30–40 (1994)
8. Maheswaran, R.T., Tambe, M., Bowring, E., Pearce, J.P., Varakantham, P.: Taking DCOP to the real world: Efficient complete solutions for distributed multi-event scheduling. In: Third International Joint Conference on Autonomous Agents and Multiagent Systems, Washington, DC, USA, July 2004, vol. 1, pp. 310–317. IEEE Computer Society, Los Alamitos (2004)
9. McCarthy, J., Anagnost, T.: Musicfx: an arbiter of group preferences for computer supported collaborative workouts. In: Maybury, M.T. (ed.) Proceedings of the 1998 ACM conference on Computer supported cooperative work - CSCW 1998, Seattle, pp. 363–372 (1998)
10. Modi, P.J., Shen, W.-M., Tambe, M., Yokoo, M.: An asynchronous complete method for distributed constraint optimization. In: Proc. of the Second International Joint Conference on Autonomous Agents and Multiagent Systems, pp. 161–168. ACM Press, New York (2003)
11. Petcu, A., Faltings, B.: A scalable method for multiagent constraint optimization. In: Kaelbling, L.P., Saffiotti, A. (eds.) Proceedings of the Nineteenth International Joint Conference on Artificial Intelligence, Edinburgh, Scotland, pp. 266–271. Professional Book Center (August 2005)
12. Resnick, P., Iacovou, N., Suchak, M., Bergstrom, P., Riedl, J.: Grouplens: An open architecture for collaborative filtering of netnews. In: Proceedings ACM Conference on Computer-Supported Cooperative Work, pp. 175–186 (1994)
13. Ricci, F., Cavada, D., Mirzadeh, N., Venturini, A.: Case-based travel recommendations. In: Fesenmaier, D.R., Woeber, K., Werthner, H. (eds.) Destination Recommendation Systems: Behavioural Foundations and Applications, pp. 67–93. CABI (2006)
14. Yokoo, M.: Distributed Constraint Satisfaction: Foundations of Cooperation in Multi-agent Systems. Springer, Berlin (2001)

Towards the Self-regulation of Personality-Based Social Exchange Processes in Multiagent Systems

Diego R. Pereira, Luciano V. Gonçalves, Graçaliz P. Dimuro, and Antônio C.R. Costa

Programa de Pó-Graduação em Informática, Universidade Católica de Pelotas
F. da Cunha 412, 96010-000 Pelotas, Brazil
{dpereira,llvarga,liz,rocha}@ucpel.tche.br

Abstract. This paper introduces an approach for the self-regulation of personality-based social exchange processes in multiagent systems, where a mechanism of social equilibrium supervision is internalized in the agents with the goal of achieving the equilibrium of the exchanges, guaranteeing the continuity of the interactions. The decision process concerning the best exchanges that an agent should propose to its partner is modeled as one Partially Observable Markov Decision Processs (POMDPs) for each agent personality trait. Each POMDP is decomposed into sub-POMDPs, according to the current balance of exchange values. Based on the relationship that may be established between POMDPs and Belief-Desire-Intention (BDI) architectures, the paper introduces an algorithm to map the policy graphs of the sub-POMDPs to BDI plans, allowing for a library of BDI plans. Simulations were carried out considering different degrees of supervision.

Keywords: self-regulation of social exchanges, multiagent systems, social simulation, Belief-Desire-Intention, Partially Observable Markov Decision Process.

1 Introdução

The regulation of agent interactions based on Piaget's theory of *social exchanges* [1] was first proposed in [2]. According to that theory, social interactions are seen as service exchanges between pairs of agents and the qualitative evaluation of those exchanges by the agents by means of *social exchange values*: the investment value for performing a service or the satisfaction value for receiving it. The exchanges also generate values of debts and credits that help to guarantee the continuity of future interactions. A multi-agent system (MAS) is said to be in *social equilibrium* if the balances of investment and satisfaction values are equilibrated for the successive exchanges occurring along the time. A (centralized) mechanism for the regulation of social exchanges in MAS, based on the concept of *equilibrium supervisor* with an associated Qualitative Interval Markov Decision Process (QI-MDP) was introduced in [2]. This approach was extended to consider *personality-based agents* in [3].

The aim of the present work is to go towards the (distributed) *self-regulation of personality-based social exchanges*, by decentralizing the equilibrium supervisor through the internalization of its decision process in the interacting agents. For simplicity, we consider that the exchanges between each pair of agents have no influence from

G. Zaverucha and A. Loureiro da Costa (Eds.): SBIA 2008, LNAI 5249, pp. 113–123, 2008.

the exchanges occurring between other pair of agents, which allowed for a simple way of distributing the decision process. However, since the agents do not necessarily have access to the internal states (balances of exchange values) of each other, the decision processes have to operate in a *partially observable* mode.

For convenience, we have chosen the BDI architecture for the agents in the social exchange simulator [3] used in this work. In the BDI architecture [4], agents have a set of beliefs about the states of the world and a set of desires, which identify those states that the agent has as goals. By a process of deliberation, the agent formulates one or more intentions (considered here as the states which the agent is committed to bringing about). The agent then builds a plan to achieve those intentions (through, e.g, some form of means-ends reasoning), and executes it. Since this is a heuristic approach, BDI agents are often outperformed by the those using *Partially Observable Markov Decision Processes* (POMDP) [5], when the POMDP solution is tractable. However, the BDI model can scale to handle problems that are beyond the scope of a POMDP solution, and can outperform approximate POMDPs for some relatively small problems [6].

POMDPs, the extension of MDPs to deal with partially observable environments [5], can be regarded as an ideal approach for the implementation of intelligent agents that have to choose optimal actions in partially observable stochastic domains. There are several algorithms for finding optimal policies for POMDPs (see, e.g., the *Witness* [5] algorithm, which yields policy graphs mapping actions to observations). However, the very nature of these algorithms, and in particular the need to establish the outcome of every possible action in every belief state, means that finding policies is intractable in many practical cases, particularly in large or complex problems [6].

Hybrid models have been proposed in order to take advantages of both POMDPs and BDI architecture (see, e.g., [6,7]). In [6], the formal relationships existing between certain components of the BDI model and those of POMDPs were discussed, showing how BDI plans and POMDP policies could be related to one another. It was shown that the plans derived from an optimal policy are those adopted by a BDI agent that selects plans with the highest utility, and an optimal reconsideration strategy.

In this paper, the decision about the best exchanges that an agent should propose to its partner in order to achieve social equilibrium, or to promote new interactions, is modeled as a global POMDP for each personality trait that its partner may assume. Considering a set of six personality traits (egoism, strong egoism, altruism, strong altruism, tolerance, equilibrium fanaticism), each global POMDP is decomposed into three sub-POMDPs, according to the current internal state (favorable, equilibrated or unfavorable balance of material exchange values) of the agent that is supervising the interaction.

Based on [6], we developed an algorithm to extract BDI plans from the policy graphs of each sub-POMDP, building a set of rules that form the BDI plans for interacting with agents of each personality model. Those plans are to be selected in each interaction according to the current balance of material values of the agents in each exchange stage. Such plans are said to “obey” optimal POMDP policies [6].

The paper is organized as follows. Section 2 briefly presents the modelling of social exchanges that is adopted in this work. In Sect. 3, we explain some personality traits that agents may assume in social exchanges. Our POMDP model for the self-regulation of personality-based exchanges is introduced in Sect. 4, and our approach for the extraction of BDI plans from those models is presented in Sect. 5. Results on simulations of personality-based exchanges are shown in Sect. 6. Section 7 is the Conclusion.

2 The Modelling of Social Exchanges

According to Piaget's approach to social interaction [1], a *social exchange* between two agents, α and β, involves two types of stages. In stages of type $\mathrm{I}_{\alpha\beta}$, α realizes an action on behalf of (a "service" for) β. The *exchange values* involved in this stage are the following: $r_{\mathrm{I}_{\alpha\beta}}$, which is the value of the *investment* done by α for the realization of a service for β (this value is always *negative*); $s_{\mathrm{I}_{\beta\alpha}}$, which is the value of β's *satisfaction* due to the receiving of the service done by α; $t_{\mathrm{I}_{\beta\alpha}}$ is the value of β's *debt*, the debt it acquired to α for its satisfaction with the service done by α; and $v_{\mathrm{I}_{\alpha\beta}}$, which is the value of the *credit* that α acquires from β for having realized the service for β. In stages of type $\mathrm{II}_{\alpha\beta}$, α asks the payment for the service previously done for β, and the values related with this exchange have similar meaning. The order in which the stages may occur is not necessarily $\mathrm{I}_{\alpha\beta} - \mathrm{II}_{\alpha\beta}$. The values $r_{\mathrm{I}_{\alpha\beta}}$, $s_{\mathrm{I}_{\beta\alpha}}$, $r_{\mathrm{II}_{\beta\alpha}}$ and $s_{\mathrm{II}_{\alpha\beta}}$ are called *material values* (investments and satisfactions), generated by the evaluation of *immediate exchanges*; the values $t_{\mathrm{I}_{\beta\alpha}}$, $v_{\mathrm{I}_{\alpha\beta}}$, $t_{\mathrm{II}_{\beta\alpha}}$ and $v_{\mathrm{II}_{\alpha\beta}}$ are the *virtual values* (credits and debts), concerning exchanges that are expected to happen in the future [1].

A *social exchange process* is composed by a sequence of stages of type $\mathrm{I}_{\alpha\beta}$ and/or $\mathrm{II}_{\alpha\beta}$ in a set of discrete instants of time. The *material results*, according to the points of view of α and β, are given by the sum of material values of each agent. The *virtual results* are defined analogously. A social exchange process is said to be in *material equilibrium* [2] if in all its duration it holds that the pair of material results of α and β encloses a given equilibrium point[1]. See also [2,3] for more details on this modelling.

3 Exchanges between Personality-Based Agents

In this paper, the material results constitute the *internal states* of an agent, while the virtual results that an agent manifests for the others constitute the *observations* that the others may make about it. Each agent is assumed to have access only to its own internal states, but it is able to make observations on what the others agents inform about their virtual results. Then, consider the set of internal states of an agent β given by $E_\beta = \{E^-_\beta, E^0_\beta, E^+_\beta\}$, where E^0_β denotes equilibrated results (e.g., around the zero), and E^-_β and E^+_β represent unfavorable results (e.g., negative results) and favorable results (e.g., positive results), respectively. The set of observable virtual results of an agent β is given by $\Omega_\beta = \{D_\beta, N_\beta, C_\beta\}$, where D_β and C_β means that the β has debts or credits, respectively, whereas N_β means that β has neither debts nor credits.

The agents may have different personality traits[2] that give rise to different state-transition functions, which specify, given their current states and the exchange proposal,

[1] Notice that Piaget's notion of equilibrium has no game-theoretic meaning, since it involves no notion of game strategy, and concerns just an algebraic sum.

[2] See [8] for a discussion on applications of personality-based agents in the context of MAS.

Table 1. State-transition function for a personality-based agent β

(a) Exchange stages of type $I_{\alpha\beta}$						
	Egoist agent			Altruist agent		
$\Pi(E_\beta)$	E^0	E^+	E^-	E^0	E^+	E^-
E^0	very-low	very-high	0.0	very-high	very-low	0.0
E^+	0.0	1.0	0.0	0.0	1.0	0.0
E^-	very-low	high	very-low	very-low	very-low	high
	Equilibrium fanatic agent			Tolerant agent		
$\Pi(E_\beta)$	E^0	E^+	E^-	E^0	E^+	E^-
E^0	very-high	very-low	0.0	very-low	very-high	0.0
E^+	0.0	1.0	0.0	0.0	1.0	0.0
E^-	high	very-low	very-low	high	high	very-low
(b) Exchange stages of type $II_{\alpha\beta}$						
	Egoist agent			Altruist agent		
$\Pi(E_\beta)$	E^0	E^+	E^-	E^0	E^+	E^-
E^0	very-high	0.0	very-low	very-low	0.0	very-high
E^+	low	high	very-low	very-low	very-low	high
E^-	0.0	0.0	1.0	0.0	0.0	1.0
	Equilibrium fanatic agent			Tolerant agent		
$\Pi(E_\beta)$	E^0	E^+	E^-	E^0	E^+	E^-
E^0	very-high	0.0	very-low	very-low	0.0	very-high
E^+	very-low	high	very-low	high	very-low	high
E^-	0.0	0.0	1.0	0.0	0.0	1.0

a probability distribution over the set of states that the interacting agents will try to achieve next. In the following, we illustrate some of those personality traits:

Egoism: the agent is mostly seeking his own benefit, with a very high probability to accept exchanges that represent transitions to favorable results (i.e., exchange stages in which when the other agent performs a service to it); if it has also a high probability to over-evaluate its credits and under-evaluate its debts, then it is a *strong egoist*;

Altruism: the agent is mostly seeking the benefit of the other, with a very high probability to accept exchanges that represent transitions toward states where the other agent has favorable results (i.e., exchange stages in which it performs a service to the other agent); if it has also a high probability to under-evaluate its credits and over-evaluate its debts, then it is a *strong altruist*;

Equilibrium fanaticism: the agent has a very high probability to accept exchanges that lead it to equilibrium state, avoiding other kinds of transitions;

Tolerance: the agent has a high probability to accept all kinds of exchanges.

Table 1(a) presents a pattern of the probability distribution $\Pi(E_\beta)$ over the set of states E_β, determined by the state-transition function that characterizes a personality-based agent β, when another agent α offers to perform a service to β (exchange stage of type $I_{\alpha\beta}$). Observe that, for an egoist agent β, transitions ending in favorable results (E^+) are the most probable, meaning that there is a very high probability that β will accept the service proposed by α. On the other hand, if β is an altruist agent then there is a high probability that it will refuse such proposal, remaining in the same state. See Table 1(b) to compare the converse probabilistic behavior of an egoist/altruist agent β

Table 2. Observation function for a personality-based agent β

	Egoist/altruist agent			Strong egoist agent			Strong altruist agent		
$\Pi(\Omega_\beta)$	N	D	C	N	D	C	N	D	C
E^0	very-high	very-low	very-low	very-low	very-low	very-high	very-low	very-high	very-low
E^+	very-low	very-high	very-low	very-low	very-low	very-high	0.0	1.0	0.0
E^-	very-low	very-low	very-high	0.0	0.0	1.0	very-low	very-high	very-low

when α requests a service from β (exchange stage of type $\mathrm{II}_{\alpha\beta}$). Now, if β is fanatic, then there is a very high probability that it will refuse transitions that would lead it far from the equilibrium state (E^0), in both types of exchange stages. However, there is a very high probability that a tolerant agent β will accept any kind of exchange proposal.

Table 2 shows a pattern of a virtual-value observation function that gives a probability distribution $\Pi(O_\beta)$ over the set of observations Ω_β that an agent α can do about the virtual values of a (strong) egoist/altruist agent β in any of β's (non-observable) states.

4 The Self-regulation of Social Exchange

The social control mechanism introduced in [3] is performed by a *centralized social equilibrium supervisor* that, at each time, decides on which actions it should recommend personality-based agents to perform in order to lead the system to the equilibrium, regarding the balance of the exchange values involved in their interactions. The supervisor takes its decisions based on a totally observable QI-MDP [2], which means that it has complete access to the internal states of the agents, through total observability.

For the self-regulation of social exchanges, it is necessary to decentralize the equilibrium supervisor, i.e., to distribute the decision process, which means to internalize the social control mechanism in the BDI agents, which is the agent architecture of the adopted social exchange simulator [3]. Although the distribution of the decision process is quite simply, since we consider that the exchanges between each pair of agents have no influence from the exchanges occurring between other pair of agents, one should assume that the internalized decision process has only *partial observability* of the states of the other agents, because one should not assume that agents necessarily have access to the internal states of each other. However, we assume that the agents are able to treat as observations the manifestations that the other agents make about the virtual results that they got from each exchange stage, and we assume that the agents necessarily make those manifestations, although not necessarily in a faithful way (see Table 2).

The decision process on the best exchanges an agent α should propose to an agent β, in order to achieve the equilibrium or to promote new interactions, is then modelled as a POMDP[3], denoted by $\mathrm{POMDP}_{\alpha\beta}$. For any pair of personality models, the full $\mathrm{POMDP}_{\alpha\beta}$ of the decision process internalized in the agent α has as states pairs of agent internal states (E_α, E_β), in which E_α is known by α, but E_β is non-observable. Then, for each personality model that β may assume, we decompose the $\mathrm{POMDP}_{\alpha\beta}$

[3] Decentralized POMDP's [9] were not considered here, since the agents do not perform joint actions; they perform a sequence of alternating individual actions in each exchange stage, with an agent deciding the action to perform after knowing the action previously done by its partner.

into three sub-POMDPs, one for each α's current internal state, denote by $\text{POMDP}^{-}_{\alpha\beta}$, $\text{POMDP}^{0}_{\alpha\beta}$ and $\text{POMDP}^{+}_{\alpha\beta}$, whenever α is in an unfavorable, equilibrated or favorable state, respectively.[4] Then, for each personality trait discussed in Sect. 3, we define:

Definition 1. *Let α be a personality-based agent, with an internalized equilibrium supervisor, and $E^*_\alpha \in \{E^-_\alpha, E^+_\alpha, E^0_\alpha\}$ be the current state of α. Let β be the personality-based agent that interacts with α. The POMDP for α's internalized equilibrium supervisor is defined as the tuple $POMDP^*_{\alpha\beta} = (E_\beta, A_\alpha, T_\beta, \Omega_\beta, O_\beta, R_\alpha)$, where:*

(i) $E_\beta = \{E^-_\beta, E^0_\beta, E^+_\beta\}$ *is the set of internal states of β;*
(ii) $A_\alpha = \{$**do-service**, **ask-service**$\}$ *is the set of actions available for α, with* **do-service** *meaning that α should offer a service to β (exchange stage of type $\mathrm{I}_{\alpha\beta}$), and* **ask-service** *meaning that α should request a service to β (exchange stage $\mathrm{II}_{\alpha\beta}$);*
(iii) $T_\beta = E_\beta \times A_\alpha \rightarrow \Pi(E_\beta)$ *is the state-transition function based on β's personality model, which, given β's current state and the action performed by α, states a probability distribution over E_β (i.e., the probabilities that β will accept or refuse the kind of exchange proposed by α);*
(iv) $\Omega_\beta = \{N_\beta, D_\beta, C_\beta\}$ *is the set of observations that may be realized by α about the information given by β about its virtual values;*
(v) $O_\beta = E_\beta \times A_\alpha \rightarrow \Pi(\Omega)$ *is the observation function based on the personality model of the agent β, which, given the action performed by α and β's resulting state, indicates a probability distribution over the set of possible observations Ω_β (i.e., the probability that β will inform that it has debts, credits or none of them);*
(vi) $R_\alpha = E_\beta \times A_\alpha \rightarrow \mathbb{R}$ *is the reward function for the agent α, given the expected immediate reward gained by α for each action taken.*

The solution of a $\text{POMDP}^*_{\alpha\beta}$, its *optimal policy*, helps the agent α to elaborate exchange proposals that may lead both agents towards the equilibrium. We use the *Witness* algorithm [5] to compute the policy graphs for the resulting eighteen models, although any other algorithm for POMDPs can be used.

5 Building BDI Plans from the $\text{POMDP}^*_{\alpha\beta}$ Policy Graphs

In [6], it was presented a formal discussion on the relationship between the *MDP description* and the *BDI description* of an agent's decision procedure. For a given agent in a given environment, the *state space*, the *set of actions* and the *state-transition function* may be considered to be the same for both descriptions. Then, since rewards are a means for determining policies, and desires are a step on the way to determine intentions, these two instrumental components can be effectively ignored. Indeed, the relation that should be considered in detail is that between *policies*, in the MDP description, and *plans* to achieve intentions, in the BDI description.

In fact, [6] showed that it is possible to obtain a BDI plan for a determined intension in which the actions prescribed by the plan are the same as those prescribed by the MDP

[4] The decomposition of the full $\text{POMDP}_{\alpha\beta}$ allowed to reduce the state space and to obtain directly the BDI plans for each α's current state, which may change after each exchange stage.

policy through the plan's intermediate states. Such plan is said to *obey* the MDP policy. They proved that the plans derived from an optimal policy are those adopted by a BDI agent that selects plans with the highest utility, and an optimal reconsideration strategy.

This correspondence holds under rather restrictive assumptions, in particular the optimality requirements, but ensures that the BDI plans generated reflect an optimal policy. If we are willing to relax the requirement for a set of plans to correspond to an optimal policy, we can create plans under fewer restrictions.

Based on the considerations above and in the pseudocode for mapping MDP policies into intentions presented in [6], we developed the algorithm **policyToBDIplans** to extract BDI plans from the $\text{POMDP}^*_{\alpha\beta}$ policy graphs (Algorithm 1). This algorithm was used to build a library of BDI plans for the internalized equilibrium supervisor. It takes as input a policy graph and produces as result a BDI plan that follows the policy represented in that policy graph. In the transformation from a policy graph to a set of BDI plans, nodes of the policy graph of a POMDP agent become achievement goals for the BDI agent, and the arcs of the policy graph, representing observation-based sequencing of actions in the POMDP agent, become plan rules for the BDI agent.

Algorithm 1. Algorithm for mapping $\text{POMDP}^*_{\alpha\beta}$ policy graphs into BDI plans

```
policyToBDIplan(Policygraph pg)
Policygraphline pl
BDIplan plan
BDIrule ruleN, ruleD, ruleC
set pl to firstLine(pg)
repeat
    set ruleN to BDIrule(head={Node(pl)}, context={True}, body={act(Action(pl)), aGoal(Node(pl'))})
    set ruleN to BDIrule(head={Node(pl')}, context={ObsN}, body={aGoal(ObsN(pl))})
    set ruleD to BDIrule(head={Node(pl')}, context={ObsD}, body={aGoal(ObsD(pl))})
    set ruleC to BDIrule(head={Node(pl')}, context={ObsC}, body={aGoal(ObsC(pl))})
    addBDIrules {ruleN, ruleD, ruleC} to plan
    advance ln
until ln beyond lastLine(pg)
return Plan
```

For example, if the agent α (with an internalized equilibrium supervisor) is interacting with an egoist agent β, and its current state is E^0_α, the policy graph of the related $\text{POMDP}^0_{\alpha\beta}$, which is the input for the algorithm **policyToBDIplans**, is presented in Table 3. Table 4 shows the rule for the goal State(01) of the BDI plan, in the AgentSpeak language [4], extracted from those policy graph by the algorithm **policyToBDIplans**.

6 Simulation of Personality-Based Social Exchanges

The implementation of the personality-based social exchange simulator was done in *AgentSpeak* using *Jason* [4]. For each personality trait, we performed 20 simulations of unsupervised and 20 simulations of supervised exchange processes of 500 interactions each, which allowed us to calculate the average balance of exchange values and the average number of proposed and accepted/rejected exchanges. Each simulation generated a graphic of the evolution of the material results of the agents in 500 interactions. The equilibrium state was considered as the interval $[-2, 2]$.

Table 3. Policy graph of the $\text{POMDP}^0_{\alpha\beta}$ for an egoist agent β

Node	Action	Obs. N	Obs. D	Obs. C
01	00	25	19	01
15	00	23	19	19
19	01	26	19	01
22	01	28	19	01
23	01	30	19	01
25	01	31	22	01
26	01	32	22	01
28	01	34	23	01
30	01	34	25	01
31	01	34	26	01
32	01	34	28	01
33	01	34	14	01
34	01	34	15	01

Table 4. Rule for achievement goal State(01), extracted from the policy graph of Table 3

```
+!State(01): True ->
          act(do-service),
          !State(01').
+!State(01'): obs==N ->
          !State(25).
+!State(01'): obs==D ->
          !State(19).
+!State(01'): obs==C ->
          !State(01).
```

Encoding: 00=action **do-service**, 01=action **ask-service**

6.1 Unsupervised Personality-Based Social Exchanges

We have built a *test* agent α that randomly proposed exchanges of types $\text{I}_{\alpha\beta}$ or $\text{II}_{\alpha\beta}$ to a personality-based agent β, in order to analyze the behaviour of the personality model in unsupervised interactions. The average balance of material exchange values of a tolerant agent β was -67.2 and it refused only 20% of α's exchange proposals. On the other hand, an equilibrium fanatic agent β accepted only 16.9% of α's exchange proposals, staying in the equilibrium zone in the most part of the time with an average balance of material exchange values equal to -0.04, obstructing the continuity of interactions. The average balance of material exchange values of an altruist agent β was -950.5, showing its preference by performing services for α instead of accepting services from him (it accepted 80.6% of exchange stages of type $\text{II}_{\alpha\beta}$ and 29.5% of exchange stages of type $\text{I}_{\alpha\beta}$). On the other hand, an egoist agent β presented average balance of material exchange values of 1,473.4, which showed its preference for receiving services (it accepted 89.5% of exchange stages of type $\text{I}_{\alpha\beta}$ and 10.4% of stages of type $\text{II}_{\alpha\beta}$).

Figure 1 shows examples of simulations of unsupervised exchanges between the test agent α, and tolerant (b) and altruist (d) agents β.

6.2 Supervised Personality-Based Social Exchanges

In supervised exchanges, with a supervision degree of 100%, the simulations showed that the agent α (with the internalized equilibrium supervisor) was really effective in its interactions with a tolerant agent β, with an average of 80% accepted exchanges, where β's average balance of material exchange values achieved -0.37 (i.e., controlled in the equilibrium zone). Even a supervision degree of 50% is sufficient to get the equilibrium, with β's average balance of material exchange values of -0.67.

Considering supervised exchanges, with a supervision degree of 100%, between α (with the internalized equilibrium supervisor) and an altruist agent β, β's average balance of material exchange values was -1.84 (i.e., in the equilibrium zone). The agent α generated more exchanges stages of type $\text{I}_{\alpha\beta}$ (79.8%) than of type $\text{II}_{\alpha\beta}$, in order to promote exchanges in which β could increase its material results. Supervision degrees around 90%, although not effective in leading β to the equilibrium, might be able to

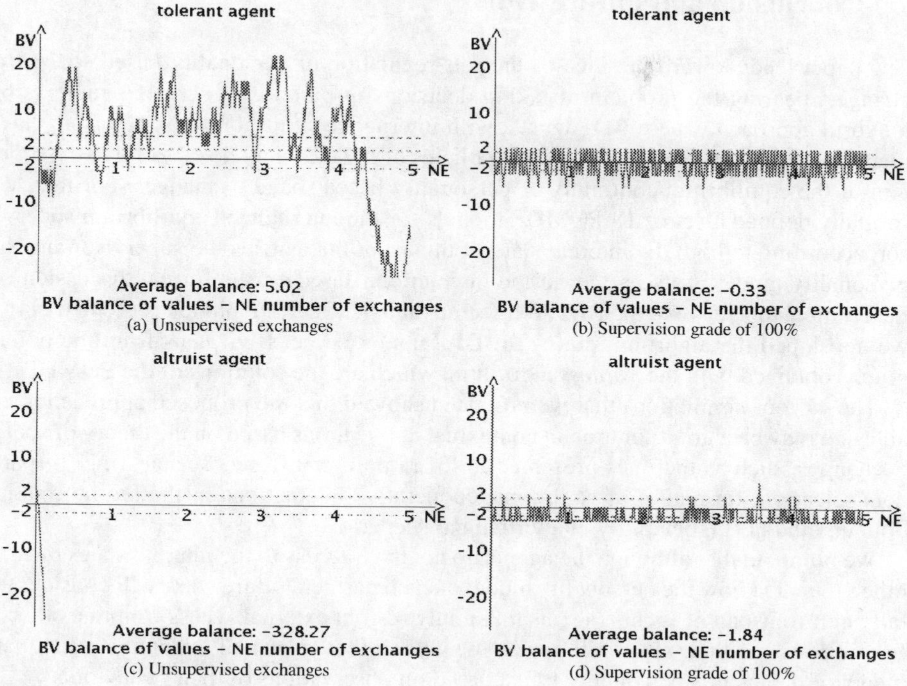

(a) Unsupervised exchanges

(b) Supervision grade of 100%

(c) Unsupervised exchanges

(d) Supervision grade of 100%

Fig. 1. Examples of simulations of the behaviour of a personality-based agent β in 500 exchanges

increase its results (in this case, β's average balance of material values increased 99.5%, in comparison with unsupervised exchanges).

We obtained analogous results for supervised exchanges, with a supervision degree of 100%, for an egoist agent β, where α generated more exchanges stages of type $II_{\alpha\beta}$ (71.7%) than of type $I_{\alpha\beta}$ and β's average balance of material values was 2.0. Using a supervision degree of 90%, β's average balance of material exchange values was 3.5, representing a decreasing of 99.8%, in comparison with unsupervised exchanges.

However, simulations showed that it was difficult to lead a strong altruist agent β to the equilibrium, do to its non-realistic self-evaluation of its virtual results. However, α succeeded in increasing β's average balance of material exchange values to -5.2 (99.5% in comparison with unsupervised exchanges). Analogous results were obtained for a strong egoist agent β, where β's average balance of material exchange values was 7.6 (a decreasing of 99.5% in comparison with unsupervised exchanges).

Figure 1 shows examples of simulations of supervised exchanges between agent α (with the internalized equilibrium supervisor), and tolerant (a) and altruist (c) agents β.

The simulations showed that the proposed approach succeeded in helping the agents to control, by themselves, the results of their personality-based exchanges, trying to maintain the material equilibrium. For a comparison with a centralized approach, see [3].

7 Conclusion and Future Work

The paper leads toward the idea of the self-regulation of personality-based social exchanges, bearing the problem of taking decisions in partially observable contexts by a hybrid approach, where BDI agents, with internalized equilibrium supervisors, have plans constructed according to optimal policies of POMDPs modeled to solve the problem of the equilibrium/continuity of personality-based social exchanges. For that, we formally defined a set of 18 POMDP models for the internalized equilibrium supervisor, according to both the internal state of the agent that carries the supervisor and the personality model of its partner in the interaction. Based on the formal discussion on the relationship between the BDI architecture and the POMDP model, presented in [6], we developed the algorithm **policyToBDIplans** to extract BDI plans from the policy graphs obtained with the *Witness* algorihtm, which are the solutions of the POMDPs.

The various simulations that we produced showed that the proposed approach is viable and may be a good solution in contextual applications based on the theory of social exchanges, such as the ones presented in [3] (a politician/voters scenario), [10] (partners selection, formation of coalitions, cooperative interactions in the bioinformatics domain) and [11] (sociability in a virtual bar scenario).

We point out that although the agents do not have access to the internal states of each other, they do know the personality model of each partner. Future work will consider totally non-transparent agents (agents that totaly restrict external access from the others). In this case, the supervisor will have to be able to recognize and maintain an adequate model of the agent's personality traits, based on observations of their behaviors.

Acknowledgements. This work was supported by Petrobrás (joint project with PENO/COPPE/UFRJ), CAPES, FAPERGS and CNPq. We are grateful to the referees for their comments.

References

1. Piaget, J.: Sociological Studies. Routlege, London (1995)
2. Dimuro, G.P., Costa, A.C.R., Palazzo, L.A.M.: Systems of exchange values as tools for multi-agent organizations. Journal of the Brazilian Computer Society 11(1), 31–50 (2005)
3. Dimuro, G.P., Costa, A.C.R., Gonçalves, L.V., Hübner, A.: Centralized regulation of social exchanges between personality-based agents. In: Noriega, P., Vázquez-Salceda, J., Boella, G., Boissier, O., Dignum, V., Fornara, N., Matson, E. (eds.) COIN 2006. LNCS (LNAI), vol. 4386, pp. 338–355. Springer, Heidelberg (2007)
4. Bordini, R.H., Hübner, J.F., Wooldrige, M.: Programming Multi-agent Systems in AgentSpeak Using Jason. Wiley Series in Agent Technology. John Wiley & Sons, Chichester (2007)
5. Kaelbling, L.P., Littman, M.L., Cassandra, A.R.: Planning and acting in partially observable stochastic domains. Artificial Intelligence 101(1-2), 99–134 (1998)
6. Simari, G.I., Parsons, S.: On the relationship between MDPs and the BDI architecture. In: Nakashima, H., Wellman, M.P., Weiss, G., Stone, P. (eds.) Proc. of the 5th Intl. Joint Conf. on Autonomous Agents and Multiagent Systems, Hakodate, pp. 1041–1048. ACM, New York (2006)
7. Nair, R., Tambe, M.: Hybrid BDI-POMDP framework for multiagent teaming. Journal of Artificial Intelligence Research 23, 367–420 (2005)

8. Castelfranchi, C., Rosis, F., Falcone, R., Pizzutilo, S.: A testbed for investigating personality-based multiagent cooperation. In: Proc. of the Symposium on Logical Approaches to Agent Modeling and Design, Aix-en-Provence (1997)
9. Bernstein, D.S., Givan, R., Immerman, N., Zilberstein, S.: The complexity of decentralized control of MDPs. Mathematics of Operations Research 27(4), 819–840 (2002)
10. Rodrigues, M.R., Luck, M.: Cooperative interactions: an exchange values model. In: Noriega, P., Vázquez-Salceda, J., Boella, G., Boissier, O., Dignum, V., Fornara, N., Matson, E. (eds.) COIN 2006. LNCS (LNAI), vol. 4386, pp. 356–371. Springer, Heidelberg (2007)
11. Grimaldo, F., Lozano, M., Barber, F.: Coordination and sociability for intelligent virtual agents. In: Sichman, J.S., Padget, J., Ossowski, S., Noriega, P. (eds.) COIN 2007. LNCS (LNAI), vol. 4870, pp. 58–70. Springer, Heidelberg (2008)

Probabilistic Multiagent Patrolling

Tiago Sak, Jacques Wainer, and Siome Klein Goldenstein

Instituto de Computação, Universidade Estadual de Campinas,
CEP 13084-851, Campinas, SP - Brasil
{tiago.sak,wainer,siome}@ic.unicamp.br

Abstract. Patrolling refers to the act of walking around an area, with some regularity, in order to protect or supervise it. A group of agents is usually required to perform this task efficiently. Previous works in this field, using a metric that minimizes the period between visits to the same position, proposed static solutions that repeats a cycle over and over. But an efficient patrolling scheme requires unpredictability, so that the intruder cannot infer when the next visitation to a position will happen. This work presents various strategies to partition the sites among the agents, and to compute the visiting sequence. We evaluate these strategies using three metrics which approximates the probability of averting three types of intrusion - a random intruder, an intruder that waits until the guard leaves the site to initiate the attack, and an intruder that uses statistics to forecast how long the next visit to the site will be. We present the best strategies for each of these metrics, based on 500 simulations.

1 Introduction

Patrolling an environment can be seen as finding efficient ways of performing visits to all the important points of a given area. This task can be considered inherently multiagent-like since in most cases the process will be started in a distributed manner, by a group of agents. Research on multiagent patrolling, however, is not limited to patrolling real-world problem, but they can find applications on several domains, such as network security systems and games. In other words, patrolling can be useful in any domain characterized by the need of systematically visiting a set of predefined points. Additionally, it is important to keep the visit order secret, since an intruder could use it to plan a path that avoids being seen by the patrols.

The existing works in the area give priority to shorten the time spent between visits to all the regions in the environment, proposing methodologies without any kind of variation in the order which the points are visited. Although these methods achieve an efficient way of performing the visits, the order with which the nodes are visited can be easily deducted by any external intruder. By knowing the visit-order and the period between visits, an intruder can mount a successful intrusion. For example, if the intruder knows that a security guard visits a safe with a combination lock every 50 minutes regularly, he may successfully open

G. Zaverucha and A. Loureiro da Costa (Eds.): SBIA 2008, LNAI 5249, pp. 124–133, 2008.

the lock, even if the whole process takes longer than 50 minutes. By interrupting his attempt just before the guard arrives and hiding, and because his attempt does not leave a traces in the lock that can be recognized by the guard, the intruder will have time to try all lock combinations until he can open the safe. Of course if the intruder is trying to open the safe by cutting it with a welding torch, the interruption trick will not work. If cutting the safe takes 60 minutes, then a guard that returns every 50 minutes guarantees that the intrusion will be stop. But if cutting the safe takes 40 minutes, and the intruder can wait until the guard leaves the safe to start the attack, then a fixed period of 50 between visits will not detain the intrusion. Thus, depending on the type of intrusions, unpredictability of the period of the visits, or at least variability may be more important than shortening the period.

In this work, we will model the problem of patrolling as a problem of visiting vertices in a graph. The patrol will move from vertex to vertex, transversing the edges. Each edge has a value that corresponds to the time needed to traverse it. Each vertex is a place that needs to be observed for intruders, but we will assume that such observation are instantaneous. Each patrol will be called an "agent". The intruder will choose a vertex, and will stay some time at that location to achieve the intrusion or attack. If a patrol visits the vertex while the intruder is there, the attack has been averted. If not, the attack is successful and the patrolling task failed. Intruders do not move in the graph.

The multiagent patrolling task can be partitioned into two almost orthogonal decisions: if and how to divide the work, and how to select the visiting sequence for each agent. The first decision is whether all agents will patrol the whole graph or if each one will have a different subgraph to patrol. We call the second alternative the decision to *partition* the graph, and for the lack of better name, we call the alternative a *non-partitioning* decision. If the graph will be partitioned, then one has to decide on how to select the subgraphs that will be assigned to each agent. We call it the *partition algorithm*.

Independent of the partitioning decision, one has to select the sequence of vertex visitation for each agent. We call this the *sequencing algorithm*. In this paper we will only deal with *homogeneous* agents, that is, they all use the same sequencing algorithm. The combination of the partitioning decision, the partition algorithm (if needed) and the sequencing algorithm is known as a *strategy*.

This paper is organized as follows: the next section briefly describes previous works on multiagent patrolling; section 3 introduces our approach in details; section 4 presents an experimental evaluation; and, in section 5, we discuss the results and directions for the future.

2 Related Work

The use of multiples agents, acting cooperatively or not, to perform search of intruders or a patrolling tasks, have been studied by many researchers. [1] proposed the combination of map-learning and pursuit of invaders in a single problem, using a *greedy* policy that directs the pursuers at each instant to the location that maximize the probability of finding an invader. [2] studied the problem of generating near-optimal trajectories that will be followed by several agents to co-operatively search for targets in an environment with some a priori information about the target distribution. [3] investigated multiagent pursuit of targets that become observable for short time periods, this is used by each agent to estimate the next position of any intruder. [4] studied the use of stochastic rules to guide agents motions in order to perform the surveillance task. Their main interest was to define the rate at which the Markov chain converges to its steady state distribution. Thus they have not defined any evaluation criteria to precisely measure effectiveness of their strategies, and have not made any comparison with other possible architectures.

Very relevant to this work, is the work of Machado et al[5], which was followed by others [6,7,8,9]. These works also model the problem of patrolling as graph traversal. Their goal was to reduce the period between two consecutive visits to any vertex. In [5] they explore different alternatives for a local decision (by each agent) on the next vertex to visit, based on properties of the agents and communication abilities. [8] considered the use of reinforcement learning to discover plans for traversing the graph. All these are variations to the sequencing algorithm. [9] explores alternatives for the partitioning algorithm based on different alternative of negotiations protocols among the agents.

[6] uses the well-know *travelling salesman problem* (TSP) as a sequencing algorithm. Consider the problem of minimize the period between two consecutive visits to a vertex. This problem can be seen as finding the minimal cycle that contains all the vertexes in the graph, which is exactly the TSP definition. Although this is a *NP-hard* problem, there techniques to solve TSP can now determine the optimal result on very large instances of the problem. The standard TSP solution is designed for one agent traveling the graph. Two multiagent extensions were proposed: the first dispose the agents on the TSP-cycle, keeping an approximate distance between them. The second strategy partitions the graph into disjoint regions, each agent is assigned to patrol one region. Except for specific cases, the cyclic strategy achieved better result. As the expected, the TSP-based solution achieved the best results of all other alternatives for most cases.

3 The Probabilistic Patrolling Problem

3.1 Evaluation Criteria

We define three evaluation criteria for any solution to the problem. These criteria are approximations to detecting an attack by three different kind of intruder.

The **random intruder** will start an attack on a random node at a random time. The attack must last A to be successful. If within the time interval A, a agent visits the attacked vertex, the intrusion is averted.

The **waiting intruder** will wait (hidden) until the agent leaves a (random) vertex and start the attack then.

The **statistical intruder** will collect statistics on the period between visits to a random node and will initiate the attack whenever the data assures that it is safe to attack. The statistical intruder keeps a visit history, and computer the correlation ρ between the period of two consecutive visits to its goal node. Given that the last period to visit the node was x, the attacker will search the visit history to the node and find out the closest period to x in the record, and discover the following visit period y. If $y \times \rho > A$ the intruder will initiate the attack. ρ is a measure of how certain the intruder is that the next period will be y (because it was so in the past).

These intruders are in an increasing order of sophistication. The metrics are approximations to the probability of averting each of these intruders. PRI(A) is the probability of catching a random intruder with attack of length A. PWI(A) is the probability of catching a waiting intruder, and PSI(A) is the probability of catching a statistical intruder.

3.2 Sequencing Algorithms

We discuss two main alternatives to the sequencing algorithm. The random walk alternatives, have high variability and unpredictability, while the TSP-based solutions achieve shorter visiting periods, but have lower variability.

Random walk based: The intuitive idea of the *random walk* is to take successive steps, each one in a random direction. In graphs, random walks are discussed in [10]. There are theoretical results that with enough time every vertex in a graph will be reached, and there are lower and upper bounds to cover completely a graph. Since, every vertex will be visited in a random walk, one can consider using random-based solutions as the sequence algorithm. We propose two random based algorithms:

- **Local random algorithm:** The agent will choose randomly one vertex among all the adjacent nodes.
- **Global random algorithm:** The agent will choose randomly any vertex of the graph to be the next objective-node. To reach this vertex in case there is no direct edge between the current vertex and the objective-node, the agent will use the *shortest path* between them.

TSP-Based: As discussed above, a solution to the *travelling salesman problem* (TSP) on the graph is clearly a sequence that will minimize the period between visits to each node in the graph. But this solution has no variability or unpredictability, and thus will not stop some of the intruders discussed above. But given the TSP solution to a graph, there are alternatives to add variability to the sequence of visits.

- **Original TSP:** This is the sequence generated by the TSP solution of the graph. The TSP cycle is repeated continuously until the end of the task.
- **TSP with local visits:** The TSP cycle is covered over and over, but for each node visited, the agent decides randomly with probability $LV_{\%}$ if a *local*

visit will be performed or not. A *local visit* consists in visiting one neighbor of the current node, after this the agent returns to the original node and the next node in the cycle will be visited.
- **TSP with local changes:** This algorithm proposes that for each cycle the agent will perform a random local change in the TSP-cycle. Two nodes are chosen randomly and the order in which they appear in the TSP-cycle is exchanged. For example, if the original TSP-cycle was $\{v_1, v_2, v_3, v_4, v_5, v_6\}$, and the two nodes chosen were v_2 and v_5 then the new cycle would be: $\{v_1, v_5, v_3, v_4, v_2, v_6\}$.
- **TSP rank of solutions:** This algorithm is based on the fact that although there may be only one optimal solution, sub-optimal solutions to the TSP problem may also be efficient, and will provide variability. It is possible to some branch and bound TSP solvers to generate also other, sub-optimal solutions. Every time a feasible solution is found, it is stored in a priority queue. The TSP solver will return not only the optimal solution (which is the first element in the queue), but the K first elements of the heap, provided the cost of each solution is less than twice the cost of the optimal solution. The TSP rank of solutions algorithm will for each cycle, choose randomly one of the K solutions returned by the altered TSP solver.

3.3 Partitioning

Non-partitioning Cyclic Strategy: In this strategy all the agents will act with the same set of nodes. Although the vertices which the agents will have access are the same, they will act independently making their decisions without any influence from others agents. Nevertheless, it is very important to defined a way in which the agents will be distributed initially in the graph. Two alternatives are presented:

- **Random cycle:** in this alternative the agents will be distributed randomly through all the vertices of the graph, the only criteria followed was that one vertex can not be assigned to more than one agent.
- **Approximate equal distance cycle:** the second alternative distribute the agents equidistantly in the TSP-cycle, that is, from one agent to the next in the cycle there exists a constant distance, that will be repeated to all the agents. This constant distance is equal to the optimal TSP-solution size divided by the number of agents. In many cases, however, the perfect distribution of the agents will not be possible. Hence, this strategy finds the best, but not necessarily optimal solution. This is achieved by verifying all the possibilities and choosing the one which less distortion of the ideal equidistant distribution.

Partition algorithms: The problem of partitioning a graph is well known and important in a wide range of applications. There are, however, many possible approaches to perform this task, all of them based on different premises and goals. Three possibilities were explored in this work.

- **Multilevel graph partitioning:** This partitioning scheme gives priority to the size of the resulting partitions. Although, for the multiagent patrolling problem a fair partitioning of the graph is one where the distances travelled by each agent are equal, we approximate this requirement to determining subgraphs with approximately the same number of vertexes.
- **K-Means:** We use the K-means clustering algorithm [11] to partition the graph in K clusters based on the euclidean distance of each vertex to the prototype of each cluster (see section 4.1 for the discussion on "euclidean distance" of vertexes).
- **Agglomerative hierarchical clustering:** Agglomerative hierarchical clustering [11] is also a clustering algorithm, and in particular using the *single linkage* alternative for metric for the algorithm seems to closely correspond to the idea that close by vertexes should be joined into the same cluster, and that each agent should be responsible for one of the K clusters.

After the partitioning process, each agent will be assigned to each partition. And without any influence or knowledge about another agents, the patrol task will be performed in its partition.

4 Results

4.1 Experimental Scenario

To evaluate the different alternatives, we developed a *graph generator* that generates instances of a patrolling problem. We will now describe the way graphs are generated. Parameters of the generation process will be written in uppercase, and are defined manually.

1. Randomly decide the number of vertices $n \leq MAX_n$ that the graph will contain. The vertices will occupy a 2D square of size MAX_c
2. Each vertex $v \in V$ have their 2D coordinates chosen randomly provided that different vertexes are sufficiently distant from one another, that is $\forall (v_1 \neq v_2) \in V, dist(v_1, v_2) \geq MIN_{dist}$ where $dist(v_1, v_2)$ is the euclidean distance.
3. Create edges between each pair of vertexes and randomly select $elim_v < CH_{elim}$ of those edges to be eliminated, but keeping the graph connected.
4. For each edge e that remains in E, calculate the euclidean distance between vertexes $W - e$. With probability $1 - CH_d$ the weight of the edge will be W_e and with probability CH_d the weight will be the distance multiplied by a distortion D_e randomly selected from 1 to MAX_d.

4.2 Simulation Details

In our experiments we used 500 graphs, the minimal distance between vertexes (MIN_{dist}) was 2, the size of the 2D square (MAX_c) was defined as 100×100. Each graph had between (MIN_n) 80 and (MAX_n) 100 vertexes. The chance of an endge has its weight distorted (CH_d) was 15% and the distortion (MAX_d) was at most 20%. The chance of an edge been removed (CH_{elim}) was 5%. Each experiment were run with 2, 4 and 6 agents. For each of the 500 evaluation scenarios created, we ran all strategies. Each simulation had a total time execution limited by $100 \times TSP_{cycle}$, where TSP_{cycle} is the time needed to travel the TSP-cycle of the graph. Each time we used the k-means to partition a graph, the algorithm was executed five times, with different seeds, and the best solution was chosen.

For the TSP based solutions, we used TSP solvers based on 1-tree relaxation of the branch and bound algorithm (see [12]). For the multilevel graph partitioning, we used the software METIS [13].

The PRI, PWI, and PSI are evaluated on just one randomly selected vertex of the graph. To calculate the PRI we use a Monte Carlo approach - given all visit times for the vertex, we generate 50 random attacks for each simulation, and count the number of failed attacks. The PWI and PSI are calculated deterministically - for each simulation we compute the number of failed attacks, and the number of tries. We used 5 different attack intervals based on the TSP-cycle divided by the number of agents, which we abbreviate as T/n: $\frac{1}{8} \times T/n$, $\frac{1}{4} \times T/n$, $\frac{1}{2} \times T/n$, $1 \times T/n$ and $2 \times T/n$.

4.3 Results

Tables 1, 4.3 and 3 are the main results of this work. They list for all attack intervals, the average rate with which the intrusion was averted in the five best solutions found considering separately each evaluation criterion. We calculated this average rate considering only experiments that have at least one attack try. The tables are ordered by decreasing value of general effectiveness (the average for all values). Table 1 is the data for PRI, table 4.3 for PWI, and table 3 for PSI.

Table 1. PRI

Number of Agents	**Partition Scheme**	**Sequencing Algorithm**	**PRI(%)** $\frac{1}{8} \times \frac{T}{n}$	$\frac{1}{4} \times \frac{T}{n}$	$\frac{1}{2} \times \frac{T}{n}$	$1 \times \frac{T}{n}$	$2 \times \frac{T}{n}$
2	K-Means	Original TSP	12	24	47	91	100
	K-Means	TSP local visits	11	22	44	84	99
	Equal distance cycle	Original TSP	12	23	43	76	99
	K-Means	TSP rank	11	22	43	79	98
	Random cycle	Original TSP	12	23	43	74	99
4	K-Means	Original TSP	12	23	47	87	100
	K-Means	TSP local visits	11	22	44	83	99
	K-Means	TSP rank	11	22	42	77	98
	Hierarchical	Original TSP	12	24	44	68	93
	Equal distance cycle	Original TSP	12	23	41	69	94
6	K-Means	Original TSP	13	25	49	87	100
	K-Means	TSP local visits	12	24	48	85	100
	K-Means	TSP rank	12	23	45	78	98
	Hierarchical	Original TSP	12	24	46	76	96
	Hierarchical	TSP local visits	12	23	45	73	94

Table 2. PWI

Number of Agents	Partition Scheme	Sequencing Algorithm	PWI(%)				
			$\frac{1}{8} \times \frac{T}{n}$	$\frac{1}{4} \times \frac{T}{n}$	$\frac{1}{2} \times \frac{T}{n}$	$1 \times \frac{T}{n}$	$2 \times \frac{T}{n}$
2	Random cycle	Original TSP	7	12	25	50	100
	Equal distance cycle	Original TSP	6	11	25	50	100
	Equal distance cycle	TSP rank	7	14	26	50	89
	Random cycle	TSP rank	7	13	26	50	89
	Random cycle	TSP local visits	6	12	24	46	89
4	Random cycle	Original TSP	8	17	32	57	88
	Equal distance cycle	Original TSP	8	16	32	57	88
	Random cycle	TSP rank	9	17	32	56	85
	Equal distance cycle	TSP rank	9	17	32	56	85
	Equal distance cycle	TSP local visits	9	16	31	54	83
6	Equal distance cycle	Original TSP	9	19	35	60	87
	Random cycle	Original TSP	9	19	35	59	87
	Random cycle	TSP rank	10	19	34	58	85
	Equal distance cycle	TSP rank	10	19	34	58	84
	Equal distance cycle	TSP local visits	9	18	33	56	83

Table 3. PSI

Number of Agents	Partition Scheme	Sequencing Algorithm	PSI(%)				
			$\frac{1}{8} \times \frac{T}{n}$	$\frac{1}{4} \times \frac{T}{n}$	$\frac{1}{2} \times \frac{T}{n}$	$1 \times \frac{T}{n}$	$2 \times \frac{T}{n}$
2	Random cycle	TSP rank	6	10	17	28	81
	Equal distance cycle	TSP rank	6	11	18	28	72
	Equal distance cycle	TSP local visits	6	12	22	29	44
	Random cycle	TSP local visits	5	11	20	27	43
	Multilevel partition	TSP local visits	1	4	7	10	56
4	Random cycle	TSP rank	8	16	29	51	100
	Equal distance cycle	TSP rank	9	16	29	50	92
	Random cycle	TSP local visits	8	15	26	50	88
	Equal distance cycle	TSP local visits	8	14	26	47	80
	Equal distance cycle	TSP local changes	7	13	24	44	85
6	Random cycle	TSP rank	10	18	33	59	100
	Equal distance cycle	TSP rank	10	18	33	57	100
	Random cycle	TSP local visits	9	16	31	55	100
	Equal distance cycle	TSP local changes	8	15	28	48	94
	Random cycle	TSP local changes	8	15	27	48	90

5 Discussion and Conclusions

The reliable patrolling is a complex problem, requiring solutions that integrate efficiency and unpredictability. The main contribution of this paper consists into presenting 3 new metrics to evaluate the patrolling problem. Each metric considers different kinds of invaders. Based on this metrics we proposed and compared various partition schemes and sequencing algorithms.

Our results point out that for invaders that act randomly - the attack is made without any knowledge about the agents - the traditional partitioning schemes are more effective than use non-partitioning strategies. However, when the attacker have some information about the agents, like an historical of visits to a specific place, or even if the invader can only perceives when an agent visit some place, the non-partitioning strategies perform better. It is also important to notice that in the non-partitioning strategies the best equal distance distribution of the agents not necessarily guarantee a better solution, since a random distribution can also contribute to the unpredictability of the strategy. Another important result is that for invaders that use statistical information to plan his actions the sequencing algorithm that achieve the best results were the TSP rank, which is not a static solution, like the original TSP strategy. This corroborate our assumption that unpredictability is an essential characteristic to the patrolling task. And as a general rule, random walk based sequencing algorithms, although very unpredictable, have so long periods between visits, that they are not usefull to avert any of the three attackers modeled.

When the attacker acts randomly or with very restricted information, perform the TSP cycle over and over was the best solution found. This happens because this approach finds the cycle that covers all the nodes with the minimal time needed, so it maximizes the number of times that all the vertexes will be visited, raising the chances that an agent averts an invasion from a random or almost-random invader.

The solutions proposed in this paper are mainly centralized solutions. For example, the TSP with local visits selection depends on calculating the TSP solution to the graph, which requires not only global knowledge of the graph but enough computational power. But once the solution is computed and distributed to the agents, they each perform based on local decisions, regardless of the other agent's decisions. Other architectures, for example, the global random architecture are also based on local decisions, and require global knowledge of the graph, but are not that computationally expensive.

From one point of view, if the terrain being patrolled is static, and if the patrolling will last for a long time, then it is probably worth to compute the "best" solution to the problem in a centralized way. In situations in which a global knowledge is not possible, or it is not worth to gather all the local knowledge into a global one, and spend time computing the best solution, a distributed way of achieving the solution could be interesting, and our results can be seen as an upper bound to what can be achieved with the distributed problem solving.

Extensions to this work include, for instance, analysis of more other strategies, modifications on the scenario generator to include new specific cases; inclusion of new characteristics to the patrolling task, for example, prioritized regions and dynamic topologies. Another important extension is to study the scenario where there is redundancy in the number of agents that visits a place, so that the compromise of one/few patrolling agent does not compromise the system as a whole.

References

1. Hespanha, J., Kim, H.J., Sastry, S.: Multiple-agent probabilistic pursuit-evasion games. In: Proceedings of the 38th IEEE Conference on Decision and Control, 1999, vol. 3, pp. 2432–2437 (1999)
2. Flint, M., Polycarpou, M., Fernandez-Gaucherand, E.: Cooperative control for multiple autonomous uav's searching for targets. In: Proceedings of the 41st IEEE Conference on Decision and Control, 2002, vol. 3, pp. 2823–2828 (2002)
3. Subramanian, S., Cruz, J.: Adaptive models of pop-up threats for multi-agent persistent area denial. In: Proceedings. 42nd IEEE Conference on Decision and Control, 2003, December 9-12, vol. 1, pp. 510–515 (2003)
4. Grace, J., Baillieul, J.: Stochastic strategies for autonomous robotic surveillance. In: 44th IEEE Conference on Decision and Control, 2005 and 2005 European Control Conference. CDC-ECC 2005, pp. 2200–2205 (2005)
5. Machado, A., Ramalho, G., Zucker, J.D., Drogoul, A.: Multi-agent patrolling: An empirical analysis of alternative architectures. In: Sichman, J.S., Bousquet, F., Davidsson, P. (eds.) MABS 2002. LNCS (LNAI), vol. 2581, pp. 155–170. Springer, Heidelberg (2003)
6. Chevaleyre, Y., Sempe, F., Ramalho, G.: A theoretical analysis of multi-agent patrolling strategies. In: AAMAS 2004: Proceedings of the 3rd International Conference on Autonomous Agents and Multiagent Systems, vol. 3, pp. 1524–1525. IEEE Computer Society, Los Alamitos (2004)
7. Almeida, A., Ramalho, G., Santana, H., Tedesco, P.A., Menezes, T., Corruble, V., Chevaleyre, Y.: Recent advances on multi-agent patrolling. In: Bazzan, A.L.C., Labidi, S. (eds.) SBIA 2004. LNCS (LNAI), vol. 3171, pp. 474–483. Springer, Heidelberg (2004)
8. Santana, H., Ramalho, G., Corruble, V., Ratitch, B.: Multi-agent patrolling with reinforcement learning. In: AAMAS 2004: Proceedings of the 3rd International Conference on Autonomous Agents and Multiagent Systems, vol. 3, pp. 1122–1129. IEEE Computer Society, Los Alamitos (2004)
9. Menezes, T., Tedesco, P., Ramalho, G.: Negotiator agents for the patrolling task. In: Sichman, J.S., Coelho, H., Rezende, S.O. (eds.) IBERAMIA 2006 and SBIA 2006. LNCS (LNAI), vol. 4140, pp. 48–57. Springer, Heidelberg (2006)
10. Barnes, G., Feige, U.: Short random walks on graphs. SIAM Journal on Discrete Mathematics 9(1), 19–28 (1996)
11. Jain, A.K., Murty, M.N., Flynn, P.J.: Data clustering: a review. ACM Computing Surveys 31(3), 264–323 (1999)
12. Volgenant, A., Jonker, R.: A branch and bound algorithm for the symmetric traveling salesman problem based on the 1-tree relaxation. European Journal of Operational Research 9, 83–89 (1982)
13. Karypis, G., Kumar, V.: A fast and high quality multilevel scheme for partitioning irregular graphs. SIAM J. Sci. Comput. 20(1), 359–392 (1998)

Proving Epistemic and Temporal Properties from Knowledge Based Programs

Mario Benevides, Carla Delgado, and Michel Carlini*

Universidade Federal do Rio de Janeiro
{mario,delgado,mcar}@cos.ufrj.br

Abstract. In this work we investigate two approaches for representing and reasoning about knowledge evolution in Multi-Agent System MAS. We use the language of the Logic of Time and Knowledge TKL [5] as a specification language for the whole system, and for each agent, we use Knowledge Based Programs KBP, [4]. We propose a method to translate a a KBP system into a set of TKL formulas. The translation method presented provides an strategy to model and prove properties of MAS without being attached to local or global representations, giving the alternative to switch from one representation to another. We also present a formal model of computation to our KBP system and prove the correctness our translation function w.r.t. this model.In order to illustrate usefulness of the translation method two examples are presented: the Muddy Children Puzzle and the Bit Exchange Protocol.

Keywords: Temporal Logic, Epistemic Logic, Knowledge Based Programs.

1 Introduction

When representing a Multi-Agent System(MAS) we have two main approaches. We can look at the system by means of agents' individual behavior or as a global unity reacting to the surrounding environment.

For the global approach, we use the Language of the Logic of Time and Knowledge TKL [5], to specify the global behavior of MAS's. TKL is a language capable of representing knowledge evolution in a group of agents.

For the individual agent's approach, we use Knowledge Based Programs (KBP) [4], which aims to describe the activities of a MAS considering agent's knowledge and abstracting from implementation details. In this approach a MAS is a collection of KBP's, one for each agent, that executes in accordance with knowledge tests applied to the agent's local state.

Using TKL formulas, it is possible to formally verify interesting global properties of the system, as deadlock, agreement, global stability, distributed knowledge and everybody's knowledge. On the other hand, there is a great variety of problems that can be represented by a KBP. Whenever we obtain a specification of

* The authors acknowledge support from Research Agencies CNPq and CAPES.

G. Zaverucha and A. Loureiro da Costa (Eds.): SBIA 2008, LNAI 5249, pp. 134–144, 2008.

a KBP expressed in TKL from the KBP system, we would get both a concise global representation of the whole system and a framework for checking global properties using theorem provers available [1]. Our approach differs from [7,8] because they use model checking techniques to verify temporal and epistemic formulas while, in our case, we extract formulas from a MAS and, using some proof system, we can prove some desirable properties.

In this work, we present a translation method to generate a specification in TKL from a set of KBP's composing a MAS. The translation is made by a translation function that receives as input a KBP system and outputs a set of formulas in TKL language. We also propose a execution model for the KBP system and show that the formulas generated hold in this model.

This method offers an strategy to model and verify MAS's in two levels: local and global, allowing one to switch from one representation to another. Sections 2 and 3 present TKL and the KBP formalisms. Section 4, defines the translation methods. Section 5 proposes the computation model and proves the correctness of the translation function. Finally, in section 6 we state our conclusions.

2 A Language for Knowledge and Time

The language is a propositional multi-modal language [5,4]. For each agent i, there is a modality K_i representing knowledge from agent's i point of view, $K_i\varphi$ means that "agent i knows φ". We also include modality B_i, the dual of K_i. Finally, we have temporal operators $\bigcirc$ (next), $\Box$ (always) and *Until*.

Formulas of TKL are given by the following grammar:

$\varphi ::= p | \perp | \neg\varphi | \varphi \wedge \psi | \varphi \vee \psi | \varphi \rightarrow \psi | K_i\varphi | B_i\varphi | \bigcirc\varphi | \varphi\, Until \psi | \Box\varphi$, where p ranges over the set *Prop* of propositional symbols and i over the set of agents.

3 Knowledge-Based Programs

Most models for distributed systems suggest that each entity is a program that executes its actions in accordance with tests applied to agent's local state [6]. Each test together with its associated actions is called a *guard*. Knowledge Based Programs were defined in [4], it contains knowledge tests to agent's local state.

Standard Program	KBP
Program $Prog_i$	**Program** KBP_i
initial: *initial conditions*	**initial:** *initial conditions, initial knowledge*
repeat	**repeat**
case of	**case of**
(a) *test_1* **do** *action_1*	**(a)** *test_1* $\wedge$ *knowledge_test_1* **do** *action_1*
(b) *test_2* **do** *action_2*	**(b)** *test_2* $\wedge$ *knowledge_test_2* **do** *action_2*
...	
end case	**end case**
until *termination_test*	**until** *knowledge_termination_test*
end	**end**

The Muddy Children Puzzle offers an interesting study about the knowledge involved in a MAS. It consists of a system formed by n children and their father.

We start with n children playing together. The father tell them that they should not get dirty. It happens that some of the children get mud on their foreheads. Each child can see the mud on the others forehead, but not on his own. After some time, the father returns and says: "At least one of you has mud on your forehead". Then, he keeps asking the following question, over and over: "Does any of you know whether you have mud on your own forehead?" Suppose that all children are perceptive, intelligent, truthful, and they answer simultaneously. If there are k muddy children, in the first $k-1$ times, all children will answer 'NO', but in the kth time the muddy children will all answer 'YES'.

Each child on the Puzzle can be modelled as a KBP. KBP, where

program MC_i %% version 2
  initial: $K_i p_j$, for $j \in$ {% set of muddy children agent i sees%}
    $K_i(\bigvee_{x=1..n} p_x)$ %% *At least one child is muddy*
  repeat
    case of
    (a) initial$_i$ $\wedge$ childheard$_i$ $\wedge$ $K_i p_i$ **do** say 'YES'
    (b) initial$_i$ $\wedge$ childheard$_i$ $\wedge$ $\neg K_i p_i$ **do** say 'NO'
    (c) childheard_yes$_i$ $\wedge$ $\neg K_i p_i$ **do** learn $K_i \neg p_i$
    (d) childheard$_i$ $\wedge \neg$ childheard_yes$_i \wedge$
      $B_i(\neg p_i \wedge B_1(\neg p_1 \wedge \cdots \wedge B_{k-2}(\neg p_{k-2} \wedge B_{k-1}(\neg p_{k-1} \wedge K_k p_k))\cdots)))$ **do**
        learn $B_i(\neg p_i \wedge B_1(\neg p_1 \wedge \cdots \wedge B_{k-2}(\neg p_{k-2} \wedge K_{k-1}(p_{k-1} \wedge K_k p_k))\cdots)))$
        say 'NO'
    (e) childheard$_i$ $\wedge \neg$ childheard_yes$_i$ $\wedge B_i(\neg p_i \wedge K_j p_j)$, for $j <> i$ **do**
        learn $K_i p_i$, say 'YES'
    end case
  until $K_i p_i \vee K_i \neg p_i$

Proposition childheard$_i$ stands for "Child i just heard father's question", and proposition p_i stands for "child i is muddy", and propositions initial$_i$ and childheard_yes$_i$ stands for "Program MC$_i$ is in the first round" and "Child i heard answer Yes from any other child" respectively. The command *learn* denotes the action of acquiring a piece of knowledge. It would be interesting to provide some explanation about the formula in guard (d), when the father asks if somebody already knows if there is mud on its forehead, the state of knowledge of each agent changes from:

$B_i(\neg p_i \wedge B_1(\neg p_1 \wedge B_2(\neg p_2 \wedge \cdots \wedge B_{k-2}(\neg p_{k-2} \wedge B_{k-1}(\neg p_{k-1} \wedge K_k p_k))\cdots)))$
to: $B_i(\neg p_i \wedge B_1(\neg p_1 \wedge B_2(\neg p_2 \wedge \cdots \wedge B_{k-2}(\neg p_{k-2} \wedge K_{k-1}(p_{k-1} \wedge K_k p_k))\cdots)))$
which means that as agents learn more and decrease their uncertainty, some of their beliefs are confirmed and turned into knowledge.

The complete specification of the MAS as a KBP system can be obtained by the composition of all MC$_i$ programs.

4 Translating KBPs into TKL formulas

The translation process consists on applying some rules of transformation to each KBP that belongs to the system. To formalize the translation method, we define these rules as a function from KBP systems into a set of TKL formulas. First, we give an intuitive overview of the complete process.

Formulas at the initial section of a KBP are put in the set of formulas corresponding to the MAS specification. They represent facts that are true at the beginning of the system. If a formula is preceded by operator K_i, for some i, then it represents local knowledge for agent i, and does not interfere with other agents' knowledge. On the other hand, formulas that are not preceded by a modal operator for knowledge represent global information.

Example: **Bit Sender KBP**
keeps sending a bit to its partner, Bit Receiver KBP, until convinced the bit was correctly received
program *BtS*
  initial: $K_s(bit = 0)$, assuming that zero is the initial value of the bit
    $K_s(bit = 0 \vee bit = 1)$
  repeat
    case of
    (a) $\neg K_s K_r(Bit)$ **do** sendbit
    end case
  until $K_s K_r(Bit)$
end

The formula $K_r Bit$ stands for $K_r(bit = 0) \vee K_r(bit = 1)$.From the initial section, we can extract formulas $K_s(bit = 0)$ and $K_s(bit = 0 \vee bit = 1)$. The guards of the KBP correspond to the preconditions of a section of actions (or commands). Actions affects the system and consequently other agents, so when representing state change rules in a MAS, we generate, for each guard of the program, a rule of the kind: *cause* $\rightarrow$ *consequence*; where *cause* corresponds to the precondition implied by the guard, and *consequence* to the effect of the actions associated to that guard. The test of a guard is already TKL formula, but actions have to be translated; it is necessary to identify the effect caused by each action and translate this effect into *TKL*.

Example: **Bit Sender KBP**
From guard (a) we identify: **cause:** $\neg K_s K_r(Bit)$ **consequence:** sendbit
cause is already a TKL formula, but the action sendbit needs translation.

Propositions, evaluated at knowledge tests on the KBPs, represent important information about state changes from the point of view of each agent. Actions are the only way to change the truth value of propositions at knowledge tests. So, the execution of an action at a KBP is translated into a set of TKL formulas that change their values as consequence of action execution. Besides, actions in the KBP representing agent i can affect agent i itself or other agents in the system; actions that affect the agent itself have their results reflected at the right moment they are performed, but actions that affect other agents will only be perceived in the next turn of the system.

Example: **Bit Sender KBP**
Action sendbit is a communication action. When talking about a system where messages are safely delivered, its consequence is that agent receiver will learn the bit value. As it is an action made by one agent that concerns another, it must be preceded by the temporal operator: $\neg K_s K_r(Bit) \rightarrow \bigcirc K_r(Bit)$

Finally, we tackle the translation of the knowledge termination test. A KBP finishes its execution when a termination condition is achieved, what corresponds to say that the agent being modelled by this KBP is not capable of doing any other action or perceiving anything else. Whenever all KBPs of the system has achieved its termination condition, it should finish its computation and whole system must halt. This can also be represented by a rule *cause* $\rightarrow$ *consequence*, where *cause* corresponds to the conjunction of all KBPs termination condition formulas ($\bigwedge_{\forall i}$ *knowledge_termination_test*$_i$), and *consequence* to the formula $\Box \perp$.

Example: **Bit Sender KBP**
Termination condition will be made from the conjunction of Bit Sender KBP termination test and the other KBPs involved, i. e., the Receiver's.

$K_s K_r(Bit) \wedge$ $knowledge\_termination\_test_{receiver} \rightarrow \Box \perp$

The formula $effect\_of\_action_i^n$ represents the fact that becomes true as consequence of the action being performed. $effect\_of\_action_i^n$ is a TKL formula involving propositional operators and operators for individual knowledge, if the action affects only the agent who performed it; if the action affects other agents, it will turn into a formula preceded by temporal operator $\bigcirc$, due to the fact that it will only be perceived at the next turn of the system.

These rules can be formalized by a translation function Ψ_{MAS} that maps KBP systems into a set of TKL formula. The function $\Psi_{MAS} : P \rightarrow 2^{\mathcal{L}}$, where P is a set of KBP's and $\mathcal{L}$ is the set of all TKL formulas. Ψ_{MAS} can be defined in terms of a subfunction ψ_{mas} from KBP items into formulas in TKL, based on the rules explained above.

For each KBP of the system:
ψ_{mas}: initial section translation
For each knowledge logic formula φ at initial section: $\psi_{mas}(\varphi) = \Box\varphi$
ψ_{mas}: guards translation
For each guard G of the form:
$test_i \wedge knowledge\_test_i$ **do** $action_i^1, action_i^2 \cdots action_i^n$
$\psi_{mas}(G) = \Box(test_i \wedge knowledge\_test_i \rightarrow effect\_of\_action_i^h)$, if $action_i^h$ affects only i
or
$\psi_{mas}(G) = \Box(test_i \wedge knowledge\_test_i \rightarrow \bigcirc effect\_of\_action_i^h)$, otherwise
(for all $action_i^h$ and $1 \leq h \leq n$)
ψ_{mas}: termination condition translation
For *knowledge_termination_test* from all KBPs
$\psi_{mas}(knowledge\_termination\_test = \Box(\bigwedge_{\forall i} knowledge\_termination\_test_i) \rightarrow \Box \perp$

In order to illustrate the use of translation function Ψ_{MAS} we apply it to the KBP of second example for the Muddy Children Puzzle, MC_i version 2.

$\psi_{mas}(K_i p_j)$: $\Box K_i p_j$, for j={% set of muddy children agent i sees%}
$\psi_{mas}(K_i(\bigvee_{x=1..n} p_x))$: $\Box K_i(\bigvee_{x=1..n} p_x)$
$\psi_{mas}(a)$: $\Box(\ (\text{initial}_i \wedge \text{childheard}_i \wedge K_i p_i) \rightarrow \bigcirc \bigwedge_{j \in G} \text{childheard\_yes}_j)$
$\psi_{mas}(b)$: $\Box((\text{initial}_i \wedge \text{childheard}_i \wedge \neg K_i p_i) \rightarrow \bigcirc \bigwedge_{j \in G} \text{childheard\_no}_j)$
$\psi_{mas}(c)$: $\Box((\text{childheard\_yes}_i \wedge \neg K_i p_i) \rightarrow K_i \neg p_i)$
$\psi_{mas}(d)$: $\Box((\text{childheard}_i \wedge \neg\ \text{childheard\_yes}_i \wedge$
$B_i(\neg p_i \wedge B_1(\neg p_1 \wedge ... \wedge B_{k-2}(\neg p_{k-2} \wedge B_{k-1}(\neg p_{k-1} \wedge K_k p_k))...)))) \rightarrow$
$B_i(\neg p_i \wedge B_1(\neg p_1 \wedge ... \wedge B_{k-2}(\neg p_{k-2} \wedge K_{k-1}(p_{k-1} \wedge K_k p_k))...))) \wedge$
$\bigcirc \bigwedge_{j \in G} \text{childheard\_no}_j)$
$\psi_{mas}(e)$: $\Box((\text{childheard}_i \wedge \neg\ \text{childheard\_yes}_i \wedge B_i(\neg p_i \wedge K_j p_j)$, for $j <> i) \rightarrow$
$K_i p_i \wedge \bigcirc \bigwedge_{j \in G} \text{childheard\_yes}_j)$
$\psi_{mas}(K_i p_i \vee K_i \neg p_i)$: $\Box(\bigwedge_{\forall i}(K_i p_i \vee K_i \neg p_i) \rightarrow \Box \bot)$

With this specification in hands and with a proof method [5,1] in hands we could prove some desirable properties:

For a group of n children where k have muddy foreheads:

For any turn $m \leq k$, no children hears any affirmative answer.

$$\bigwedge_i \underbrace{\bigcirc \bigcirc \cdots \bigcirc}_{m\ times} \neg childheard\_yes_i,\ for\ i = 1\ to\ n$$

After k turns if child j is dirty, she knows it.

$$p_j \rightarrow \underbrace{\bigcirc \bigcirc \cdots \bigcirc}_{k\ times} K_j p_j$$

A child that is not dirty will only became aware of it after she hears an affirmative answer.

$$\neg p_i \rightarrow (\neg K_i(p_i \vee \neg p_i)\ until\ childheard\_yes_i)$$

5 Relation between KBP and TKL Formulas Specification

In this section we introduce a model of computation for our KBP system and, using this model, we prove the correctness of our translations function.

5.1 Model of Computation

Definition 1. *We call each KBP program as an agent and the collection of agents is a KBP system. Given a KBP system $\mathcal{A}$, consider the following model to compute $\mathcal{A}$:*

1. *A network with m agents ($m \geq 2$) connected by reliable* fifo *channels;*
2. *A set R of asynchronous runs (definition 3) of algorithm $\mathcal{A}$;*
3. *A set E of events (definition 2) for all the runs of $\mathcal{A}$;*
4. *A set C of consistent cuts (definition 7) for all the runs of $\mathcal{A}$.*

Definition 2. *The events reflect messages' exchanging among the agents. The event is the basic element in the definitions of time for asynchronous systems. An event is defined as a tuple* $e = [a_i, t_i, s_i, s'_i, M, \mathcal{M}]$*, where:*

1. a_i *- the agent i for which the event e happens*
2. t_i *- the agent's local time when the event e happens*
3. s_i *- the agent's local state that precedes the event e*
4. s'_i *- the agent's local state that succeeds the event e*
5. M *- the message received by the agent i associated to the event e*
6. $\mathcal{M}$ *- the messages sent by the agent i due to the event e*

Definition 3. *Given an KBP system* $\mathcal{A}$*, each run r of it is a set* E_r *of events that describe a distributed computation of it. As the delivering time of the messages can vary, different orders of arrival define different events and hence distinct runs.*

In this model, a run r of a KBP system $\mathcal{A}$ is illustrated by a *graph of events' precedence*. Consider the diagram of the figure 1 as one possible run of the KBP system. Each vertex in the precedence graph of the figure 1 represents an event and the arrows represent the order the events happened.

Definition 4. *Let e and e′ be the following causality relations:*

- *"e happened immediately before e′",* $e \rightarrow e'$*, if and only if one of the two conditions happened:*
  1. *e and e′ happened for the same agent i and, for times* $t_i < t'_i$*, there is no other event e″ for i at time* t''_i *such that* $t_i < t''_i < t'_i$*;*
  2. *e and e′ happened for different agents i and j, and the message received from j in the event e′ was sent from i in the event e, that is,* $e = [a_i, t_i, s_i, s'_i, M_e, \mathcal{M}_e]$, $e' = [a_j, t_j, s_j, s'_j, M_{e'}, \mathcal{M}_{e'}]$, $M_{e'} \in \mathcal{M}_e$.
- *"e happened before e′",* $e \rightarrow^+ e'$*, if the relation* $\rightarrow^+$ *is the transitive closure under the relation* $\rightarrow$*, that is, when it happened one of the two conditions:*
  1. $e \rightarrow e'$
  2. $\exists e_1, \ldots, e_k$ *for some* $k > 0$ *such that* $e \rightarrow e_1, \ldots, e_{k-1} \rightarrow e_k \rightarrow e'$.

Note that the relation *happens before* ($\rightarrow^+$) is a partial order on the events.

Definition 5. *Let* E_r *be the set of events in the run r. Consider the* Past *and the* Future *of* $e \in E_r$ *as:*
$Past(e) = \{e' \in E_r | e' \rightarrow^+ e\}$;
$Future(e) = \{e' \in E_r | e \rightarrow^+ e'\}$.

Definition 6. *Two events e and e′ are* concurrent *if and only if e did not happen before e′ and e′ did not happen before e:*
$\neg(e' \rightarrow^+ e) \wedge \neg(e \rightarrow^+ e')$.

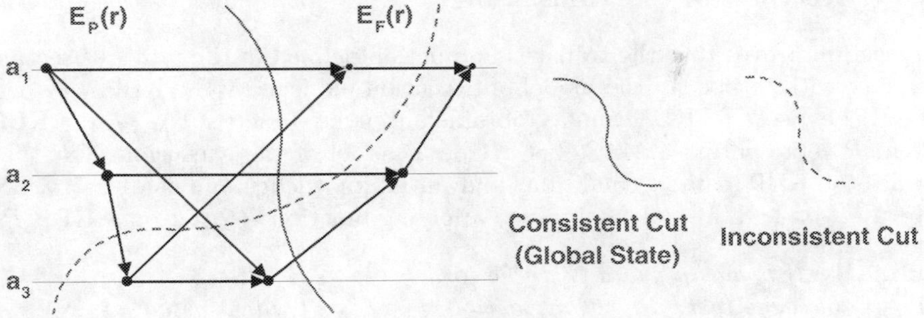

Fig. 1. Graph of Events' Precedence and Consistent Cut

Definition 7. *Let E_r be the set of events in the run r. A global state or a* consistent cut *c in r is a partition of E in two sets E_P and E_F such that, if $e \in E_P$ then $Past(e) \subseteq E_P$.*

Therefore, a consistent cut divides the graph of events in two sets of events, E_P and E_F, called, respectively, past and future with respect to the cut c. The graph of the figure 1 illustrates two partitions or two states of the system, but just one of them is a global state, that is, just one is a consistent cut.

The relation $\rightarrow^+$ over the set of events E can induce a partial order relation $\leq$ over the set of global states (consistent cuts). Each run induces a sequence (possibly infinite) of global states which we call a history.

Now we are ready to give a semantic to the TKL language presented in section 2.

A model for TKL is a tuple $\mathcal{M} = \langle S, H, l, R_1, R_2, \cdots \rangle$ where S is set of global states, H is the set of all possible histories over S ($H \subseteq S^{\mathbb{N}}$), l is a labelling function $l : S \rightarrow 2^{Prop}$, and for every agent i, R_i is an infinite sequence of equivalence relations, i. e. for any $k \in \mathbb{N}$, on R_i^k is an equivalence relation on $S_k^{\mathbb{N}}$, the set of all possible histories at instant k. The intuitive meaning for R_i^k is that any two histories ω and τ are equivalent with respect to i and k if the knowledge that agent i has gathered about the world and its history up to instant k is not enough for her to distinguish between ω and τ.

The notion of satisfaction of a formula φ in model $\mathcal{M}$ for a history σ and state $k \in \mathbb{N}$ is defined as follows:

1. $\mathcal{M}, k, \sigma \models p$ iff $p \in l(\sigma_k)$;
2. $\mathcal{M}, k, \sigma \models \neg\varphi$ iff not $\mathcal{M}, k, \sigma \models \varphi$;
3. $\mathcal{M}, k, \sigma \models \varphi \wedge \psi$ iff $\mathcal{M}, k, \sigma \models \varphi$ and $\mathcal{M}, k, \sigma \models \psi$;
4. $\mathcal{M}, k, \sigma \models K_i\varphi$ iff $\mathcal{M}, k, \tau \models \varphi$ for all histories τ such that $(\sigma, \tau) \in R_i^k$;
5. $\mathcal{M}, k, \sigma \models B_i\varphi$ iff $\mathcal{M}, k, \tau \models \varphi$ for some history τ such that $(\sigma, \tau) \in R_i^k$;
6. $\mathcal{M}, k, \sigma \models \bigcirc\varphi$ iff $\mathcal{M}, k+1, \sigma \models \varphi$;
7. $\mathcal{M}, k, \sigma \models \Box\varphi$ iff for all $n \geq k$ $\mathcal{M}, n, \sigma \models \varphi$;
8. $\mathcal{M}, k, \sigma \models \varphi$ *Until* ψ iff for all $n \geq k$ such that $\mathcal{M}, n, \sigma \not\models \varphi$ there is some m, $k \leq m \leq n$ such that $\mathcal{M}, m, \sigma \models \psi$.

5.2 Correctness of the Translation

This section proves that the translation function defined in the previous section is correct with respect to the model of execution induced by the KBP.

Let S be a set of TKL formulas, obtained applying function Ψ_{MAS} to a KBP system P representing a MAS. Let $\mathcal{M}$ be a model of the execution of P that satisfies the KBP's initial conditions and initial knowledge and σ a history. Let S_i be a set of TKL formulas, obtained applying function Ψ_{MAS} to one KBP P_i.

Lemma 1. *If P_i has a guard G of the form* $test_i \wedge knowledge\_test_i$ **do** $action_i^1, action_i^2 \cdots action_i^n$*, then for all $k \in \mathbb{N}$* $\mathcal{M}, k, \sigma \models test_i \wedge knowledge\_test_i \rightarrow effect\_of\_action_i^h$*, if $action_i^h$ affects only agent i or* $\mathcal{M}, k, \sigma \models test_i \wedge knowledge\_test_i \rightarrow \bigcirc effect\_of\_action_i^h$ *otherwise, for $1 \leq h \leq n$.*

Proof. We have three cases.

- If $\mathcal{M}, k, \sigma \models test_i \wedge knowledge\_test_i$ and $action_i^h$ only affects agents i, then $action_i^h$ is performed by P_i and the global state of the system does not change and the $effect\_of\_action_i^h$ is true in k and thus $\mathcal{M}, k, \sigma \models test_i \wedge knowledge\_test_i \rightarrow effect\_of\_action_i^h$.
- If $\mathcal{M}, k, \sigma \models test_i \wedge knowledge\_test_i$ and $action_i^h$ affects other agents, then $action_i^h$ is performed by P_i and the global state of the system does change and the $effect\_of\_action_i^h$ is true in the next global state $k+1$ and $\mathcal{M}, k+1, \sigma \models effect\_of\_action_i^h$. Thus, $\mathcal{M}, k, \sigma \models test_i \wedge knowledge\_test_i \rightarrow \bigcirc effect\_of\_action_i^h$.
- If $\mathcal{M}, k, \sigma \not\models test_i \wedge knowledge\_test_i$, then the lemma holds.

Corollary 1. *Suppose P_i has a guard G, if $\varphi \in \psi_{mas}(G)$, then $\mathcal{M}, k, \sigma \models \varphi$, for all $k \in \mathbb{N}$.*

Proof. It follows straightforward from lemma 1 and by the satisfability condition for the □

Lemma 2. *Let S_i be a set of TKL formulas, obtained applying function Ψ_{MAS} to one KBP P_i. If $\varphi \in S_i$, then for all $k \in \mathbb{N}$ $\mathcal{M}, k, \sigma \models \varphi$.*

Proof. The proof is by induction on k.

- Base $k = 0$, we have three cases
  1. Initial section: by hypothesis $\mathcal{M}, 0, \sigma \models \varphi$ for all formulas φ in the initial conditions and initial Knowledge;
  2. Guards: from corollary 1 $\mathcal{M}, 0, \sigma \models \varphi$, for all $\varphi \in \psi_{mas}(G)$;
  3. Termination conditions: if the termination condition is satisfied at the initial state the P is will not perform any action and consequently no event will take place and there is no next state, therefore $\mathcal{M}, 0, \sigma \models \Box\bot$.

– Induction Hypothesis: Suppose holds for $k = m$,
  All formulas generated by the Ψ_{MAS} are of the form $\Box\varphi$. By the definition o satisfaction $\mathcal{M}, m, \sigma \models \Box\varphi$ iff for all $n \geq m$ $M, n, \sigma \models \varphi$ iff for all $n \geq m+1$ $M, n, \sigma \models \varphi$ iff $\mathcal{M}, m+1, \sigma \models \Box\varphi$.

Theorem 1. *Let S be a set of TKL formulas, obtained applying function Ψ_{MAS} to a KBP system P. If $\varphi \in S$, then for all $k \in \mathbb{N}$ $\mathcal{M}, k, \sigma \models \varphi$.*

Proof. As $S = \bigcup S_i$ it follows straightforward from lemma 2.

6 Conclusions

In this work we investigate two different approaches for representing and reasoning about knowledge in a MAS. We use the language of the Logic of Time and Knowledge [5] as specification language to represent global properties of the system, and for each individual agent, we use Knowledge Based Programs [4]. The latter is more suitable for modelling the behavior of each agent in the system as an autonomous process and observe how interaction takes place; the former has the advantage of providing a concise representation of the whole system and the possibility of checking global properties using theorem provers available [1].

We define a translation function that receives as input a KBP systems that represents a MAS and gives as output a set of formulas in TKL language, which can be used as a specification of the MAS to prove properties about the system. We also propose a computation model for our KPB system and prove its correctness w.r.t this model.

Two examples of MAS were presented in order to illustrate the translation method: the Muddy Children Puzzle and the Bit Exchange Protocol. The Muddy Children Puzzle illustrates the relations between knowledge and time, and the Bit Exchange Protocol evidences the role played by interaction in MAS's. This approach could be used to model many practical problems related to MAS's, as for example, the implementation of protocols of communication and games.

As future works, first we could define another translation method to accomplish the inverse process, i.e., the construction of a set of KBPs from a set of formulas corresponding to a specification of a MAS written in TKL. Second, we would like to use larger application to validate the method. Finally, we need to establish complexity results for the method itself and for the length of the formulas generated by it.

References

1. Dixon, C., Nalon, C., Fisher, M.: Tableaux for Temporal Logics of Knowledge: Synchronous Systems of Perfect Recall Or No Learning. In: Proc. of 10th Int. Symp. on Temporal Representation and Reasoning, Australia (2003)
2. Fagin, R., Vardi, M.: Knowledge and Implicit Knowledge in a Distributed Environment. In: Conf. on Theoretical Aspects of Reasoning about Knowledge, pp. 187–206 (1986)

3. Fagin, R., Halpern, J., Moses, Y., Vardi, M.: Reasoning About Knowledge. MIT Press, Cambridge (1995)
4. Fagin, R., Halpern, J., Moses, Y., Vardi, M.: Knowledge-Based Programs. Distributed Computing 10(4), 199–225 (1997)
5. Lehmann, D.: Knowledge, common knowledge, and related puzzles. In: Proc. 3rd Ann. ACM Conf. on Principles of Distributed Computing, pp. 62–67 (1984)
6. Lamport, L.: Time, Clocks and the Ordering of Events in a Distributed System. Communications of the ACM 21(7), 558–565 (1978)
7. Meyden, R., Shilov, N.: Model Checking Knowledge and Time in Syst. with Perfect Recall. Found. of Soft. Techn. and Theoret. Comp. Science, 432–445 (1999)
8. Lomuscio, A., Raimondi, F.: Model Checking Knowledge, Strategies and Games in Multi-agent Systems. In: AAMAS 2006: 5th Int. Joint Conf. on Autonomous Agents and Multi-agent Systems, pp. 161–168 (2006)

Detecting Code Evolution in Programming Learning

Thais Castro[1,2], Hugo Fuks[1], and Alberto Castro[2]

[1] Department of Informatics (PUC-Rio)
R. Marquês de S. Vicente, 225 RDC – Gávea – 22453-900 – Rio de Janeiro – RJ – BR
[2] Computer Science Department (UFAM)
Av. Gal. R. O. J. Ramos, 3000 – 69077-300 – Manaus – AM – BR
{tcastro,hugo}@inf.puc-rio.br, albertoc@dcc.ufam.edu.br

Abstract. This article describes a research in code evolution for programming learning that has as its main components the identification of the specific demands for detecting code evolution in the context of programming learning, and the identification of an initial set of programming learning supporting strategies. The findings within the knowledge represented in the codes allowed us to carry out a deeper analysis concerning the inferences internal to the code strategy, which is especially useful for exploring the supporting strategies and the techniques students normally use for their knowledge acquisition in programming.

Keywords: Supporting Strategies for Code Evolution, Programming Learning, Knowledge Representation.

1 Introduction

The search for knowledge about programming skills is recurrent in this research area, starting with Dijkstra [5] in his work "On the Teaching of Programming" and in Weinberg [13], which addresses the psychological aspect of programming learning. Although there is a lot of research on the subject, the milestones of programming learning are not fully elicited yet and that is why it is so difficult to understand which elements contribute to the knowledge acquisition in programming.

In the context of programming learning, in which the students are starting to acquire knowledge in programming, a possible way to be aware of this process is by tracking students' code evolution and categorizing the changes made from one version to the next.

This paper aims at finding out how to join simple techniques and tools from knowledge representation, in order to reduce teacher's work load, improving the feedback and support to programming learners, making the programming learning process more explicit.

In order to address the issues aforementioned, the following subsection contextualizes this work. Section 2 presents a bibliographical review on knowledge acquisition in programming. As a result of that review and its analysis in the context of programming learning, Section 3 presents a deeper view in the inferences internal to the code, concerning code evolution. Finally, Section 4 explorers the analysis made in the previous section, relating it to the development and use of a tool to support code evolution.

G. Zaverucha and A. Loureiro da Costa (Eds.): SBIA 2008, LNAI 5249, pp. 145–156, 2008.

1.1 Programming Learning Supporting Strategies

In the context of programming learning, we have been conducting a series of case studies, since 2002 [2], to find out what kind of knowledge first year undergraduate students need for acquiring programming skills. The introductory course where these studies were conducted is based on a methodology in which the focus is in problem solving and coding in Haskell.

Based on these previous results, literature (Section 2) and interviews with experts in programming learning, we have identified some supporting strategies tailored to help the teacher in her search for coding improvement evidences.

In order to find out what milestones are most relevant in programming learning, it is important to extract information from every activity proposed during the introductory course, inferring the knowledge acquired in each programming topic. The coding improvement evidences will depend on the methodology adopted. Following, the improvement evidences are presented based on the assumption that the introductory course where this study takes place proposes activities as coding and conversation records.

1. Code evolution
   a. A quantitative analysis for each solution
   b. Inferences internal to the code
   c. Inferences external to the code
2. Conversation follow-up
3. Conversation report
4. Use of search and inference resources to follow the transition from individual to group solution

This work is part of a research project, which aims at investigating and refining the coding improvement evidences mentioned above. This article deals with the code evolution supporting strategy (1b), specifically with the inferences internal to the code. The other supporting strategies will be discussed in future works, based mainly on [6] and [11].

2 Bibliographical Review

By understanding which cognitive processes are involved in knowledge acquisition in programming, the practices used by each learner become evident and the knowledge generated can be reused in programming education for beginners. There are some works, summarized in the paragraphs below, which use different techniques and methods to elicit elements that may indicate an increase in the knowledge acquisition in programming.

In the article described in [9], a survey on the cognitive aspects involved in the acquisition of the skills needed for programming is carried out. Based on this survey the authors discuss the sequencing of the curriculum in introductory courses as well as the interdependence among the concepts. Such reflection does a better job of organizing introductory courses but does not guarantee that the students will learn more, once the aspects found were not elicited by means of an experiment.

In another work, described in [10], a study is conducted which objective is to elicit the manner in which expert programmers organize their knowledge on programming in comparison with beginners. The results were reported in quantitative terms, the most experienced students getting higher marks using the card sorting technique to measure their programming skills. However, there is no evidence that this technique helps in the knowledge acquisition.

Most of the articles, such as [4] and [8], attempt to elicit the skills needed for learning programming either through the discussion on models for programming understanding by relating them to models in the domain of reading [4], or through the assessment of the cognitive capability of storing new information in memory [8].

Problem solving is related to some investigations, such as those described in [1] and [7], that use the clinical method proposed by Piaget and explained in [3] to elicit aspects relevant for knowledge acquisition in programming. This approach is closer to what we propose, but in our work collaboration is a means for learning.

3 Inferences Internal to the Code Strategy

Observing code, we could infer three categories of changes found in code evolution. Along with examples in Haskell, we provide Prolog fragments that capture the example's code changes.

3.1 Syntactic

E.g. indentation, insertion and characters deletion. It aims at capturing changes made to make it correctly interpreted by the Haskell interpreter Hugs, suggesting exhaustive trial and error attempts in order to find a solution.

3.1.1 Indentation

E.g. indentation of the function's second line, positioning the `+ x/y` code after the equality symbol, in order to be viewed as a single line by the Hugs. The following example shows a needed adjustment (from Program A to Program B) to Haskell's specificities, without changing program representation:

Program A	**Program B**
`f x y = x*y`	`f x y = x*y`
`+ x/y`	`          + x/y`

A way to automatically detect the extra spaces added to Program B is transforming terms (every character in the program) into list elements. Then, it is possible to compare each list element, and infer what type of change occurred. There is a Prolog code example below, which illustrates how to compare any two programs.

```
sameF(T1,T2) :-
              T1 =.. [FunctionName|BodyList],
              T2 =.. [FunctionName|BodyList].
```

3.1.2 Insertion, Change or Character Deletion

E.g. misuse of , instead of ; or vice versa and also the absence of some connective symbol, as an arithmetic operator. The following example shows a needed alteration in the solution as well as a needed adjustment to Haskell's specificities:

Program A	**Program B**
`f (x,y) = (x;1) + (2,y)`	`f (x,y) = (x,1) + (2,y)`

In Program B, it was necessary to change the symbol used in the tuple (x,1).

A way to automatically detect such changes involves the implementation of a DCG (Definite Clause Grammar), capable of understanding Haskell code. In the Prolog program presented below there is a fragment of a Haskell grammar, based on the predicate pattern/1, which detects the aforementioned syntactic errors. The other predicates namely, expr/1, variable/1 and pattern_seq/1, are an integral part of Programs A and B. The predicates sp/0 and optsp/0 match for necessary and optional spaces in the code.

```
decls(Z) --> pattern(X), optsp, "=", optsp, expr(Y),
{Z=..[f,head(X),body(Y)]}.

decls(Z) --> pattern(X), optsp, "=", optsp, decl_seq(Y),
{Z=..[f,head(X),Y]}.

decls(Z) --> variable(X1), sp, patternseq(X2), optsp, "=",
optsp, expr(Y),  {X=..[head,X1|X2], Z=..[f,X,body(Y)]}.

decls(Z) --> variable(X1), sp, pattern_seq(X2), optsp, "=",
optsp, decl_seq(Y), {X=..[head,X1|X2], Z=..[f,X,Y]}.

decl_seq(Z) --> declseq(X), {Z=..[body|X]}.

declseq(Z) --> expr(X), {Z=[X]}.

declseq(Z) --> expr(X), ";", declseq(Y), {Z=[X|Y]}.
```

3.1.3 Inclusion of a New Function

E.g. adding a new function to the program, in order to incrementally develop and test the whole solution. This is normally found in student's codes and indicates the use of a good programming practice, emphasized in the introductory course, based on the divide and conquer strategy.

A way to automatically detect such changes involves the implementation of the same type of rule as the one presented for the previous change, shown in 3.1.2.

3.2 Semantics

E.g. data structures modification; changing from tuples to lists; inclusion of a recursive function; and bugs correction. These changes directly affect function evaluation, resulting in a wrong output.

3.2.1 Changing from Independent Variables to Tuples

E.g. a second degree equation could be calculated in two different ways: using independent roots or using tuples. Both representation methods use the same arguments, although in the former there is a need for duplicating the function definition. Example:

Program A

```
r x a b c = (-b) + e / 2*a
    where
      e = sqrt(b^2-4*a*c)
r y a b c = (-b) - e / 2*a
    where
      e = sqrt(b^2-4*a*c)
```

Program B

```
rs a b c = (x,y)
    where
    x = ((-b) + e)/2*a
    y = ((-b) - e)/2*a
    e = sqrt(b^2-4*a*c)
```

In Program A two functions are used to solve the second degree equation. Besides the programmer extra effort in repeating the functions, if these functions were too large, code readability would be affected. Program B presents a more elegant and accurate solution to the problem. By directing the output to a tuple (x,y) and defining locally the formula it becomes more readable, saving the programmer from an extra effort.

A way to automatically detect such changes also involves the implementation of a DCG. This DCG consists of transforming the terms into lists and comparing each set in both programs. As illustrated in the code below, it is possible to identify what structure has changed between code versions. There is a Prolog fragment example below, which suits both the semantic changes presented in this subsection and in 3.2.2.

```
compare2(T1,T2,Subst) :-
        T1 =.. [f,H1,B1|[]],
        T2 =.. [f,H2,B2|[]],
        H1 =.. [head|Head1],
        B1 =.. [body|Body1],
        H2 =.. [head|Head2],
        B2 =.. [body|Body2],
        comp2(Head1,Head2,[],Subst1),
        comp2(Body1,Body2,Subst1,Subst).
```

In the above fragment, a substitution list has been constructed, making it possible the comparison between T2 and T1, T2 = subst(T1).

3.2.2 Changing from Tuples to Lists

E.g. if a solution could be represented using tuples or lists, it would be better to use lists instead of tuples for handling large collections. In the following example, it is shown the change from tuples (Program A) to lists (Program B), which is a desirable technique to improve data handling:

Program A

```
composed a b = (a, b, b+a)
```

Program B

```
composed a b = [a, b, b+a]
```

A way to automatically detect such changes also involves the implementation of a DCG. This directly identifies changes in structure, between two code versions.

Moreover, using the Prolog built-in predicate ':=/2', changes such as the exemplified above are easily detected, as in the Prolog fragment shown in 3.1.1.

3.2.3 Inclusion of a Recursive Function

E.g. using a recursive function instead of an iterative one. In most cases, in functional programming, recursion suits better and it is often more accurate. That is why usually the teacher would like to know whether recursion has been correctly understood. The following example shows the representation of the solution for the factorial function, where Program A presents an iterative solution and Program B a recursive one:

Program A

```
fat n = if n==0 then 1
        else product [1..n]
```

Program B

```
fat n = if n==0 then 1
        else n * fat(n - 1)
```

A way to automatically detect such changes involves the implementation of a rule for detecting whether a program has been written using a recursive function. This involves the identification of specific terms in a code version.

3.2.4 Bugs Correction

E.g. small changes made to formulae' or functions' definitions, consisting of simple bug correction in a trial and error fashion. When a function does not work, students usually make consecutive changes without stopping to reason, ignoring this way the process used in problem solving techniques. The following example shows the inclusion of an IF-THEN-ELSE conditional structure:

Program A

```
fat n = n * fat (n - 1)
```

Program B

```
fat n = if n==0 then 1
        else n * fat(n - 1)
```

A way to automatically detect such change involves the implementation of a pattern matching predicate, which checks whether certain desired structure, as the one shown above, is part of the next code version. This is especially useful for checking whether the students are taking some time reasoning about the problem, before start coding.

3.3 Refactoring

E.g. changes which objective is to improve the code, according to known software engineering quality metrics. The changes presented below were extracted from the Haskell refactoring catalog [12]. We have adapted the catalog in order to fit it into two categories: (i) data structure, which affects data representation, and consequently all functions involved by the refactoring (e.g. algebraic or existential type, concrete to abstract data type, constructor or constructive function, adding a constructor); and (ii) naming, which implies that the binding structure of the program remains unchanged after the refactoring (e.g. adding or removing an argument, deleting or adding a definition, renaming). Most refactoring are too complex for beginners, so when that is the case, it is not followed by an example.

3.3.1 Algebraic or Existential Type

E.g. a data type, that could be recursive, is transformed into an existential type, in the same way as an object oriented data representation. This transformation is unified applying binary functions to the data type.

3.3.2 Concrete to Abstract Data Type

E.g. it converts a concrete data type into an abstract one with hidden constructors. Concrete data types provide direct representations for a variety of data types, and pattern matching syntax in data types constitutes an intuitive style of programming. The disadvantage of it is the resulting rigidity: the data representation is visible to all type clients. On the other hand, that is especially useful in order to turn the representation into an abstract one, by a modification or data type extension.

3.3.3 Constructor or Constructor Function

E.g. a data type constructor must be defined in terms of its constructors. There are complementary advantages in keeping the constructor and in eliminating it for its representation.

3.3.4 Adding a Constructor

E.g. new patterns are aggregated to case analysis that recurs over the data type in question. The positioning of the new pattern is allocated, preserving the order in which the new data type was defined.

3.3.5 Adding or Remove an Argument

E.g. adding a new argument to a function or constant definition. The new argument default value is defined at the same level as the definition. The position where the argument is added is not accidental: inserting the argument at the beginning of the argument list, implies that it can only be added to partial function applications. The following example shows the inclusion of an undefined function:

Program A	**Program B**
`f x = x + 17`	`f y x = x + 17`
	`f_y = undefined`
`g z = z + f x`	`g z = z + f f_y x`

A way to automatically detect such change is by comparing two functions and checking whether the function names or function parameters have changed. The following Prolog fragment shows a simple change in the function parameters, suggesting a solution refinement. It exemplifies a pattern matching, keeping the data in a substitution list. The code presented below is similar to the one presented in 3.2.1, with a few simplifications.

```
compare1(T1,T2,Subst) :-
        T1 =.. [FunctionName|List1],
        T2 =.. [FunctionName|List2],
        comp1(List1,List2,[],Subst).
```

3.3.6 Deleting or Adding a Definition

E.g. deleting a definition that has not been used. The following example shows the deletion of the function `table`:

Program A

```
showAll  = ...
format   = ...
table    = ...
```

Program B

```
showAll  = ...
format   = ...
```

A way to automatically detect such change is by using the same Prolog fragment presented in 3.2.1, `compare2/3`, which also checks whether a function name has been modified or deleted.

3.3.7 Renaming

E.g. renaming a program identifier, which could be a valued variable, type variable, a data constructor, a type constructor, a field name, a class name or a class instance name. The following example shows the change in the function named `format` to `fmt`:

Program A

```
format = ... format ...
entry  = ... format ...
table  = ...
         where
         format = ...
```

Program B

```
Fmt    = ... fmt    ...
entry  = ... fmt    ...
table  = ...
         where
         format = ...
```

A way to automatically detect such change is by using a predicate as `compare/2` (3.2.1). The potential of this change, besides readability, is especially useful for the teacher for detecting plagiarism. If the teacher is aware of such change and notices no improvement in the software quality metrics, she may look for a similar code in someone else's code versions and look for plagiarism.

4 Exploring the Supporting Strategies

The supporting strategies described above were implemented in a tool, named AcKnow, which aims at analyzing and categorizing student's code evolution. AcKnow takes as input codes from students that were indicated by a quantitative analysis made by a version control tool named AAEP [1]. These indications are based on the number of versions that each student did. When this number significantly differs from the normal distribution, i.e., the student is identified as a special case and her codes are submitted to AcKnow. Then, based on AcKnow's analysis of each pair of versions, the teacher may do a qualitative analysis, providing feedback relative to that student's current milestone. Figure 1 shows the AcKnow's architecture, including its subsidiary parts.

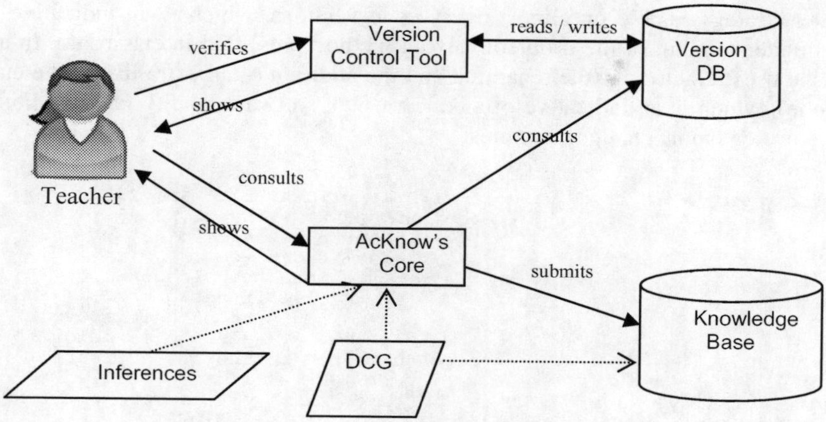

Fig. 1. AcKnow's Architecture

AcKnow classifies as syntactic changes modifications such as indentation, addition or removal of comma or semi-colon, spacing between characters and addition or removal of parentheses. The semantic changes are those related to function evaluation. For example, changes in data structures and operations in functions are classified in this category. Furthermore, refactoring, in this context, are special cases of semantic changes aimed at improving code quality according to software quality metrics. Currently, AcKnow partially uses the Haskell catalogue of refactoring [12].

Table 1. Jane Doe's History

Version	**Interval**	**Category**
1	0	Syntactic
2	Same minute	Syntactic
3	1 minute	Syntactic
4	1 minute	Syntactic
5	8 minutes	Refactoring
6	171 minutes	Semantic
7	44 hours	Refactoring

AcKnow has been used to analyze the development history of the students of the introductory programming course. That will be exemplified here by Jane Doe's code evolution. In Table 1 the categories of changes in Jane Doe's versions are presented, associated with the time intervals between them.

Table 2 shows an instantiation of the rules, applied to Jane Doe's code versions for the following problem: Given a line segment r, through two points, a and b, located in the Cartesian plane's first quarter, and any point p, determine whether p belongs to the segment or to the continuity of r or is located above or under the line. The texts inside the boxes indicate the changes made from one version to the next. After each version there is a description of Jane Doe did to the code.

Based on the code version's analysis presented next, the teacher observes that Jane Doe correctly uses the incremental development of solutions, using the divide and conquer strategy visible on almost every code versions, which is an indicative of a sound understanding of the problem solving method presented in classroom. In addition, she uses refactoring (the renaming and the addition of an extra line at the end of the code), which is an indicative of a concern with software quality metrics. Following we present some change examples:

(1)
```
pertseg x1 y1 x2 y2 x3 y3 = if x1>0 && y1>0 && x2>0 && y2>0
                             then if seg x1 y1 x2 y2 x3 y3
                                  then "p belongs to AB"
                                  else if det==0
                                       then "p belongs to r"
                                       else if det>0
                                            then "p is above r"
                                            else "p is below r"
                             else "p is not in the 1st quarter"

 seg x1 y1 x2 y2 x3 y3 = (cresc x1 y1 x2 y2 x3 y3||decresc x1 y1 x2 y2
x3 y3||conth x1 y1 x2 y2 x3 y3||contv x1 y1 x2 y2 x3 y3)
```

(2)
```
det x1 y1 x2 y2 x3 y3 = x1*y2+x2*y3+x3*y1-(x3*y2)-(x2*y1)-(x1*y3)
```

As it is shown in the above codes (1) and (2), in the first version (1), she partially solves the problem, but it is not accurate enough. Then, in the second version (2), she keeps the previous code and adds a new line, describing another function.

(3)
```
cresc x1 y1 x2 y2 x3 y3 = x3>x1 && x3<x2 && y3>y1&& y3<y2
decresc x1 y1 x2 y2 x3 y3 = x3>x1 && x3<x2 && y3<y1 && y3>y2
conth x1 y1 x2 y2 x3 y3 = x3>x1 && x3<x2 && y3==y1 && y3==y2
contv x1 y1 x2 y2 x3 y3 = x3==x1 && x3==x2 && y3>y1 && y3<y2
```

As it is shown in the above code (3), in the third version, she adds the functions needed to establish the line's condition. In the forth version, she attempts to modify the last code line, but she gives up after adding and removing a space. In the fifth version, she renames `pertseg` with `segment`.

(4)
```
 segm (x1,y1) (x2,y2) (x3,y3) = (cresc (x1,y1) (x2,y2) (x3,y3)||decresc
(x1,y1) (x2,y2) (x3,y3)||conth (x1,y1) (x2,y2) (x3,y3)||contv (x1,y1)
(x2,y2) (x3,y3))
 det (x1,y1) (x2,y2) (x3,y3) = x1*y2+x2*y3+x3*y1-(x3*y2)-(x2*y1)-
(x1*y3)
```

As it is shown in the above code (4), in the sixth version, she has a cognitive jump, changing `seg x1 y1 x2 y2 x3 y3` for `segm (x1,y1) (x2,y2) (x3,y3)` and `det x1 y1 x2 y2 x3 y3` for `det (x1,y1) (x2,y2) (x3,y3)` in order to use tuples. In the seventh version, she adds a new line at the end of the code.

Another relevant piece of information related to Table 1 and the above described code's evolution is that between versions 5 and 6 Jane Doe solved another 5 problems from the same list of exercises. Some of these exercises asked for the use of tuples in their solution. After that, she returned to the problem in question and submitted version 6 using tuples in its solution. So, Jane Doe had the winning insight when solving another problem.

The course where AcKnow was used had 100 students enrolled. The statistical method used by AAEP indicated an average of 3 students per problem as special cases. Besides Jane Doe, the other students presented a similar pattern of changes, carrying out initially many syntactic changes within a short time interval followed by one or two semantic changes within a longer time interval; only few of them (4 in 2 different exercises) carried out refactoring. When refactoring occurred, it was only identified in the very last versions.

5 Conclusion

This article proposes an initial set of programming supporting strategies which inspects students' code evolution in order to detect the following situations: expertise levels, difficulties, cognitive jumps in programming; and bottlenecks in solution planning.

We understand that having access to more code from future case studies, it will be possible to identify other strategies being used by these students. This way we will be moving towards a more complete set of programming supporting strategies and, consequently, their formal representation for the sake of knowledge acquisition research in programming.

Finally, in order to provide the teacher with a tool to follow up the programming learning development of her students, AcKnow was implemented. At this moment of time it does not support all the supporting strategies. For the sake of proof of concept the programming learning development of Jane Doe, a fictitious name for a real student, was analyzed.

Acknowledgement

This project is partially sponsored by the Brazilian Ministry of Science and Technology through the CTInfo project grant nº 550865/2007-1. Hugo Fuks is sponsored by CNPq individual grant nº 301917/2005-1 and he also receives grant from the FAPERJ project "Cientistas do Nosso Estado". Thais Castro receives a CNPq PhD grant. This research also receives resources from project ColabWeb – Proc.553329/2005-7, CNPq/CT-Amazônia n.27/2005.

References

1. Almeida Neto, F.A., Castro, T., Castro, A.N.: Utilizando o Método Clínico Piagetiano para Acompanhar a Aprendizagem de Programação. In: XVII Simpósio Brasileiro de Informática na Educação, 2006, Brasília. Simpósio Brasileiro de Informática na Educação, vol. 17, pp. 184–193. Gráfica e Editora Positiva Ltda, Brasília (2006)
2. Castro, T., Castro, A., Menezes, C., Boeres, M., Rauber, M.: Utilizando Programação Funcional em Disciplinas Introdutórias de Computação. In: XXII Congresso da Sociedade Brasileira de Computação / X Workshop sobre Educação em Computação, 2002, Workshop sobre Educação em Computação, vol. 4, pp. 157–168. SBC, Porto Alegre (2002)
3. Delval, J.: Introdução à Prática do Método Clínico. Artmed Publisher (2002) ISBN 8536300132

4. Détienne, F.: What Model(s) for Program Understanding? In: The Proceedings of the Colloque Using Complex Information, UCIS 1996 (1996)
5. Dijkstra, E.: On the Teaching of Programming, i.e. on the Teaching of Thinking. In: Selected Writings on Computing: A Personal Perspective. Springer, Heidelberg (1982)
6. Gerosa, M.A., Pimentel, M., Fuks, H., Lucena, C.J.P.: Development of Groupware based on the 3C Collaboration Model and Component Technology. In: Dimitriadis, Y.A., Zigurs, I., Gómez-Sánchez, E. (eds.) CRIWG 2006. LNCS, vol. 4154, pp. 302–309. Springer, Heidelberg (2006)
7. Gomes, A., Mendes, A.J.: Problem Solving in Programming. In: The Proceedings of PPIG as a Work in Progress Report (2007)
8. Mancy, R., Reid, N.: Aspects of Cognitive Style and Programming. In: The Proceedings of the 16th Workshop of the Psychology of Programming Interest Group, Carlow, Ireland (2004)
9. Mead, J., Gray, S., Hamer, J., James, R., Sorva, J., St. Clair, C., Thomas, L.: A Cognitive Approach to Identify Measurable Milestones for Programming Skill Acquisition. In: The Proceedings of ITiCSE. ACM, New York (2006)
10. Murphy, L., McCauley, R., Westbrook, S., Fossum, T., Haller, S., Morrison, B., Richards, B., Sanders, K., Zander, C., Anderson, R.E.: A Multi-Institutional Investigation of Computer Science Seniors' Knowledge of Programming Concepts. In: The Proceedings of SIGCSE. ACM, Missouri (2005)
11. Pimentel, M., Escovedo, T., Fuks, H., Lucena, C.J.P.: Investigating the assessment of learners' participation in asynchronous conference of an online course. In: 22nd ICDE - World Conference on Distance Education: Promoting Quality in On-line, Flexible and Distance Education (CD-ROM), September 3-6. ABED, Rio de Janeiro (2006)
12. Thompson, S., Reinke, C., Li, H.: Refactoring Functional Programs. Final Report GR/R75052/01 (2006), http://www.cs.kent.ac.uk/projects/refactor-fp
13. Weinberg, G.M.: The Psychology of Computer Programming. Computer Science Series, F9264-000-4. Litton Educational Publishing, USA (1971)

Revising Specifications with CTL Properties Using Bounded Model Checking

Marcelo Finger and Renata Wassermann

Department of Computer Science
Institute of Mathematics and Statistics
University of São Paulo, Brazil
{mfinger,renata}@ime.usp.br

Abstract. During the process of software development, it is very common that inconsistencies arise between the formal specification and some desired property. Belief Revision deals with the problem of accommodating new information that may be inconsistent with an existing knowledge base.

In this paper, we propose the use of belief revision techniques in order to deal with inconsistencies in formal specifications. The main problem to be solved is that the most well known results for belief revision only hold for logics which are monotonic and compact, while most discrete-time temporal logics used to express system properties – and in particular, CTL — are not compact. We suggest the use of bounded model-checking, transforming the problem from CTL into classical propositional logic and then transforming back the results to suggest revisions to the user.

Keywords: Model-checking, belief revision, formal specification, CTL.

1 Introduction

A system specification evolves when there is a conflict between the actual properties of the system and its intended new behaviour. In terms of formal specifications, this can be modelled by the existence of an inconsistency between an existing specification and a desired property.

Handling inconsistencies in specifications is a critical activity in the software development process. A variety of techniques has been developed for checking specifications for inconsistencies. These include formal techniques such as those based on model checking or theorem proving [1,2,3,4,5]. Model checking of specifications, in particular, is usually performed in CTL logic or in some related formalism.

While many of these approaches provide rigorous, and often automated, analysis of software specifications to reveal inconsistencies, they often also do not support the system developer in solving these inconsistencies after they have been discovered.

To address this issue, a first proposal of evolving system specifications in CTL was presented in [6] based on techniques of belief revision to suggest ways for

G. Zaverucha and A. Loureiro da Costa (Eds.): SBIA 2008, LNAI 5249, pp. 157–166, 2008.

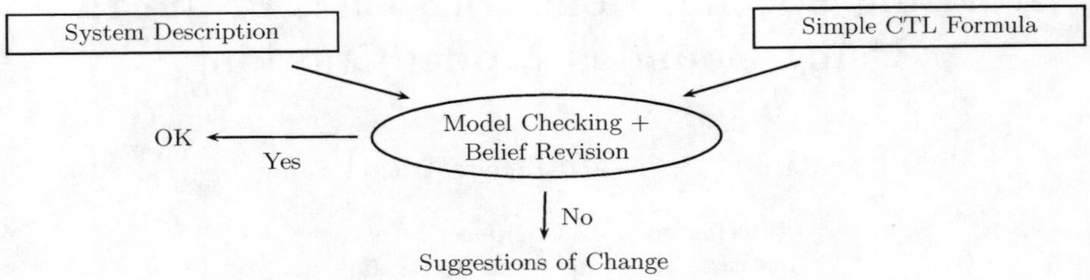

Fig. 1. Model-checking with belief revision

changing system specifications, as illustrated in Figure 1. Belief revision [7,8] is a sub-area of artificial intelligence whose main focus is to keep the consistency of a set of beliefs when new beliefs are incorporated.

Belief revision is normally done in terms of *revision operations*. However, it is not always the case that such operations exist for all formalisms. The method presented in [6] could only deal with simple CTL expressions, without embedded temporal operators, but typically one would like to check complex system properties. In the case of revising a specification with an arbitrarily complex property, it is very important to determine if a specification can be revised at all.

In this work, we study CTL specification and model checking techniques with respect to obtaining a method that guarantees that revised specifications always exist. In particular, we study the existence of *contraction* operations over specifications, which is the belief revision operation that performs the removal of some currently held belief, that is, some current system property that follows from the specification. The addition of new properties can be done by first contracting its negation, and then simply adding the new property to the contracted specification.

The starting point of this study is a quite general result on the existence of contraction operations for formal systems [9]. It turns out that CTL (and any other temporal logic used for model checking such as LTL, CTL and CTL*) *does not fulfil the required conditions* to guarantee the existence of a contracted theory. This means that, in general, one cannot guarantee that a revised specification exists using the traditional tools of belief revision.

The main contribution of this work is an interaction between model checking and bounded model checking that helps us solve this problem. Our proposal is illustrated is Figure 2.

The idea is that, once an inconsistency has been detected in a specification by model checking, one can determine a *bound* for *bounded model checking* [10]. The technique of bounded model checking transforms the model checking problem into a classical satisfiability problem. As classical logic satisfies the conditions of existence of a revised specification [9], one then performs a contraction over the transformed specification, and finally one has to work the contraction back to obtain the suggestions on how to revise the original specification.

A second advantage of this approach is that, with the translation into classical logic, CTL statements with arbitrary complexity can be accepted for model

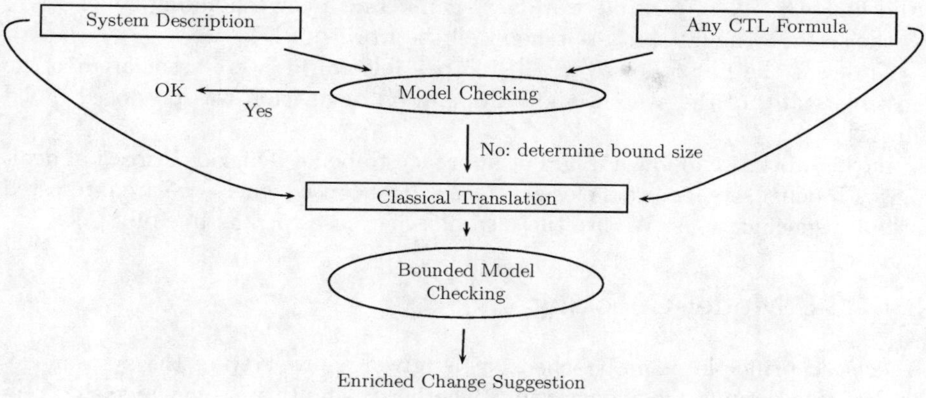

Fig. 2. Specification Revision via Bounded Model Checking

checking and specification revision, so a new kind of change suggestions can be obtained, enriching the capabilities of the method.

The paper is organised as follows: in the next two sections, we briefly introduce the areas of Belief Revision and CTL Model Checking. In Section 4 we discuss the applicability of standard belief revision to CTL. Next in Section 5 we present our method. Then we conclude, pointing towards future work.

2 Belief Revision

The necessity to model the behaviour of dynamic knowledge bases formed the basis of belief revision theory [7,8]. Most of the literature in the area is based on the seminal work of Alchourrón, Gärdenfors and Makinson [11], that have proposed some postulates that describe the formal properties that a revision process should obey. They have also proposed some constructions that satisfy the postulates. The work is usually known as the *AGM paradigm*, due to the initials of the three authors.

In AGM theory, the beliefs of an agent are represented by a belief set, a set of formulas closed under logical consequence ($K = Cn(K)$, where Cn is a supraclassical consequence operator). There are three types of change operators for a belief set (K). In *expansion* ($K + \alpha$), a consistent information α, together with its logical consequences, is added to the belief set K. In *contraction* ($K - \alpha$), the information α is abandoned. Since the set K is logically closed, it could be necessary to abandon other beliefs that would imply α. In *revision* ($K * \alpha$), an information α is added to K and to keep consistency it may be necessary to abandon other beliefs of K.

Besides belief sets, one can use the idea of possible worlds in order to represent the beliefs of the agent. Possible worlds can be thought of as possible states of the world (or, in propositional logics, propositional valuations). Given a belief set K, $[K]$ is the set of the possible worlds where all formulas of K are true.

And for a set W_k of possible worlds we can define a corresponding belief set K as the set of formulas that are true in all the worlds of W_k.

This was the line followed in [6]: the possible worlds were associated to the possible states of the system and an operation of revision was proposed based on [12].

In this work, we follow a different approach to belief revision. Instead of dealing with belief sets, i.e., sets closed under logical consequences, we are interested in finite specifications. We use the idea of belief bases, in the line of [8].

3 CTL Model-Checking

The basic principle of model-checking is: given a property of the system, described in a temporal logic, determine whether the finite state machine that represents the described system satisfies such property. In other words, one wants to verify whether a formula f is true in a graph G of states.

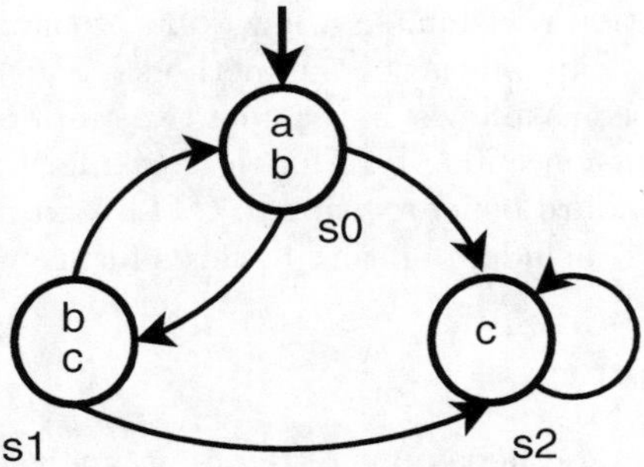

Fig. 3. Kripke Structure

The finite state machine that represents the system can be described as a specific Kripke structure, that is defined as: a set S of states, a set R of transitions between states (where each state must have a successor), a set I of initial states and a function L that associates to each state a set of propositions that hold in that state. To model a state in *deadlock* (state without successors) it suffices to create a transition from the state to itself. Figure 3 represents a Kripke structure where $P = \{a, b, c\}$, $S = \{s0, s1, s2\}$, $R = \{(s0, s1), (s0, s2), (s1, s0), (s1, s2), (s2, s2)\}$, $I = \{s0\}$ with $L(s0) = \{a, b\}$, $L(s1) = \{b, c\}$ and $L(s2) = \{c\}$.

3.1 Computation Tree Logic

The properties of the system being checked are described in temporal logic. Temporal logic is a type of modal logic where it is possible to represent and to reason about propositions related to time. Through temporal logic it is possible to express sentences of the type "I am *ALWAYS* hungry" or "I will be hungry *UNTIL* I eat in a all-you-can-eat restaurant".

Clarke and Emerson [13] proposed CTL (Computation Tree Logic), a logic capable to consider different possible futures, through the notion of branching time. The idea of this logic is to quantify over the possible runs of a program through the notion of paths that exist in the space of states of the system. The properties can be evaluated with respect to all the runs or some run. This logic is used in some model-checkers, such as NuSMV, which we have used in our implementation. The syntax of CTL is given by the following definition:

$$\phi ::= p|\neg\phi|\phi \wedge \phi|(AX\phi)|(AG\phi)|(AF\phi)|(EX\phi)|(EG\phi)|(EF\phi)|E(\phi\ U\phi)|A(\phi\ U\phi)$$

where p is a propositional atom, $\neg$, $\wedge$ are the usual logical connectives and the other are temporal operators. Each temporal operator is composed of a path quantifier (E, "there exists a path", or A, "for all paths") followed by a state operator (X, next state in the path, U, until, G, globally, or F, finally). CTL has the following semantic definition:

Definition 1. *Let M be a Kripke structure and $\pi(i)$ the i-th state of a path. We say that $M, s \models \phi$ if and only if ϕ is true in the state s of M. Thus we have:*

1. $M, s \models p$ *iff* $p \in L(s)$
2. $M, s \models \neg\phi$ *iff* $M, s \not\models \phi$
3. $M, s \models \phi_1 \wedge \phi_2$ *iff* $M, s \models \phi_1$ *and* $M, s \models \phi_2$
4. $M, s \models EX\ \phi$ *iff there is a state* s' *of* M *such that* $(s, s') \in R$ *and* $M, s' \models \phi$
5. $M, s \models EG\ \phi$ *iff there is a path* π *of* M *such that* $\pi(1) = s$ *and* $\forall i \geq 1 \bullet M, \pi(i) \models \phi$
6. $M, s \models E(\phi_1\ \ U\ \phi_2)$ *iff there is a path* π *of* M *such that* $\pi(1) = s$ *and* $\exists i \geq 1 \bullet (M, \pi(i) \models \phi_2 \wedge \forall j, i > j \geq 1 \bullet M, \pi(j) \models \phi_1)$

The other temporal operators can be derived from EX, EG and EU:

$\mathrm{AX}\ \phi = \neg\ \mathrm{EX}\ \neg\phi$
$\mathrm{AG}\ \phi = \neg\ \mathrm{EF}\ \neg\phi$
$\mathrm{AF}\ \phi = \neg\ \mathrm{EG}\ \neg\phi$
$\mathrm{EF}\ \phi = \mathrm{E}\ [\mathrm{true}\ \mathrm{U}\ \phi]$
$\mathrm{A}[\phi\ \mathrm{U}\ \beta] = \neg\ \mathrm{E}[\neg\beta\ \mathrm{U}\ \neg\phi \wedge \neg\beta] \wedge \neg\ \mathrm{EG}\ \neg\beta$

We say that a Kripke structure M satisfies a formula CTL ϕ if $M, s_0 \models \phi$, where s_0 is an initial state.

The main approach for automatic formula verification in CTL is based on the fixed point theory to characterize its operators. For details the reader is referred to [14].

4 Guaranteeing the Existence of Specification Revisions

The AGM framework for belief revision was originally formulated having classical logic in mind. There have been some attempts to apply the operations for different logics, such as modal or description logics, but in [15] it was shown that AGM can only be applied to logics that are decomposable, i.e., logics for which it holds that for any sets of formulas X, K if $Cn(\emptyset) \subset Cn(X) \subset Cn(K)$, then there exists a set Y such that $Cn(Y) \subset Cn(K)$ and $Cn(X \cup Y) = Cn(K)$. Flouris has shown that some important description logics are not decomposable and therefore, do not admit an AGM style contraction operation.

CTL (and modal logics in general) are decomposable. This means that AGM-style operations can be applied. However, the use of logically closed sets leads to problems from the computational point of view, since they are typically infinite.

For belief bases, [9] has shown that if the logic is compact and monotonic, the typical constructions can be applied. But CTL and other temporal logics based on discrete time are not compact. This means that it is not possible to directly inherit all the results, as the theorems in [9] do not say anything about the case where the logic is not compact.

The approach we follow here is to avoid the issue by first translating the problem into propositional logic and then applying standard methods of belief revision.

5 Revising Via Bounded Model Checking

A solution to the problem of ensuring the applicability of belief revision techniques is proposed here. This is achieved by means of Bounded Model Checking and its associated translation of both specification and CTL property to propositional logic. The process is illustrated in Figure 2 and detailed next.

5.1 Bounded Model Checking

The idea of Bounded Model Checking [10,16] is to fix a bound k for the maximal path in a Kripke Model that can be traversed. This allows a given Kripke Model K to be translated into a propositional classical formula K^k.

The semantic of CTL formulas is then altered to guarantee that the size of paths traversed in the evaluation of a formula are never larger than k. A special model-theoretic construct is needed for the semantics of the EG operator, namely the k-loop, that is a path of size at most k containing a loop. The k-bounded semantic of CTL formulas, $M, s \models^k \phi$, $k \geq 0$, is defined as

1. $M, s \models^k p$ iff $p \in L(s)$
2. $M, s \models^k \neg\phi$ iff $M, s \not\models^k \phi$
3. $M, s \models^k \phi_1 \wedge \phi_2$ iff $M, s \models^k \phi_1$ and $M, s \models^k \phi_2$
4. $M, s \models^k \mathrm{EX}\,\phi$ iff there is a state s' of M such that $(s, s') \in R$ and $M, s' \models^{k-1} \phi$
5. $M, s \models^k \mathrm{EG}\ \phi$ iff there is a k-loop π of M such that $\pi(1) = s$ and $\forall i \geq 1 \bullet M, \pi(i) \models^{k-i} \phi$
6. $M, s \models^k \mathrm{E}(\phi_1\ \mathrm{U}\ \phi_2)$ iff there is a path π of M such that $\pi(1) = s$ and $\exists i \geq 1 \bullet (M, \pi(i) \models^{k-i} \phi_2 \wedge \forall j, i > j \geq 1 \bullet M, \pi(j) \models^{k-j} \phi_1)$

With a fixed bound k and this k-bounded semantics, one can translate a bounded CTL formula ϕ into a classical formula A^k_ϕ. One then submits the formula $K^k \wedge A^k_\phi$ to some efficient SAT solver (that is, a classical propositional theorem prover, such as Chaff[17] or Berkmin [18]) to obtain a verdict. If the formula is unsatisfiable, this means that $\neg\phi$ holds at the Kripke Model K up to the given bound. Otherwise a classical valuation v is presented by the SAT solver, which represents a path in traversed the model such that ϕ holds at the Kripke model.

5.2 The Method of BMC Revisions

By checking a formula ϕ against a model, one is trying to force the property $\neg\phi$ over the revised model. In fact, if K and ϕ are inconsistent, this means that K "implies" $\neg\phi$. So if $K^k \wedge A^k_\phi$ is inconsistent, nothing needs to be done, and this is our ideal case.

The method for obtaining change suggestions from bounded model checking is triggered in the situation where the SAT solver outputs a valuation that satisfies the formula $K^k \wedge A^k_\phi$, as represented in Figure 4.

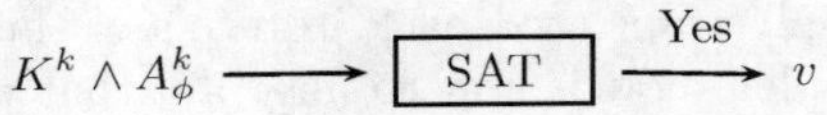

Fig. 4. Bounded Model Checking Triggering Belief Revision

Any classical valuation v can be directly transformed into a formula $A_v = \bigwedge_{v(ell_i)=1} \ell_i$ that conjoins all literals satisfied by v.

The basic idea of the method is to revise the specification represented by $K^k \wedge A^k_\phi$ so that valuation v is not allowed. However, it may be the case that $K^k \wedge A^k_\phi \wedge \neg A_v$ may be a satisfiable formula, satisfied by valuation v_2, so that $\neg A_{v_2}$ is added to the theory. This process is iterated m times until an inconsistent theory K' is obtained:

$$K' = K^k \wedge A^k_\phi \wedge \neg A_{v_1} \wedge \ldots \neg A_{v_m}$$

As the number of possible valuations is finite, the process always terminates. The revised specification K_{rev} is obtained by

$$K_{\text{rev}} = K' * \neg A^k_\phi = (K' - \neg A^k_\phi) + \neg A^k_\phi$$

One last step is still missing, namely, the translation back from propositional classical K_{rev} into change suggestions in the original specification K.

Theorem 1. *The revised specification K_{rev} always exists.*

This means that the initial specification can be revised with respect to any CTL formula *provided a large enough bound k is used.* One possible way of determining such bound is to run a usual (unbounded) model checking with respect to the desired property, $\neg\phi$.

This method is apparently trading a PSPACE-complete problem (model checking) for an NP-complete one (SAT). Is there a price to pay for it? In fact:

- No upper limit has been established over m, the number of times that SAT is executed; in the worst case, it is an exponential number, in terms of the number of propositional variables.
- The bound k may be hard to find. In particular, an approximate approach may be used via bounded model checking.

5.3 Generating Change Suggestions

To generate change suggestions, one has to compare the initial translation K^k with the revised theory K_{rev} and propagate back the differences to the specification. The translation back is not a problem because:

- The atomic formulas of K all represent clear facts about the specification such as "p holds at state S" and "S_0 accesses S_1".
- Belief revision techniques usually do not introduce new atomic symbols, so K_{rev} is a boolean combination of the same literals present in K.

As K_{rev} can be simply translated back in terms of the original specification elements, the only problem now is how to interpret that revised specification.

In general, the initial specification K is a *definite description*, presented as a conjunction of literals, such as "p holds at state S_0 which accesses state S_1, where p is false". In contrast, the revised model K_{rev} will very likely have parts of it constituted of the disjunction of several options, eg. statements of the form "state S_0 accesses either S_i or S_j and p either holds in both S_i and S_j, or in none of them".

Once the differences between K and K_{rev} have been detected, one may present them to the user as *the* output solution. However, what one needs in the end of the revision process is a definite model again, so one is faced with the following possibilities:

- Present the possible definite models in some (perhaps arbitrary) order, such that all presented definite models are the only possible ones entailed by the specification; there may be an exponentially large number of such models.
- Choose a single definite model to present as a suggestion. The preference of one model over the others may be built-in in the revision method, so that K_{rev} is in fact a definite description.

We believe that end-users must have a say on which kind of suggestion they find more useful.

6 Conclusions

We have shown how belief revision can be used to revise specifications even when the properties to introduce are specified in CTL. The technical problem that arises is that one cannot guarantee the existence of a revised model in CTL and similar discrete-time temporal logics, such as LTL and CTL*.

The solution comes in using Bounded Model Checking and a translation of both specification and CTL formulas to propositional logic, such that a SAT solver is applicable. A method was proposed to iterate through these theories until belief revision is applicable, and the revised version is guaranteed to exist, and easily translatable back to the specification level.

Future works include studying the formal properties of the proposed method and implementing such specification revision operations. In particular, a process of stepwise refinements of a specification is to be submitted to such a process. The feedback from such a process will help us decide on the best way to present to the specifier the change suggestions provided by the method.

References

1. Winter, K.: Model checking for abstract state machines. Journal of Universal Computer Science 3(5), 689–701 (1997)
2. Büessow, R.: Model Checking Combined Z and Statechart Specifications. PhD thesis, Technical University of Berlin, Faculty of Computer Science (November 2003), `http://edocs.tu-berlin.de/diss/2003/buessow_robert.pdf`
3. Leuschel, M., Butler, M.: ProB: A model checker for B. In: Araki, K., Gnesi, S., Mandrioli, D. (eds.) FME 2003. LNCS, vol. 2805, pp. 855–874. Springer, Heidelberg (2003)
4. Kolyang, S.T., Wolff, B.: A structure preserving encoding of Z in Isabelle/HOL. In: von Wright, J., Harrison, J., Grundy, J. (eds.) TPHOLs 1996. LNCS, vol. 1125, pp. 283–298. Springer, Heidelberg (1996)
5. Clarke, E.M., Grumberg, O., Hamaguchi, K.: Another look at LTL model checking. In: Dill, D.L. (ed.) CAV 1994. LNCS, vol. 818, pp. 415–427. Springer, Heidelberg (1994)
6. de Sousa, T.C., Wassermann, R.: Handling inconsistencies in CTL model-checking using belief revision. In: Proceedings of the Brazilian Symposium on Formal Methods (SBMF) (2007)
7. Gärdenfors, P.: Knowledge in Flux: Modeling the Dynamics of Epistemic States. MIT Press, Cambridge (1988)
8. Hansson, S.O.: A Textbook of Belief Dynamics. Kluwer Academic Publishers, Dordrecht (1997)
9. Hansson, S.O., Wassermann, R.: Local change. Studia Logica 70(1), 49–76 (2002)
10. Biere, A., Cimatti, A., Clarke, E.M., Zhu, Y.: Symbolic model checking without bdds. In: Cleaveland, W.R. (ed.) TACAS 1999. LNCS, vol. 1579, pp. 193–207. Springer, Heidelberg (1999)
11. Alchourrón, C., Gärdenfors, P., Makinson, D.: On the logic of theory change: Partial meet contraction and revision functions. Journal of Symbolic Logic 50(2), 510–530 (1985)
12. Grove, A.: Two modellings for theory change. Journal of Philosophical Logic 17(2), 157–170 (1988)
13. Clarke, E.M., Emerson, E.A.: Design and synthesis of synchronization skeletons using branching time temporal logic. In: Kozen, D. (ed.) Logic of Programs 1981. LNCS, vol. 131, pp. 52–71. Springer, Heidelberg (1982)
14. Huth, M., Ryan, M.: Logic in Computer Science: Modelling and Reasoning about Systems. Cambridge University Press, Cambridge (2000)

15. Flouris, G.: On Belief Change and Ontology Evolution. PhD thesis, University of Crete (2006)
16. Penczek, W., Wozna, B., Zbrzezny, A.: Bounded model checking for the universal fragment of CTl. Fundamenta Informaticae 51(1), 135–156 (2002)
17. Moskewicz, M.W., Madigan, C.F., Zhao, Y., Zhang, L., Malik, S.: Chaff: Engineering an Efficient SAT Solver. In: Proceedings of the 38th Design Automation Conference (DAC 2001), pp. 530–535 (2001)
18. Goldberg, E., Novikov, Y.: Berkmin: A Fast and Robust SAT Solver. In: Design Automation and Test in Europe (DATE 2002), pp. 142–149 (2002)

Toward Short and Structural $\mathcal{ALC}$-Reasoning Explanations: A Sequent Calculus Approach

Alexandre Rademaker and Edward Hermann Haeusler

Department of Informatics, PUC-Rio, Rio de Janeiro, Brazil
{arademaker,hermann}@inf.puc-rio.br
http://www.tecmf.inf.puc-rio.br

Abstract. This article presents labelled sequent calculi $\mathcal{S}_{\mathcal{ALC}}$ and $\mathcal{S}^{[]}_{\mathcal{ALC}}$ for the basic Description Logic (DL) $\mathcal{ALC}$. Proposing Sequent Calculus (SC) for dealing with DL reasoning aims to provide a more structural way to generated explanations, from proofs as well as counter-models, in the context of Knowledge Base and Ontologies authoring tools. The ability of providing short (Polynomial) proofs is also considered as an advantage of *SC*-based explanations with regard to the well-known Tableaux-based reasoners. Both, $\mathcal{S}_{\mathcal{ALC}}$ and $\mathcal{S}^{[]}_{\mathcal{ALC}}$ satisfy cut-elimination, while $\mathcal{S}^{[]}_{\mathcal{ALC}}$ also provides $\mathcal{ALC}$ counter-example from unsuccessful proof-trees. Some suggestions for extracting explanations from proofs in the presented systems is also discussed.

Keywords: description logics, sequent calculus, proof theory.

1 Introduction

Logic languages have been frequently used as a basis to build tools and frameworks for Knowledge Representation (KR) and artificial intelligence (AI). Automatic Theorem Provers offered the mechanical support to achieve the derivation of meaningful utterances from already known statements. One could consider First-Order Logic (FOL) as a basis for *KR*-system, however, it is undecidable and hence, would not provide answers to any query. The use of decidable fragments of FOL and other logics have been considered instead. **DATALOG** is among the logics historically considered. In general after getting to a decidable logic to KR, one starts extending it in order to accommodate more powerful features, recursive **DATALOG** is an example. Other possibilities include extending by adding temporal, deontic or even action modalities to the core logic. However, authoring a Knowledge Base (KB) using any *KR*-Logic/Framework, is far from an easy task. It is a conceptually hard task, indeed. The underlying semantics of the specific domain (SD) has to be formalized by means of linguistic components as *Predicates*, *Individual-Designators*, *Rules or Axioms*, even *Modalities* and so on. The Logic Language should not provide any glue concerning this task, since it is expected to be neutral. Anyway, two main questions raise up when one designs a KB: (1) Is the KB consistent? (2) Is the KB representing the right Knowledge at all? Again, with the help of an Automatic Proof Procedure[1] for

[1] A Sat-Solver Procedure can be also considered.

G. Zaverucha and A. Loureiro da Costa (Eds.): SBIA 2008, LNAI 5249, pp. 167–176, 2008.

the underlying logic it is possible to design authoring environments to support the KB correct definition. It is recommendable to use a decidable underlying logic, any case. The authoring environment ability on providing glues on why the current KB is inconsistent is as important as the consistency test itself. It is also important the ability to *explain* how and why a known piece of knowledge is supported (or not) by the current KB. To either ability we simply refer to *explanation*, according to our terminology.

Description Logics (DL) are quite well-established as underlying logics for KR. The core of the DL is the $\mathcal{ALC}$ description logic. In a broader sense, a KB specified in any description logic having $\mathcal{ALC}$ as core is called an Ontology. We will not take any discussion on the concept just named as well as we will not discuss also the technological concerns around Ontologies and the **Web**. A DL theory presentation is a set of axioms in the DL logical language as well as an OWL-DL file. DL have implemented reasoners, for example Pellet and Racer,[2] only to mention two. There are also quite mature Ontology Editors. Protegé is the more popular and used free editor. However, the Reasoners (theorem provers) as well as the Editors do not have a good, if any, support for explanations. As far as we know, the existing *DL*-Reasoners implement the Tableaux proof procedure first published in [1]. Concerning approaches on explanation in DL, the papers [2,3] describe methods to extract explanations from *DL*-Tableaux proofs. In [4] it is described the explanation extraction in quite few details, making impossible a feasible comparison with the approach followed in our article.

Simple Tableaux[3] cannot produce short proofs (polynomially lengthy proofs). This follows from the theorem that asserts that simple Tableaux cannot produce short proofs for the Pigeonhole Principle (PHP). PHP is easily expressed in propositional logic, and hence, is also easily expressed in $\mathcal{ALC}$. On the other hand, Sequent Calculus (SC) (with the cut rule) has short proofs for PHP. In [5,6] it is shown, distinct SC proof procedures that incorporate mechanisms that are somehow equivalent to introducing cut-rules in a proof. Anyway, both articles show how to obtain short proofs, in SC, for the PHP. We believe that super-polynomial proofs, like the ones generated by simple Tableaux, cannot be considered as good sources for text generation. The reader might want to consider that only the reading of the proof itself is a super-polynomial task regarding time complexity.

The generation of an explanatory text from a formal proof is still under investigation by the community, at least if one consider *good* explanations. The use of Endophoras[4] in producing explanations is a must. However, the produced text should not contain unstructured nesting of endophoras, a text like this is hard to follow. Some structure is need relating the endophoras. We believe that

[2] Racer is an industrial product, and currently must be bought.

[3] A simple Tableaux is not able to implement analytical cuts. The Tableaux used for $\mathcal{ALC}$ is simple.

[4] An anaphora is a linguistic reference to an antecedent piece of text. A Cataphora is a linguistic reference to a posterior piece of text in a phrase. Endophora refer to both, anaphora and cataphora.

as more structured the proof is, as easier the generation of a better text, at least concerning the use of endophoras.

In section 2 we present $\mathcal{S}_{\mathcal{ALC}}$ and its main proof-theoretical features. Section 3 shows how the structural subsumption algorithm described in [7] for a fragment of $\mathcal{ALC}$ is subsumed by $\mathcal{S}_{\mathcal{ALC}}$. Section 4 shows how to obtain a counter-model from an unsuccessful proof-tree, providing an explanatory power for negative answers similar to Tableaux, from $\mathcal{S}^{[]}_{\mathcal{ALC}}$, an implementation-driven conservative extension of $\mathcal{S}_{\mathcal{ALC}}$. Section 5 presents an example of explanation extracted from a proof in $\mathcal{S}_{\mathcal{ALC}}$ focusing on comparing our system with the well-known work described in [2] as well as more recent works on $\mathcal{ALC}$-explanation. Our main contribution, besides the proof theoretical one, is an step towards short and structured explanations in $\mathcal{ALC}$ theorem proving.

2 The $\mathcal{ALC}$ Sequent Calculus

$\mathcal{ALC}$ is a basic description language [7] and its syntax of concept descriptions is described as follows:

$$\phi_c ::= \bot \mid A \mid \neg\phi_c \mid \phi_c \sqcap \phi_c \mid \phi_c \sqcup \phi_c \mid \exists R.\phi_c \mid \forall R.\phi_c$$

where A stands for atomic concepts and R for atomic roles.

The Sequent Calculus for $\mathcal{ALC}$ ($\mathcal{S}_{\mathcal{ALC}}$) that it is shown in Figures 1 and 2 was first presented in [8] where it was proved to be sound and complete. It is based on the extension of the language ϕ_c. Labels are lists of (possibly skolemized) role symbols. Its syntax is as follows:

$$L ::= R, L \mid R(L), L \mid \emptyset$$
$$\phi_{lc} ::= {}^{L}\phi_c{}^{L}$$

where R stands for roles, L for list of roles and $R(L)$ is an skolemized role expression.

Each *consistent* labeled $\mathcal{ALC}$ concept has an $\mathcal{ALC}$ concept equivalent. For instance, ${}^{Q_2,Q_1}\alpha^{R_1(Q_2),R_2}$ is equivalent to $\exists R_2.\forall Q_2.\exists R_1.\forall Q_1.\alpha$.

Let α be an ϕ_{lc} formula; the function $\sigma : \phi_{lc} \rightarrow \phi_c$ transforms a labeled $\mathcal{ALC}$ concept into an $\mathcal{ALC}$ concept. $\mathcal{ALC}$ sequents are expressions of the form $\Delta \Rightarrow \Gamma$ where Δ and Γ are finite *sequences* of labeled concepts. The natural interpretation of the sequent $\Delta \Rightarrow \Gamma$ is the $\mathcal{ALC}$ formula $\bigsqcap_{\delta\in\Delta} \sigma(\delta) \sqsubseteq \bigsqcup_{\gamma\in\Gamma} \sigma(\gamma)$.

In Figures 1 and 2 the lists of labels are omitted whenever it is clear that a rule does not take into account their specific form. This is the case for the structural rules. In all rules α, β stands for $\mathcal{ALC}$ concepts (formulas without labels), γ, δ stands for labeled concepts, Γ, Δ for list of labeled concepts. L, M, N stands for list of roles. All of this letters may have indexes whenever necessary for distinction. If ${}^{L_1}\alpha^{L_2}$ is a consistently labeled formula then $\mathcal{D}(L_2)$ is the set of role symbols that occur inside the *skolemized role expressions* in L_2. Note that $\mathcal{D}(L_2) \subseteq L_1$ always holds.

Consider ${}^{L_1}\alpha^{L_2}$; the notation ${}^{\frac{L_2}{L_1}}\alpha^{\frac{L_1}{L_2}}$ denotes the exchanging of the universal roles occurring in L_1 for the existential roles occurring in L_2 in a consistent way

$$\overline{\alpha \Rightarrow \alpha}$$

$$\frac{\Delta \Rightarrow \Gamma}{\Delta, \delta \Rightarrow \Gamma}\ \text{weak-l} \qquad \frac{\Delta \Rightarrow \Gamma}{\Delta \Rightarrow \Gamma, \gamma}\ \text{weak-r}$$

$$\frac{\Delta, \delta, \delta \Rightarrow \Gamma}{\Delta, \delta \Rightarrow \Gamma}\ \text{contraction-l} \qquad \frac{\Delta \Rightarrow \Gamma, \gamma, \gamma}{\Delta \Rightarrow \Gamma, \gamma}\ \text{contraction-r}$$

$$\frac{\Delta_1, \delta_1, \delta_2, \Delta_2 \Rightarrow \Gamma}{\Delta_1, \delta_2, \delta_1, \Delta_2 \Rightarrow \Gamma}\ \text{perm-l} \qquad \frac{\Delta \Rightarrow \Gamma_1, \gamma_1, \gamma_2, \Gamma_2}{\Delta \Rightarrow \Gamma_1, \gamma_2, \gamma_1, \Gamma_2}\ \text{perm-r}$$

Fig. 1. The System $\mathcal{S}_{\mathcal{ALC}}$: structural rules

such that the skolemization is dually placed. This is used to express the negation of labeled concepts. If $\beta \equiv \neg\alpha$ the formula $\frac{Q}{R}\beta\frac{R}{Q(R)}$ is the negation of ${}^{R}\alpha^{Q(R)}$.

In the rules ($\forall$-r) and ($\forall$-l) $R \notin \mathcal{D}(L_2)$ must hold. In rules (prom-2), ($\forall$-r) and ($\forall$-l), the notation L_2', N_i' means the reconstructions of the skolemized expressions on those lists regarding the modification of the lists L_1 and M_i,

$$\frac{\Delta, {}^{L_1,R}\alpha^{L_2} \Rightarrow \Gamma}{\Delta, {}^{L_1}(\forall R.\alpha)^{L_2'} \Rightarrow \Gamma}\ \forall\text{-l} \qquad \frac{\Delta \Rightarrow \Gamma, {}^{L_1,R}\alpha^{L_2}}{\Delta \Rightarrow \Gamma, {}^{L_1}(\forall R.\alpha)^{L_2'}}\ \forall\text{-r}$$

$$\frac{\Delta, {}^{L_1}\alpha^{R(L_1),L_2} \Rightarrow \Gamma}{\Delta, {}^{L_1}(\exists R.\alpha)^{L_2} \Rightarrow \Gamma}\ \exists\text{-l} \qquad \frac{\Delta \Rightarrow \Gamma, {}^{L_1}\alpha^{R(L_1),L_2}}{\Delta \Rightarrow \Gamma, {}^{L_1}(\exists R.\alpha)^{L_2}}\ \exists\text{-r}$$

$$\frac{\Delta, {}^{L}\alpha^{\emptyset}, {}^{L}\beta^{\emptyset} \Rightarrow \Gamma}{\Delta, {}^{L}(\alpha \sqcap \beta)^{\emptyset} \Rightarrow \Gamma}\ \sqcap\text{-l} \qquad \frac{\Delta \Rightarrow \Gamma, {}^{L}\alpha^{\emptyset} \quad \Delta \Rightarrow \Gamma, {}^{L}\beta^{\emptyset}}{\Delta \Rightarrow \Gamma, {}^{L}(\alpha \sqcap \beta)^{\emptyset}}\ \sqcap\text{-r}$$

$$\frac{\Delta, {}^{\emptyset}\alpha^{L} \Rightarrow \Gamma \quad \Delta, {}^{\emptyset}\beta^{L} \Rightarrow \Gamma}{\Delta, {}^{\emptyset}(\alpha \sqcup \beta)^{L} \Rightarrow \Gamma}\ \sqcup\text{-l} \qquad \frac{\Delta \Rightarrow \Gamma, {}^{\emptyset}\alpha^{L}, {}^{\emptyset}\beta^{L}}{\Delta \Rightarrow \Gamma, {}^{\emptyset}(\alpha \sqcup \beta)^{L}}\ \sqcup\text{-r}$$

$$\frac{\Delta \Rightarrow \Gamma, {}^{L_1}\alpha^{L_2}}{\Delta, \frac{L_2}{L_1}\neg\alpha\frac{L_1}{L_2} \Rightarrow \Gamma}\ \neg\text{-l} \qquad \frac{\Delta, {}^{L_1}\alpha^{L_2} \Rightarrow \Gamma}{\Delta \Rightarrow \Gamma, \frac{L_2}{L_1}\neg\alpha\frac{L_1}{L_2}}\ \neg\text{-r}$$

$$\frac{{}^{L_1}\alpha^{L_2} \Rightarrow {}^{M_1}\beta_1{}^{N_1}, \ldots, {}^{M_n}\beta_n{}^{N_n}}{{}^{L_1}\alpha^{L_2,R} \Rightarrow {}^{M_1}\beta_1{}^{N_1,R}, \ldots, {}^{M_n}\beta_n{}^{N_n,R}}\ \text{prom-1}$$

$$\frac{{}^{M_1}\beta_1{}^{N_1}, \ldots, {}^{M_n}\beta_n{}^{N_n} \Rightarrow {}^{L_1}\alpha^{L_2}}{{}^{R,M_1}\beta_1{}^{N_1'}, \ldots, {}^{R,M_n}\beta_n{}^{N_n'} \Rightarrow {}^{R,L_1}\alpha^{L_2'}}\ \text{prom-2}$$

$$\frac{\Delta_1 \Rightarrow \Gamma_1, {}^{L_1}\alpha^{L_2} \quad {}^{L_1}\alpha^{L_2}, \Delta_2 \Rightarrow \Gamma_2}{\Delta_1, \Delta_2 \Rightarrow \Gamma_1, \Gamma_2}\ \text{cut}$$

Fig. 2. The System $\mathcal{S}_{\mathcal{ALC}}$: logical rules

respectively. The restrictions in the rules (∀-r) and (∀-l) means that the role R can only be removed from the left list of labels if none of the skolemized role expressions in the right list depends on it.

In [9] the cut-elimination[5] theorem is proved for $\mathcal{S}_{\mathcal{ALC}}$.

3 Comparing with the Structural Subsumption Algorithm

The *structural subsumption algorithms* (SSC), presented in [7], compare the syntactic structure of two normalized concept descriptions in order to verify if the first one is subsumed by the second one. Due to lack of space, we can only say that each step taken by a bottom-up construction of a $\mathcal{S}_{\mathcal{ALC}}$ proof corresponds to a step towards this matching by means of the *SSC* algorithms. The following construction on $\mathcal{S}_{\mathcal{ALC}}$ deals with normalized concepts (the sub-language of $\mathcal{ALC}$ required by *SSC*) [7]. It would conclude the subsumption (sequent) whenever the top-sequents ensure also their respective subsumptions. This is just what the (recursive) *SSC* does. Consider:

$$\dfrac{\dfrac{\dfrac{A_1 \Rightarrow B_1}{\forall R_1.C_1, A_1 \Rightarrow B_1}}{A_1, \forall R_1.C_1 \Rightarrow B_1} \qquad \dfrac{\dfrac{\dfrac{\dfrac{{}^{R_1}C_1 \Rightarrow {}^{S_1}D_1}{{}^{R_1}C_1 \Rightarrow \forall S_1.D_1}}{\forall R_1.C_1 \Rightarrow \forall S_1.D_1}}{A_1, \forall R_1.C_1 \Rightarrow \forall S_1.D_1}}{}}{\dfrac{A_1, \forall R_1.C_1 \Rightarrow B_1 \sqcap \forall S_1.D_1}{A_1 \sqcap \forall R_1.C_1 \Rightarrow B_1 \sqcap \forall S_1.D_1}}$$

4 Obtaining Counter-Models from Unsuccessful Proof-Trees

The structural subsumption algorithm is restricted to a quite inexpressive language and simple Tableaux based algorithms generally fails to provide short proofs. On the other hands, the later has an useful property, it returns a counter-model from an unsuccessful proof. A counter-model, that is, an interpretation that falsifies the premise, is a quite useful artifact to a knowledge-base engineer.

In this section we show how to extend $\mathcal{S}_{\mathcal{ALC}}$ system in order to be able to construct a counter-model from unsuccessful proofs too. In this way, $\mathcal{S}_{\mathcal{ALC}}$ can be compared with Tableaux algorithms, indeed. In fact $\mathcal{S}_{\mathcal{ALC}}$ is a structural-free sequent calculus designed to provide sequent proofs without considering backtracking during the proof-construction from conclusion to axioms. Concerning $\mathcal{ALC}$ the novelty is focused on the promotion and frozen rules, shown in the sequel.

Let us consider the system $\mathcal{S}^{[]}_{\mathcal{ALC}}$, a conservative extension of $\mathcal{S}_{\mathcal{ALC}}$, presented in Figure 3. $\mathcal{S}^{[]}_{\mathcal{ALC}}$ sequents are expressions of the form $\Delta \Rightarrow \Gamma$ where Δ and Γ

[5] This theorem states that any $\mathcal{S}_{\mathcal{ALC}}$-provable sequent can be proved without the cut-rule.

are *sets* of labeled concepts (possibly frozen). A frozen formula (labeled concept) α is represented as $[\alpha]$. The notation $[\Delta]$ means that each $\delta \in \Delta$ is frozen (i.e. $\{[\delta] \mid \delta \in \Delta\}$). Frozen formulas may also be indexed as $[\alpha]^n$. The notation can also be extended to a set of formulas $[\Delta]^n = \{[\delta]^n \mid \delta \in \Delta\}$.

Due to space limitation, it is only displayed the rules of $\mathcal{S}^{[]}_{\mathcal{ALC}}$ that modify the indexes of the formulas in the sequents. The remaining formulas are those of Figure 2 except *cut* (*prom-1* and *prom-2* of Figure 3 replace theirs corresponding in Figure 2) and considering that Δ and Γ range over formulas and frozen formulas (indexed or not).

Reading bottom-up the rules *prom-1* and *prom-2* freeze all formulas that do not contain the removed label. In rules *frozen-exchange*, *prom-1* and *prom-2* $[\Delta]^n$ and $[\Gamma]^n$ contains all the indexed-frozen formulas of the set of formulas in the antecedent (resp. succedent) of the sequent. The n is taken as the greatest index among all indexed-frozen formulas present in Γ and Δ. In rule *prom-2* the notation L_2' means the reconstructions of the skolemized expressions on list L_2 regarding the modification of the list L_1. The notation $\Gamma^{(R)}$ and ${}^{(R)}\Gamma$ denotes the addition of the Role R to the beginning of the existential and universal labels respectively. In rule *frozen-exchange* all formulas in Δ_2 and Γ_2 must be atomic.

$$\overline{\Delta, \alpha \Rightarrow \alpha, \Gamma} \qquad \frac{[\Delta]^n, [\Delta_1], {}^{L_1}\alpha^{L_2} \Rightarrow \Gamma_1, [\Gamma_2], [\Gamma]^n}{[\Delta]^n, \Delta_1, {}^{L_1}\alpha^{L_2,R} \Rightarrow {\Gamma_1}^{(R)}, \Gamma_2, [\Gamma]^n}\ \text{prom-1}$$

$$\frac{[\Delta]^n, [\Delta_2], \Delta_1 \Rightarrow {}^{L_1}\alpha^{L_2}, [\Gamma_1], [\Gamma]^n}{[\Delta]^n, \Delta_2, {}^{(R)}\Delta_1 \Rightarrow {}^{R,L_1}\alpha^{L_2'}, \Gamma_1, [\Gamma]^n}\ \text{prom-2}$$

$$\frac{\Delta_1, [\Delta_2]^{n+1}, [\Delta]^n \Rightarrow \Gamma_1, [\Gamma_2]^{n+1}, [\Gamma]^n}{[\Delta_1], \Delta_2, [\Delta]^n \Rightarrow [\Gamma_1], \Gamma_2, [\Gamma]^n}\ \text{frozen-exchange}$$

Fig. 3. Sub-system of $\mathcal{S}^{[]}_{\mathcal{ALC}}$ containing the rules that modify indexes

In Section 2, we stated the natural interpretation of a sequent $\Delta \Rightarrow \Gamma$ in $\mathcal{S}_{\mathcal{ALC}}$ as the $\mathcal{ALC}$-formula $\bigsqcap_{\delta\in\Delta} \sigma(\delta) \sqsubseteq \bigsqcup_{\gamma\in\Gamma} \sigma(\gamma)$. Given an interpretation function $\cdot^{\mathcal{I}}$ we write $\mathcal{I} \models \Delta \Rightarrow \Gamma$, if and only if, $\bigsqcap_{\delta\in\Delta} \sigma(\delta)^{\mathcal{I}} \sqsubseteq \bigsqcup_{\gamma\in\Gamma} \sigma(\gamma)^{\mathcal{I}}$ [8]. Now we have to extend that definition to give the *semantics* of sequents with frozen and indexed-frozen formulas.

Definition 1 (Satisfability of frozen-labeled sequents). *Let $\Delta \Rightarrow \Gamma$ be a sequent with its succedent and antecedent having formulas that range over labeled concepts, frozen labeled concepts and indexed-frozen labeled concept. This sequent has the general form $\Delta_1, [\Delta_2], [\Delta_3]^1, \ldots, [\Delta_k]^{k-2} \Rightarrow \Gamma_1, [\Gamma_2], [\Gamma_3]^1, \ldots, [\Gamma_k]^{k-2}$. Let $(\mathcal{I}, \mathcal{I}', \mathcal{I}_1 \ldots, \mathcal{I}_{k-2})$ be a tuple of interpretations. We say that this tuple satisfy $\Delta \Rightarrow \Gamma$, if and only if, one of the following clauses holds: $\mathcal{I} \models \Delta_1 \Rightarrow \Gamma_1$, $\mathcal{I}' \models \Delta_2 \Rightarrow \Gamma_2$, $\mathcal{I}_1 \models \Delta_3 \Rightarrow \Gamma_3$, …, $\mathcal{I}_{k-2} \models \Delta_k \Rightarrow \Gamma_k$.*

Obviously, $\Delta \Rightarrow \Gamma$ is not satisfiable by a tuple of interpretations, if and only if, no interpretation in the tuple satisfy the corresponding indexed (sub) sequent.

The following lemma shows that $\mathcal{S}^{[]}_{\mathcal{ALC}}$ is a conservative extension of $\mathcal{S}_{\mathcal{ALC}}$.

Lemma 1. *Consider $\Delta \Rightarrow \Gamma$ a $\mathcal{S}_{\mathcal{ALC}}$ sequent. If P is a proof of $\Delta \Rightarrow \Gamma$ in $\mathcal{S}^{[]}_{\mathcal{ALC}}$ then it is possible to construct a proof P' of $\Delta \Rightarrow \Gamma$ in $\mathcal{S}_{\mathcal{ALC}}$.*

Proof. The proof of Lemma 1 is done by simple induction over the number of applications of *prom-1* and *prom-2* occurring in a proof P. Let us consider a top most application of rule *prom-1* of the system $\mathcal{S}^{[]}_{\mathcal{ALC}}$. Due to space limitations, we present *prom-1* cases only, *prom-2* can be dealt similarly.

Below we present how we can build a new proof P' in $\mathcal{S}_{\mathcal{ALC}}$ (right) from a proof P in $\mathcal{S}^{[]}_{\mathcal{ALC}}$ (left). We have to consider two cases according to the possible outcomes from the top most application of the rule *prom-1*. The first case deals with the occurrence of a *frozen-exchange* rule application between the *prom-1* application and the initial sequent. This situation happens whenever the formulas frozen by *prom-1* rule were necessary to obtain the initial sequent, in a bottom-up construction of a $\mathcal{S}^{[]}_{\mathcal{ALC}}$ proof. Note that above the top most *prom-1* application, there might exist at most one *frozen-exchange* application.

$$
\dfrac{\dfrac{\begin{array}{c}\Delta'_1, [\Delta^*_2], \alpha \Rightarrow \alpha, [\Gamma^*_1], \Gamma'_2 \\ \Pi_1 \\ \Delta_1, [\Delta^*_2] \Rightarrow [\Gamma^*_1], \Gamma_2\end{array}}{\begin{array}{c}[\Delta_1], \Delta^*_2 \Rightarrow \Gamma^*_1, [\Gamma_2] \\ \Pi_2 \\ [\Delta_1], {}^{L_1}\beta^{L_2} \Rightarrow \Gamma_1, [\Gamma_2]\end{array}}\text{frozen-exch}}{\Delta_1, {}^{L_1}\beta^{L_2,R} \Rightarrow {\Gamma_1}^R, \Gamma_2}\text{prom-1}
\qquad
\dfrac{\begin{array}{c}\dfrac{\alpha \Rightarrow \alpha}{\Delta'_1, \alpha \Rightarrow \alpha, \Gamma'_2}\text{some weak} \\ \Pi'_1 \\ \Delta_1 \Rightarrow \Gamma_2\end{array}}{\Delta_1, {}^{L_1}\beta^{L_2,R} \Rightarrow {\Gamma_1}^R, \Gamma_2}\text{some weak}
$$

In the second case, the initial sequent is obtained (bottom-up building of the proof) without the application of *frozen-exchange* rule:

$$
\dfrac{\begin{array}{c}[\Delta_1], \Delta'_2, \alpha \Rightarrow \alpha, \Gamma'_1, [\Gamma_2] \\ \Pi_1 \\ [\Delta_1], {}^{L_1}\beta^{L_2} \Rightarrow \Gamma_1, [\Gamma_2]\end{array}}{\Delta_1, {}^{L_1}\beta^{L_2,R} \Rightarrow {\Gamma_1}^R, \Gamma_2}\text{prom-1}
\qquad
\dfrac{\dfrac{\begin{array}{c}\dfrac{\alpha \Rightarrow \alpha}{\Delta'_2, \alpha \Rightarrow \alpha, \Gamma'_1}\text{some weak} \\ \Pi'_1 \\ {}^{L_1}\beta^{L_2} \Rightarrow \Gamma_1\end{array}}{{}^{L_1}\beta^{L_2,R} \Rightarrow {\Gamma_1}^R}\text{prom-1}}{\Delta_1, {}^{L_1}\beta^{L_2,R} \Rightarrow {\Gamma_1}^R, \Gamma_2}\text{some weak}
$$

Applying recursively the transformations above from top to bottom we obtain a proof in $\mathcal{S}_{\mathcal{ALC}}$ from a proof in $\mathcal{S}^{[]}_{\mathcal{ALC}}$. □

A fully atomic $\mathcal{S}^{[]}_{\mathcal{ALC}}$ sequent has every (frozen and not frozen) formula in the antecedent as well as in the succedent as atomic concepts. A fully expanded proof-tree of $\Delta \Rightarrow \Gamma$ is a tree having $\Delta \Rightarrow \Gamma$ as root, each internal node is a premise of a valid $\mathcal{S}^{[]}_{\mathcal{ALC}}$ rule application, and each leaf is either a $\mathcal{S}^{[]}_{\mathcal{ALC}}$ axiom (initial sequent) or a fully atomic sequent. In the following lemma we are interested in fully expanded proof-trees that are not $\mathcal{S}^{[]}_{\mathcal{ALC}}$ proofs.

Lemma 2. *Let Π be a fully expanded proof-tree having $\Delta \Rightarrow \Gamma$, a sequent of $\mathcal{S}^{[]}_{\mathcal{ALC}}$, as root. From any non-initial top-sequent, one can explicitly define a counter-model for $\Delta^* \Rightarrow \Gamma^*$, where $\Delta^* \Rightarrow \Gamma^*$ is the $\mathcal{S}_{\mathcal{ALC}}$ sequent related to $\Delta \Rightarrow \Gamma$.*

Proof. Note that the construction of the counter-model (a tuple of interpretation) is made from top to bottom, starting from any top-sequent that is no initial in the fully expanded tree.

From the natural interpretation of a sequent in $\mathcal{S}_{\mathcal{ALC}}$ as an $\mathcal{ALC}$ concepts subsumption, we know that an interpretation $\mathcal{I}$ falsifies a sequent $\Delta \Rightarrow \Gamma$ if there exists and element c such that $c \in \Delta^{\mathcal{I}}$ and $c \notin \Gamma^{\mathcal{I}}$. In $\mathcal{S}^{[]}_{\mathcal{ALC}}$ a tuple of interpretation falsifies a sequent $\Delta \Rightarrow \Gamma$ if each of its elements (interpretations) falsifies the correspondent sequent (Definition 1).

We proof Lemma 2 by cases, considering each rule of the system $\mathcal{S}^{[]}_{\mathcal{ALC}}$. For each rule, we must prove that given a tuple t that falsifies its premises, one can provide a tuple t' that falsifies its conclusion. In t' each interpretation may be an extension of its correspondent in t.

It can be easily seen that from a non-initial and fully atomic top-sequent, we are able to define a tuple falsifying it.

Since rules ($\sqcap$-r), ($\sqcap$-l), ($\sqcup$-r), ($\sqcup$-l), ($\neg$-r), ($\neg$-l), ($\exists$-r), ($\exists$-l), ($\forall$-r) and ($\forall$-l) do not touch in the frozen formulas, to verify the preservation of falsity in those rules, one can consider just the second projection of the tuple. Due the lack of space, we only show the case ($\sqcup$-l). If there exist an $\mathcal{I}$ that falsifies $\Delta, {}^{\emptyset}\alpha^{L} \Rightarrow \Gamma$ then we can state that in the domain of $\mathcal{I}$ there exist an individual c such that $c \in \sigma(\Delta)^{\mathcal{I}} \cap \sigma({}^{\emptyset}\alpha^{L})^{\mathcal{I}}$ and $c \notin \sigma(\Gamma)^{\mathcal{I}}$. From the hypothesis and basic Set Theory, we can conclude that $c \in \sigma(\Delta)^{\mathcal{I}} \cap \sigma({}^{\emptyset}\alpha^{L})^{\mathcal{I}} \cup \sigma({}^{\emptyset}\beta^{L})^{\mathcal{I}}$, which from $\exists R.C \sqcup \exists R.D \equiv \exists R.(C \sqcup D)$ is equivalent to $c \in \sigma(\Delta)^{\mathcal{I}} \cap \sigma({}^{\emptyset}\alpha \sqcup \beta^{L})^{\mathcal{I}}$, falsifying the conclusion since $c \notin \sigma(\Gamma)^{\mathcal{I}}$. The same idea holds for the right premise.

For the rule *frozen-exchange* we must consider the whole tuple of interpretations, since it manipulates the frozen formulas. Suppose a tuple such that $(\mathcal{I}, \emptyset, \mathcal{I}_1, \ldots, \mathcal{I}_{n+1}) \not\models \Delta_1, [\Delta_3]^1, \ldots, [\Delta_k]^n, [\Delta_2]^{n+1} \Rightarrow \Gamma_1, [\Gamma_3]^1, \ldots, [\Gamma_k]^n, [\Gamma_2]^{n+1}$ where the second projection of the tuple is an empty interpretation since there is no frozen formulas without index in the premise. In order to falsify the conclusion of the rule one simply considers the tuple $(\mathcal{I}_{n+1}, \mathcal{I}, \mathcal{I}_1, \ldots, \mathcal{I}_n) \not\models \Delta_2, [\Delta_1], [\Delta_3]^1, \ldots, [\Delta_k]^n \Rightarrow \Gamma_2, [\Gamma_1], [\Gamma_3]^1, \ldots, [\Gamma_k]^n$.

For rule *prom-1*, suppose that $(\mathcal{I}, \mathcal{I}', \mathcal{I}_1, \ldots, \mathcal{I}_n)$ falsifies the premise. From that, we can obtain a new tuple $t' = (\mathcal{I}'', \emptyset, \mathcal{I}_1, \ldots, \mathcal{I}_n)$ that falsifies the conclusion ${}^{L_1}\alpha^{L_2,R}, \Delta_1, [\Delta_3]^1, \ldots, [\Delta_k]^n \Rightarrow {\Gamma_1}^{(R)}, \Gamma_2, [\Gamma_3]^1, \ldots, [\Gamma_k]^n$. To construct t', the unique non trivial part is the construction of $\mathcal{I}''$. This comes from $\mathcal{I}$ and $\mathcal{I}'$ from the tuple that falsifies the premise. $\mathcal{I}''$ preserve the structure of both $\mathcal{I}$ and $\mathcal{I}'$ and also the properties that make those interpretation falsify the premise. Due space limitation, the complete procedure can not be presented here. The principal idea is that whenever a *crash* of names of individuals occurs, that is, names that occurs in both, one can consistently renames the individuals from I or I'. The reader must note that this is a perfectly adequate construct.

Starting from a tuple of interpretations defined to falsify the fully atomic top-sequent we obtain by the preservation of the falsity, provide by the $\mathcal{S}^{[]}_{\mathcal{ALC}}$ rules, the falsity of the root of the proof-tree. We can see that the counter-model construction also works for $\mathcal{S}_{\mathcal{ALC}}$ sequents when submitted to the $\mathcal{S}^{[]}_{\mathcal{ALC}}$ proof-system. □

5 Providing Explanations from Proofs

Consider the proof:

$$
\begin{array}{cl}
\mathit{Doctor} \Rightarrow \mathit{Doctor} & \\ \hline
\mathit{Doctor} \Rightarrow \mathit{Rich}, \mathit{Doctor} & \text{weak-r} \\ \hline
\mathit{Doctor} \Rightarrow (\mathit{Rich} \sqcup \mathit{Doctor}) & \sqcup\text{-r} \\ \hline
{}^{\mathit{child}}\mathit{Doctor} \Rightarrow {}^{\mathit{child}}(\mathit{Rich} \sqcup \mathit{Doctor}) & \text{prom-2} \\ \hline
\top, {}^{\mathit{child}}\mathit{Doctor} \Rightarrow {}^{\mathit{child}}(\mathit{Rich} \sqcup \mathit{Doctor}) & \text{weak-l} \\ \hline
\top \Rightarrow \neg \mathit{Doctor}^{\mathit{child}}, {}^{\mathit{child}}(\mathit{Rich} \sqcup \mathit{Doctor}) & \mathit{neg}\text{-r} \\ \hline
\top \Rightarrow \neg \mathit{Doctor}^{\mathit{child}}, \mathit{Lawyer}^{\mathit{child}}, {}^{\mathit{child}}(\mathit{Rich} \sqcup \mathit{Doctor}) & \text{weak-r} \\ \hline
\top \Rightarrow \neg \mathit{Doctor}^{\mathit{child}}, \exists \mathit{child}.\mathit{Lawyer}, {}^{\mathit{child}}(\mathit{Rich} \sqcup \mathit{Doctor}) & \exists\text{-r} \\ \hline
\top \Rightarrow \exists \mathit{child}.\neg \mathit{Doctor}, \exists \mathit{child}.\mathit{Lawyer}, {}^{\mathit{child}}(\mathit{Rich} \sqcup \mathit{Doctor}) & \exists\text{-r} \\ \hline
\top \Rightarrow ((\exists \mathit{child}.\neg \mathit{Doctor}) \sqcup (\exists \mathit{child}.\mathit{Lawyer})), {}^{\mathit{child}}(\mathit{Rich} \sqcup \mathit{Doctor}) & \sqcup\text{-r} \\ \hline
\top^{\mathit{child}} \Rightarrow ((\exists \mathit{child}.\neg \mathit{Doctor}) \sqcup (\exists \mathit{child}.\mathit{Lawyer}))^{\mathit{child}}, {}^{\mathit{child}}(\mathit{Rich} \sqcup \mathit{Doctor})^{\mathit{child}} & \text{prom-1} \\ \hline
\top^{\mathit{child}}, {}^{\mathit{child}}\neg((\exists \mathit{child}.\neg \mathit{Doctor}) \sqcup (\exists \mathit{child}.\mathit{Lawyer})) \Rightarrow {}^{\mathit{child}}(\mathit{Rich} \sqcup \mathit{Doctor})^{\mathit{child}} & \neg\text{-l} \\ \hline
\top^{\mathit{child}}, {}^{\mathit{child}}\neg((\exists \mathit{child}.\neg \mathit{Doctor}) \sqcup (\exists \mathit{child}.\mathit{Lawyer})) \Rightarrow \forall \mathit{child}.(\mathit{Rich} \sqcup \mathit{Doctor})^{\mathit{child}} & \forall\text{-r} \\ \hline
\top^{\mathit{child}}, \forall \mathit{child}.\neg((\exists \mathit{child}.\neg \mathit{Doctor}) \sqcup (\exists \mathit{child}.\mathit{Lawyer})) \Rightarrow \forall \mathit{child}.(\mathit{Rich} \sqcup \mathit{Doctor})^{\mathit{child}} & \forall\text{-l} \\ \hline
\top^{\mathit{child}}, \forall \mathit{child}.\neg((\exists \mathit{child}.\neg \mathit{Doctor}) \sqcup (\exists \mathit{child}.\mathit{Lawyer})) \Rightarrow \exists \mathit{child}.\forall \mathit{child}.(\mathit{Rich} \sqcup \mathit{Doctor}) & \exists\text{-r} \\ \hline
\exists \mathit{child}.\top, \forall \mathit{child}.\neg((\exists \mathit{child}.\neg \mathit{Doctor}) \sqcup (\exists \mathit{child}.\mathit{Lawyer})) \Rightarrow \exists \mathit{child}.\forall \mathit{child}.(\mathit{Rich} \sqcup \mathit{Doctor}) & \exists\text{-l} \\ \hline
\exists \mathit{child}.\top \sqcap \forall \mathit{child}.\neg((\exists \mathit{child}.\neg \mathit{Doctor}) \sqcup (\exists \mathit{child}.\mathit{Lawyer})) \Rightarrow \exists \mathit{child}.\forall \mathit{child}.(\mathit{Rich} \sqcup \mathit{Doctor}) & \sqcap\text{-l}
\end{array}
$$

which could be explained by: "(1) Doctors are Doctors or Rich (2) So, Everyone having all children Doctors has all children Doctors or Rich. (3) Hence, everyone either has at least a child that is not a doctor or every children is a doctor or rich. (4) Moreover, everyone is of the kind above, or, alternatively, have at least one child that is a lawyer. (5) In other words, if everyone has at least one child, then it has one child that has at least one child that is a lawyer, or at least one child that is not a doctor, or have all children doctors or rich. (6) Thus, whoever has all children not having at least one child not a doctor or at least one child lawyer has at least one child having every children doctors or rich."

The above explanation was build from top to bottom (toward the conclusion of the proof), by a procedure that tries to not repeat conjunctive particles (if - then, thus, hence, henceforth, moreover etc) to put together phrases derived from each subproof. In this case, phrase (1) come from weak-r, $\sqcup$-r; phrase (2) come from prom-2; (3) is associated to weak-l, neg-r; (4) corresponds to weak-r, the two following $\exists$-r and the $\sqcap$; (5) is associated to prom-1 and finally (6) corresponds to the remaining of the proof. The reader can note the large possibility of using endophoras in the construction of texts from structured proofs as the ones obtained by either $\mathcal{S}_{\mathcal{ALC}}$ or $\mathcal{S}^{[]}_{\mathcal{ALC}}$.

6 Conclusion and Further Work

In this article it is shown two sequent calculi for $\mathcal{ALC}$. Both are cut-free, and one ($\mathcal{S}^{[]}_{\mathcal{ALC}}$) is designed for implementation without the need of any backtracking resource. Moreover $\mathcal{S}^{[]}_{\mathcal{ALC}}$ provides counter-model whenever the subsumption candidate is not valid in $\mathcal{ALC}$. We briefly suggest how to use the structural feature of sequent calculus in favour of producing explanations in natural language from proofs. As it was said at the introduction, the use of the cut-rule can provide shorter proofs. The cut-rule may not increase the complexity of the explanation, since it simply may provide more structure to the original proof. With the help of the results reported in this article one has a solid basis to build mechanisms to provide shorter and good explanation for $\mathcal{ALC}$ subsumption in the context of a KB authoring environment. The inclusion of the cut-rule, however, at the implementation level, is a hard one. Presently, there are approaches to include analytical cuts in Tableaux, as far as we know there is no research on how to extend this to $\mathcal{ALC}$ Tableaux. This puts our results in advantage when taking explanations, and the size of the proofs as well, into account. There are also other techniques, besides the use of the cut-rule, to produce short proofs in the sequent calculus, see [5] and [10], that can be used in our context.

References

1. Schmidt-Schau, M., Smolka, G.: Attributive concept descriptions with complements. Artificial Intelligence 48(1), 1–26 (1991)
2. McGuinness, D.: Explaining Reasoning in Description Logics. PhD thesis, Rutgers University (1996)
3. Liebig, T., Halfmann, M.: Explaining subsumption in $\mathcal{ALEHF}_{R+}$ tboxes. In: Horrocks, I., Sattler, U., Wolter, F. (eds.) Proc. of the 2005 International Workshop on Description Logics - DL 2005, Edinburgh, Scotland, pp. 144–151 (July 2005)
4. Deng, X., Haarslev, V., Shiri, N.: Using patterns to explain inferences in $\mathcal{ALCHI}$. Computational Intelligence 23(3), 386–406 (2007)
5. Gordeev, L., Haeusler, E., Costa, V.: Proof compressions with circuit-structured substitutions. In: Zapiski Nauchnyh Seminarov POMI (to appear, 2008)
6. Finger, M., Gabbay, D.: Equal Rights for the Cut: Computable Non-analytic Cuts in Cut-based Proofs. Logic Journal of the IGPL 15(5–6), 553–575 (2007)
7. Baader, F.: The Description Logic Handbook: theory, implementation, and applications. Cambridge University Press, Cambridge (2003)
8. Rademaker, A., do Amaral, F., Haeusler, E.: A Sequent Calculus for $\mathcal{ALC}$. Monografias em Ciência da Computação 25/07, Departamento de Informática, PUC-Rio (2007)
9. Rademaker, A., Haeusler, E., Pereira, L.: On the proof theory of $\mathcal{ALC}$. In: XV EBL-15th Brazilian Logic Conference (to appear)
10. Finger, M.: DAG sequent proofs with a substitution rule. In: Artemov, S., Barringer, H., dAvila Garcez, A., Lamb, L., Woods, J. (eds.) We will show Them – Essays in honour of Dov Gabbays 60th birthday, vol. 1, pp. 671–686. Kings College Publications, London (2005)

A Case for Numerical Taxonomy in Case-Based Reasoning*

Luís A.L. Silva[1], John A. Campbell[1], Nicholas Eastaugh[2], and Bernard F. Buxton[1]

[1] Department of Computer Science, University College London, Malet Place, London, WC1E 6BT, UK
{l.silva,j.campbell,b.buxton}@cs.ucl.ac.uk
[2] The Pigmentum Project, Research Laboratory for Archaeology and the History of Art, University of Oxford, Dyson Perrins Building, South Parks Road, Oxford, OX1 3QY, UK
nicholas.eastaugh@pigmentum.org

Abstract. There are applications of case-like knowledge where, on the one hand, no obvious best way to structure the material exists, and on the other, the number of cases is not large enough for machine learning to find regularities that can be used for structuring. Numerical taxonomy is proposed as a technique for determining degrees of similarity between cases under these conditions. Its effect is illustrated in a novel application for case-like knowledge: authentication of paintings.

1 Introduction

Case-based reasoning (CBR) relies partly on an effective structuring or indexing of a case base [1, 2] to maximise the relevance of the cases retrieved in response to a query that represents a new problem. If an application has no evident structure in itself, e.g. no structure following from an accepted underlying theory, or no obvious set of key terms or properties in the cases, one may use techniques of machine learning to discover a possible indexing. However, this is practicable or reliable only when a relatively large set of cases is available.

There are applications where the numbers of cases are small and are likely to remain so, and where at the same time no obvious indexing suggests itself. The latter condition, in particular, has occurred in science in the past, where the "base" has consisted not of cases but of physical specimens. Numerical taxonomy was developed in response to the need to find order in sets of such specimens. We have found that, regarded suitably, cases can be treated in the same way and that numerical taxonomy is a valuable tool to supplement the existing techniques of CBR. We illustrate this finding through an examination of a painting authentication application [3].

2 The Role of Numerical Taxonomy

Suppose that a set of specimens is given where each specimen has various features and where the values of the features can be noted. Suppose also that it is possible to

* This work is supported by CAPES/Brazil (grant number 2224/03-8).

G. Zaverucha and A. Loureiro da Costa (Eds.): SBIA 2008, LNAI 5249, pp. 177–186, 2008.

assign or measure directly the relative distances or differences between any pair of values of any one feature. Direct measurement of distance is appropriate, for example, if such a feature is numerical by nature. Translating symbolic features to numerical scales is needed if there is no obvious coordinate in any space, e.g. where a feature is a certain kind of texture or a specimen's preferred food. Following such an encoding of specimen's features, it is possible to form a classification of the specimens in a tree-like structure showing their relative closeness to each other. Proceeding upwards from the individuals, the closest are associated first, forming small groups, and further specimens or groups are associated into larger groups, until a single group is reached as the root of the tree. Each intermediate node between the leaves and the root represents one association between individuals or groups of them. What is necessary to make this process work?

Historically, the earliest demands for such a process came from botany and zoology. Either (usually) no strong theory which could be used to make the classification existed, or one needed to investigate what generalisations the tree (a "dendrogram") could suggest if no theory of classification in the application domain were assumed (e.g. as a way of validating theoretical classification by means of more objective and systematic methods). Numerical taxonomy is the response to this demand. A somewhat classical reference [4] is still a relevant source of information about this subject.

In brief, for features 1, 2 ... of specimens X and Y, only a distance function D of all the differences x1 – y1, x2 – y2, ... is needed. There are then several good clustering algorithms [5] which can use D in generating the dendrogram. For any application there are still questions to be answered in order to define D, e.g. normalisation of the ranges of value of the different attributes, scaling of the normalised attributes, choice of the functional form of D (Pythagorean form is a reasonable default choice), and – even before these – choice of the features to be parameterised in setting up the distance function between specimens.

For conventional cases in CBR where specimens are the cases, the same approach is applicable in organising or indexing a case base. There is no reason to suppose procedures of numerical taxonomy would be better than existing approaches of structuring cases or indexing in CBR [1, 2] where some convincing scheme of theory or generalisation is already known and trusted for the cases in an application domain. One may also not assume that numerical taxonomy would provide satisfactory results in situations where traditional approaches of structuring cases and indexing do not provide good results but the number of available cases in the case base is large enough for machine learning techniques to be applied and their outputs to be taken (at least initially) on trust. However, there are several applications of CBR in which the number of cases available is too small for machine learning approaches, while not small enough for users to be confident about mistake-free manual indexing. There are also situations when experts in the application fields tend to believe that no overall theoretical scheme exists for the organisation of the problems (or cases) by traditional structuring or indexing procedures. Where neither the indexing nor the machine learning approach is effective for any particular CBR application, the problem is the same as for the situations where numerical taxonomy has been most useful, outside CBR, in the past.

When we introduce procedures of numerical taxonomy into to the CBR scenario as in the line of research of [6-8], entire cases have to be taken first as the specimens.

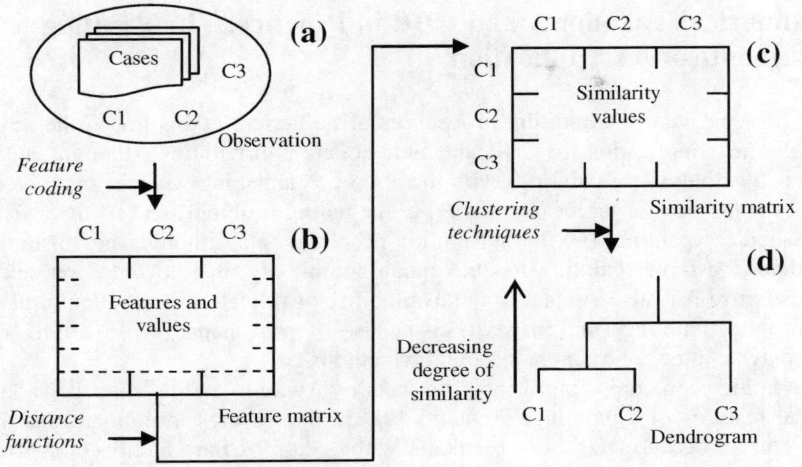

Fig. 1. Cases as taxonomic entities in numerical taxonomy

The encoding of the conventional contents (facts, actions) of cases as taxonomic features, whose values can then be regarded as the coordinates of a specimen, is the first step of numerical taxonomy (Fig. 1 – (a)).

Once taxonomic features are recorded in a feature matrix (Fig. 1 – (b)), the next step is the estimation of similarity and construction of a similarity matrix (Fig. 1 – (c)). We should emphasise that there is no formula or theory to build a distance measure for a given set of characteristics in numerical taxonomy and CBR: such measures are always found in practice by trial and error. In traditional numerical taxonomy, this starts with equal a priori weights for all the features that contribute to the estimation of similarity. Expert assessment of the quality of the dendrogram that results from the use of the inter-case similarities that can be computed can then be used to tune the weights in the underlying similarity functions.

The dendrogram is simply a clustering of taxonomic entities (Fig. 1 – (d)) (details of clustering algorithms can be found in [5]). Taking the similarity matrix as input, or starting directly from the feature matrix, the clustering divides a collection of cases into groups of similar cases according to some numerical measure D of similarity. In taxonomic studies, the best clustering algorithms for our purposes in CBR are hierarchical. The multilevel hierarchy of similarity in a typical dendrogram is shown in (Fig. 1 – (d)). In it, the overall similarity relationships between cases are evident. In our experience, such tree-like hierarchical structure is simple enough for an expert to inspect the groupings and assess their validity. This leads either to expert feedback to improve D, e.g. by adjusting the scaling of individual feature dimensions or to explanation of why the member cases of particular groupings belong together [6]. Fundamentally, the latter amounts to a bottom-up process of elicitation of knowledge previously held only tacitly by the expert.

3 Numerical Taxonomy and CBR in Practice: The Painting Authentication Application

In our project, we are exploiting procedures of numerical taxonomy in the development of a new application for CBR: the authentication of paintings. Painting authentication is the analysis of paintings with the aim of establishing whether there is reason to doubt i) the date of target paintings (i.e. the dating problem) and ii) the artist who has created the painting (i.e. the attribution problem) [3]. Although the attribution of an author to a target painting has too much complexity to be decided on scientific grounds, since it is also subject to the availability of reliable historical/cultural information about paintings and artists (i.e. the use of provenance information), dating issues may be attacked by more systematic methods.

The dating problem is approached routinely by trying to assign (ranges of) dates to physical features of a painting. Pigments [9] are one of most prominent dating features. This is because there are pigments with a date of introduction (sometimes a discovery date, for instance) in paintings and a point in time after which one can no longer expect to find them in paintings (a terminal date). Thus paintings contain some features that may be characterised in time and also space, since such features are the product of materials and techniques available to artists. Once these pigments are identified by analysts in paintings, they are taken as findings in the construction of dating conclusions in reports on the authenticity or otherwise of a painting. Among other reasons, such reports are needed since painting investors may want to secure their investment. They are also developed for the solution of legal disputes involving ownership. Basically, the very simple information of when a certain painting was created may be the key factor for deciding who wins the dispute.

3.1 Cases of Painting Authentication

Routinely, past cases of painting authentication are underused by painting analysts in the solution of dating problems. Although steps in the formalisation of painting pigments have been given in a standard Pigment Compendium reference [9], there is still a need to record these pigments systematically in cases of painting authentication to avoid ambiguities as far as possible and to help the knowledge to be re-used. Such a recording should be broad enough to involve not only the traditional artistic (e.g. stylistic) criteria but also the newer kinds of knowledge relying on studies of pigmentation through chemistry. Once such painting information is available in a computable format, one can consider methods for approaching new painting authentication cases in terms of past authentication episodes.

Although the task of painting authentication is handled naturally on a case-by-case basis and analysts' reports mention various other artists and dates as background, such information has previously been regarded as implicit. Hence, knowledge of previous painting authentication episodes, such as knowledge from the authentication of paintings dated incorrectly on purely stylistic grounds, is available essentially only as small samples. The numbers of detailed case-like reports on authentication of paintings is small and will probably remain small. In our project, 50 reports (in total: 66 paintings) involving quite distinct kinds of paintings – e.g. small, regular and large paintings, "fake" paintings, paintings from different artists and dates, etc – are being utilised.

Each such report is rich in knowledge – but in forms of knowledge that have no single underlying theoretical base. There is also no theory for how the existing cases of painting authentication should be organised and examined as a means of obtaining hypotheses for the investigation of dating in new examples of authentication. All these considerations point to the situation – small numbers of cases, no ready-made framework for generalisations about them – where numerical taxonomy is most appropriate as a technique for aiding the development of CBR applications.

In the 66 cases available, 36 contain pigment information – the pigment-based case base, while 30 contain information regarding elemental composition of painting samples only – the "EDX"-based case base (in total, 16 cases are common in these two case bases). Pigments and "EDX" compositions (i.e. elemental compositions from energy dispersive X-ray spectrometry) represent two distinct but overlapping views on the analysis of materials in a painting. "Pigments" is based on the interpreted results of sampling and analysis, expressed as compounds that have been used historically in the painting process as listed in a standardised compendium [9]. Examples might be "cadmium sulfide" and "synthetic ultramarine". A range of analytical techniques may have been used to reach the pigment identifications given, including (but not exclusively) the method known as "EDX". EDX is typically performed as an adjunct to scanning electron microscopy ("SEM") on paint samples, yielding the localised elemental composition of the pigments present in the paint sample. The EDX-based case base is thus the collection of elements identified (e.g. major amount of Zinc – Zn) during such an analysis of a painting. While it mirrors the pigments present, it does not identify them uniquely.

An important difference between the two case bases concerns the types of information about a painting that each contains. The "Pigments" characterisation aims to reflect the complete set of pigments used by an artist in a specific work. Consequently it embodies the choice an artist has made in selecting materials from those available at a particular time and place, but not the way in which they were combined. Information on aspects of painting technique involving mixtures, such as when pigments are habitually used together to achieve certain colours (information which reveals methods of painting production and aspects for painting conservation) is consequently lost. EDX features on the other hand, while not reflecting the level of analytical detail and pigment specificity contained in the "Pigments" features, whatever does still contain information associated with mixtures.

From an expert understanding of the information contained in the two case bases we would anticipate the groups of cases formed from "Pigments" characteristics might outperform the groups formed from EDX data in respect of date determination. This is because there is a greater amount of information with temporal variation encoded into the former. While the EDX features contain information on mixtures discarded in the "Pigments" features, such mixtures only contain weak temporal variations and have thus not been widely used until now for making judgments regarding painting dates. However, as the identification of some pigments may not be straightforward for non-experts, information regarding elemental compositions is worth investigation as taxonomic features of cases. In fact, input features in CBR queries are either pigments or elemental compositions, or both, since these are features obtained quickly in the process of authentication. When a query case is formed, similarity is computed against all cases in both elemental composition-based and pigment-based case bases. The hierarchical taxonomic structure involving cases is not used to reduce the search procedure since these case bases are composed of a small number of cases. Nevertheless, the final hierarchical arrangement of cases is used by the CBR application when discussing retrieved results according to the "dating taxonomies" determined. In a process that resembles the analysis of cluster results in tasks of knowledge discovery, the CBR application makes use of "templates" (previously constructed) in order to explain traditional retrieval results according to local and global groups of cases in both composition-based and pigment-based taxonomies. Fundamentally, the combined analysis of retrieval results from pigments and elemental compositions may provide the contrasting support, and cross-validation, for the construction of case-based dating hypotheses. Further details regarding the actual CBR process involving taxonomic results of this kind can be found in [6].

3.2 Selecting and Coding Case Features

We have treated each case of authentication of a painting as a tuple <problem, solution>. The problem description consists of 27 elemental compounds in cases of the EDX-based case base. The problem consists of 32 different types of pigments (where some represent different variations of an historical pigment [9]) in cases of the pigment-based case base. The solution part of the cases consists of a dating conclusion as stated by an expert in the field.

For representing EDX features in cases, one has to compute the normalised amount of each element in each painting. This is done by first coding EDX elemental measures as "trace" (which we coded as 1), "minor" (2) and "major" (3). For each element, one then sums these coded values in each painting sample, where the resulting sum is divided by the number of samples occurring in the painting. For example, the Zn element is identified in 3 of 10 samples. This element is characterised as "minor" in 2 of the samples, and "major" in 1 sample, which results in an absolute amount of Zn = 7 (i.e. 2 samples * 2 "minor" + 1 sample * 1 "major"). Then, this absolute value is divided by 10, giving the normalised amount of Zn observed. We treat each elemental component of the paintings in the same way. For representing pigments in cases, a list of pigments was selected from: a) Pigment Compendium [9] material, b) analysis of different kinds of painting authentication statements in authenticity reports (e.g. arguments regarding authentication explanations) and c) refinements from the analysis of intermediate grouping results formed. In particular, statements in authenticity reports involve different pigments, although they are not limited to pigment information. An example of a statement is: "Numerous false paintings in the style of the Russian Avant Garde movement contain titanium dioxide white pigments, for which current evidence suggests there was little general use before the 1950s". Although work is on going in the examination of these aspects of argumentation in the capture of case-based expert knowledge and the search for a better similarity assessment in the problem as proposed in [6], this example already shows the kind of information one may exploit in aiding the selection of relevant case features, e.g. the "titanium dioxide white" pigment in this application.

3.3 Similarity Computation and Clustering

Computation of similarity between cases is the next step. For cases formed by EDX compositions, these numerical values are ranged and scaled in the computation of similarity. All elemental features have the same weights in the similarity function. This amounts to the application of a traditional numerical taxonomy procedure over case facts, where a "general-purpose classification" is constructed. For cases formed by pigments, these are recorded as present, absent or unknown. Pigments present in two cases are treated as evidence for similarity between the cases. Pigments present in one case and absent in another case are treated as evidence for significant difference between them. Absent or unknown pigments in two cases being compared are neutral (not used) in the computation of similarity. EDX composition and pigment features are the dimensions for separate weighted Euclidean distance D computations, where D is equivalent (at least weakly) to various other distance functions presented in [4]. Similarities between cases are computed and stored in a similarity matrix which is then used as input in the clustering analysis. We have utilised hierarchical methods of clustering [4, 5] considering both "average" and "complete" versions of linking (since these most often give the best hierarchical representation of relationships expressed in a similarity matrix). The hierarchical methods are the most frequently used in traditional applications of numerical taxonomy. One reason why this is so is also relevant for the development of CBR applications, namely: a dendrogram structure is clear and simple enough for participants from the application field to give feedback about the organisation and relevance of the groups formed.

Fig. 2 presents the dendrogram arising from the similarities of the EDX features only. The clusters formed are reasonable in the sense that paintings with broadly similar dating ranges occur together. Fig. 3 present the dendrogram arising from the similarities of the pigment features only. Weights of pigment features were explored by means of an "extreme heuristic", e.g. using a weight of 20 for each of the pigments while the other weights were unchanged, and examining the changes of cluster membership. The weights were then adjusted proportionately, subject to expert agreement that they did

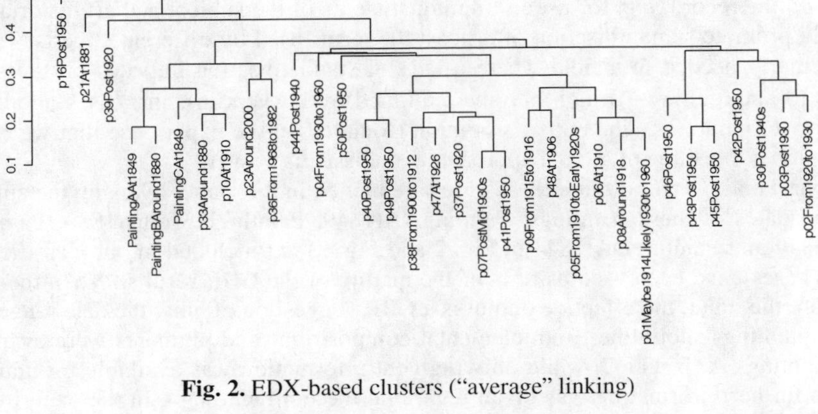

Fig. 2. EDX-based clusters ("average" linking)

not distort the relative importance of the dimensions, and values were chosen that maximised overall agreement between the dating of cases in the same clusters.

A key process in the use of numerical taxonomy in CBR is this machine-supported process of user-directed learning through construction and examination of different taxonomies. It is also highly desirable to involve the domain expert in the search for a good taxonomy (and the weights that may lead to it). In essence, this taxonomic process is a two-way task of both the successive formalisation of the domain by the expert and the improvement of the taxonomic structure (and similarity assessment) by the knowledge engineer. The most satisfactory resulting taxonomy from adjusted pigment weights is presented in Fig. 3.

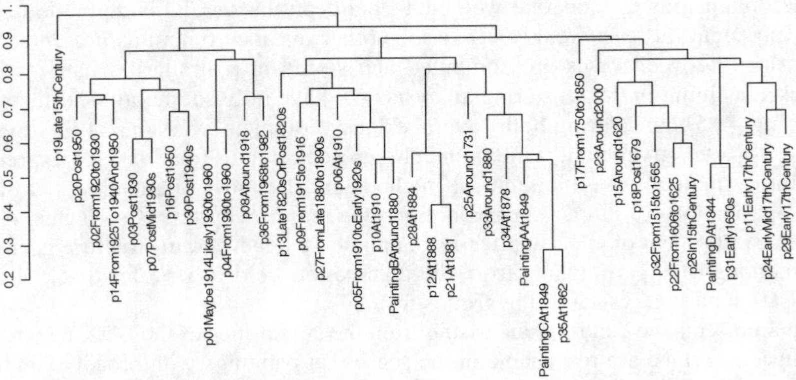

Fig. 3. Pigment-based clusters ("average" linking)

3.4 Initial Experiences: Using the Taxonomies in CBR

According to the expert in the application field, there is a strong need for systematic recording of the knowledge of painting authentication procedures as expressed in past reports on individual paintings. Besides putting a single specialist source of knowledge on the record, e.g. for use in training students of the subject and art historians, it should promote standardisation of at least the terms used by different experts. This is particularly needed in a field where a lack of standardisation impedes progress and leads to ambiguities – which may be exploited by lawyers in court cases about disputes over paintings. The expert's opinion (founded on the experience that we report here) about the value of a CBR approach to the issue is positive.

Four paintings (for clarity, these cases are named in this paper by using the authenticated dates of these paintings: PaintingAAt1849, PaintingBAround1880, PaintingCAt1849 and PaintingDAt1844 in Fig. 2 and Fig. 3) not included in our classified set of 66 cases have been used as tests of the quality of the CBR results. When their details are the input, the effective output is a CBR suggestion of what the date ranges for these paintings should be. Both elemental compositions and pigments were available for paintings A, B and C, while only pigment information was available for painting D (no further information was given regarding these 4 paintings). In the experiments involving these paintings, the explanation templates generated with the retrieval results show that painting D is significantly older than paintings A, B and C. Paintings A and C belong to the same group in both CBR results. In particular, the pigment-based group of paintings from around the 1870s allows the claim that paintings A and C come from this epoch. In terms of EDX compositions, painting B can not be distinguished from paintings A and C. When pigment characteristics are used, painting B is from around the 1910s, indicating that this painting could have been created later than A, C and D. According to a later expert verdict of these results, the dating of paintings A-D should be shifted ~30-40 years. The same expert says this is highly likely to be because most of the cases available are 20th century paintings. Paintings A and C are the closest ones indeed. According to the expert, painting D is the oldest one but there is less data from such times in the case base. Painting D is also a different kind of painting in which the pigments are not very informative for dating. That B is the latest is crucial to the discussion of these 4 paintings since the issue is to find the temporal relationship between painting B and the others (this key piece of information was revealed only after the experiment). This simple but objective experiment presents evidence that pure fact-based CBR has the potential for generating improved dating hypotheses for further investigation. Exploitation of expert arguments regarding other authentication aspects may add further knowledge and dating precision to CBR beyond what we report here.

Finally, there may be situations where an apparently "wrong" classification may be significant because it is in fact right. This is exactly what happens in case of painting forgeries where painting "p23Around2000", for instance, supposedly from the 17th to 18th century, is associated in the process of clustering with paintings from that period of time when pigment information is used (Fig. 3) – since all but one of the pigments present are indeed old, but this painting is particularly associated with "p36From1968to1982" when elemental compositions are used (Fig. 2). The expert has confirmed this outcome obtained from CBR. In effect, the classification is a knowledge elicitation tool: it draws the expert's attention to features that are unusual enough to need explanation and/or correction. Explanations are explicit versions of previously tacit knowledge and can enhance the value of a case when they are added to its previous contents.

4 Concluding Remarks

The effectiveness of numerical taxonomy has been demonstrated in this paper, even though we have not yet completed all of the treatments (e.g. capture, typing, and use of the argument parts of experts' reports on authentication of paintings) that are possible in a combined factual and argumentation approach to CBR. Numerical taxonomy is effective not only in obtaining results (here, estimates of the dates of paintings), but also for knowledge elicitation, correction, and indication of anomalies (here, forgeries; elsewhere in our cases, past mistakes in data recording), as we have remarked in section 3.4. In particular, we demonstrate the applicability of taxonomic procedures to CBR when there are relatively small case bases where the nature of an application does not allow any single useful a priori scheme of indexing to support the process of case retrieval.

References

1. Mántaras, R.L.d., McSherry, D., Bridge, D., Leake, D., Smyth, B., Craw, S., Faltings, B., Maher, M.L., Cox, M.T., Forbus, K., Keane, M., Aamodt, A., Watson, I.: Retrieval, reuse, revision and retention in CBR. Knowledge Engineering Review 20, 215–240 (2005)
2. Kolodner, J.L.: Case-based reasoning. Morgan Kaufmann Publishers, Inc., San Mateo (1993)
3. Eastaugh, N.: Scientific Dating of Paintings. Infocus Magazine, 30–39 (2006)
4. Sneath, P.H., Sokal, R.R.: Numerical taxonomy - The principles and practice of numerical classification. W. H. Freeman and Company, San Francisco (1973)
5. Jain, A.K., Murty, M.N., Flymn, P.J.: Data clustering: A review. ACM Computing Surveys 31, 264–323 (1999)
6. Silva, L.A.L., Buxton, B.F., Campbell, J.A.: Enhanced Case-Based Reasoning through Use of Argumentation and Numerical Taxonomy. In: The 20th Int. Florida Artificial Intelligence Research Society Conference (FLAIRS-20), pp. 423–428. AAAI Press, Key West (2007)
7. Campbell, J.A.: Numerical Taxonomy: A Missing Link for Case-Based Reasoning and Autonomous Agents. Review of the Royal Academy of Exact, Physical and Natural Sciences, Serie A: Matemáticas 98, 85–94 (2004)
8. Popple, J.: SHYSTER: A pragmatic legal expert system. Department of Computer Science, p. 432. Australian National University, Canberra (1993)
9. Eastaugh, N., Walsh, V., Chaplin, T., Siddall, R.: The pigment compendium: Optical microscopy of historical pigments. Elsevier, Butterworth-Heinemann (2004)

Methodical Construction of Symbolic Diagrams*

Paulo A.S. Veloso[1] and Sheila R.M. Veloso[2]

[1] Progr. Engenharia de Sistemas e Computação, COPPE, Universidade Federal do Rio de Janeiro; Caixa Postal 68511, 21945-970, Rio de Janeiro, RJ, Brasil
[2] Depto. Engenharia de Sistemas e Computação, Fac. Eng., Universidade Estadual do Rio de Janeiro; Rio de Janeiro, RJ, Brasil

Abstract. We introduce a stepwise approach for computing symbolic diagrams. Such logical diagrams display formulas and connections between them and they are useful tools for visualizing connections between formulas as well as for reasoning about them. This incremental approach is modular: it starts from small diagrams and expands them. We explain the method and justify it. We also comment on the application of these ideas and illustrate them with some examples.

Keywords: Knowledge representation, symbolic reasoning, diagrams.

1 Introduction

In this paper, we introduce a method for the construction of symbolic diagrams. Diagrams are employed in many branches of science as useful tools for visualization. Symbolic diagrams display term-like objects and connections between them. They are useful tools for many purposes: visualizing connections, reasoning and experimentation. So, we provide a method for constructing tools with a wide spectrum of possible uses.

Logical diagrams display logical objects and connections between them. Some simple examples are modality diagrams [6] with modalities and their connections (implications, equivalences, etc.). Other examples are diagrams displaying derivations and reductions or formulas connected by inference rules [12].

Logics have been used in many areas of Artificial Intelligence and Computer Science, mainly because they provide insights into reasoning problems without implementation details. In fact, one can distinguish three uses of logics in this context: as tools for analysis, as bases for knowledge representation and as programming languages [10]. Examples of logics with important impact are: feature logics (to encode linguistic information [14]), description logics (to describe ontologies [1], [2]), and modal logics, such as epistemic logics (to reason about knowledge of agents as well as distributed and common knowledge [7], [9]) and dynamic logics [8].

* Research partly sponsored by CNPq (Brazilian National Research Council) and FAPERJ (Rio de Janeiro State Research Foundation).

G. Zaverucha and A. Loureiro da Costa (Eds.): SBIA 2008, LNAI 5249, pp. 187–196, 2008.

Logical diagrams share some structural features in that they consist of two basic entities: the nodes are (representatives of) equivalence classes of formulas and the edges display some logical connections. The formulas are generated by some formula-building operations, like negation and modalities (or quantifiers). Since the connections are logical, a first step in constructing such a diagram amounts to determining its nodes. In the case of few formula-building operations with simple behavior, the task is relatively simple, but not so otherwise.

We propose a stepwise approach for the construction of symbolic diagrams. This incremental approach is modular: the basic idea is starting from diagrams with few operations and combining them or adding one operation at a time. The method may also join diagrams for sublanguages and then adapt the result. The adaptions are of two kinds: coalescing existing nodes and adding new nodes. We explain the method and justify it. The justification rests on regarding a (perhaps partial) diagram as a kind of algebra. We will introduce and illustrate the ideas for the case of modality-like diagrams, where the operations are unary (in Sections 2 to 6), and then we will indicate how to extend them to the general case (in Section 7).

The structure of this paper is as follows. In Section 2 we examine the basic ideas on logical diagrams consisting of nodes and arcs. In Section 3 we introduce our approach to diagram construction: first nodes, then arcs. In Section 4 we examine some ideas underlying modality-like diagrams, namely normal form and equivalence. In Section 5 we take a closer look at the construction of modality-like diagrams by stepwise addition of modalities. In Section 6 we present our method for the construction of logical diagrams by expanding partial diagrams. In Section 7 we present some variations and extensions of our method. Finally, we present some closing remarks about our ideas in Section 8.

2 Basic Ideas

We will introduce the idea of logical diagrams consisting of nodes and arcs, by means of a classical example: the square of oppositions [11].

The classical case of the square of oppositions is obtained as follows. The assertions are classified as either affirmative or negative and as either universal or particular. This classification leads to four kinds of assertions as follows.

(A) Affirmative universal: $\forall x\, P(x)$ (or, equivalently, $\neg\,\exists x\,\neg P(x)$).
(I) Affirmative particular $\exists x\, P(x)$ (or, equivalently, $\neg\,\forall x\,\neg P(x)$).
(E) Negative universal $\forall x\,\neg P(x)$ (or, equivalently, $\neg\,\exists x\, P(x)$).
(O) Negative particular $\exists x\,\neg P(x)$ (or, equivalently, $\neg\,\forall x\, P(x)$).

The classical square of oppositions displays logical connections involving these four kinds of assertions: being contrary, sub-contrary, contradictory, etc.[1]. It is as follows:

[1] Two formulas are *contrary* when they cannot be both true, *sub-contrary* when they cannot be both false and *contradictory* when they are both contrary and sub-contrary.

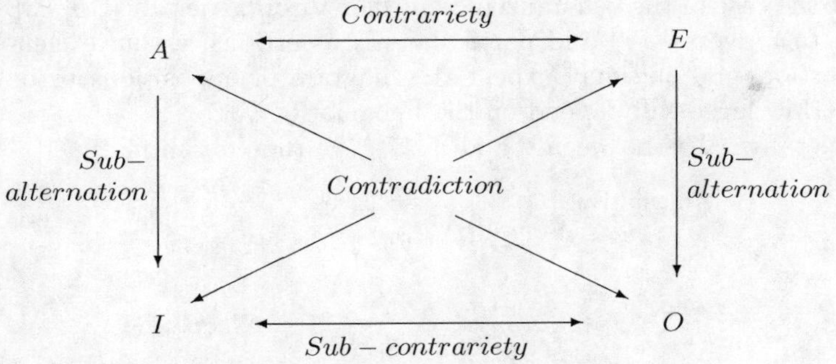

The nodes are actually representatives of logical equivalence classes:

	Affirmative	Negative
Universal	$A : \{\forall x\, P(x), \neg\, \exists x\, \neg P(x)\}$	$E : \{\forall x\, \neg P(x), \neg\, \exists x\, P(x)\}$
Particular	$I : \{\exists x\, P(x), \neg\, \forall x\, \neg P(x)\}$	O: $\{\exists x\, \neg\, P(x), \neg\, \forall x\, P(x)\}$

The arcs display logical relationships between nodes, for instance

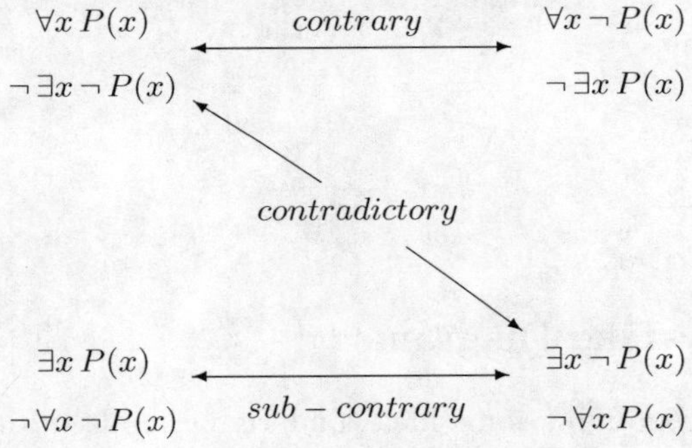

These logical relationships between nodes are independent of representatives, so can choose suitable representatives, e.g. $\forall x\, P(x)$ contradicts $\neg\, \forall x\, P(x)$.

3 Diagram Construction

One can construct a logical diagram in two steps as follows.

1. Obtain nodes as (representatives) of logical equivalence classes generated by formula building operations.
2. Add arcs displaying the logical relationships between these nodes.

Here, we will focus on the construction of node diagrams. This task may very well be far from being trivial. This can be seen by considering other cases of the square of oppositions. The classical case has 4 nodes (cf. Section 2); what about other cases (e.g. intuitionistic logic or logics with additional quantifiers)?

Let us consider the case of the generalized quantifier ∇ for 'generally' [15], [3]. Now, in addition to universal ($\forall$) and particular ($\exists$) assertions, we have 'generalized' (∇) assertions. So one may expect the diagram of oppositions to be somewhat larger. How large will depend on the behavior of ∇.

For proper 'generally', ∇ is between $\forall$ and $\exists$ [17]. We then obtain

$$
\begin{array}{ccccccc}
 & & \text{Affirmative} & & \text{Negative} & & \\
\neg\exists v\neg\varphi & \Leftrightarrow & \forall v\varphi & \overset{contrar.}{\leftrightarrow} & \forall v\neg\varphi & \Leftrightarrow & \neg\exists v\varphi \\
 & & \Downarrow & & & & \Downarrow \\
 & & \nabla v\varphi & \overset{contradict.}{\leftrightarrow} & & & \neg\nabla v\varphi \\
\Downarrow & & & & \Downarrow & & \Downarrow \\
\neg\nabla v\neg\varphi & & & \overset{contradict.}{\leftrightarrow} & \nabla v\neg\varphi & & \\
\Downarrow & & \Downarrow & & \Downarrow & & \\
\neg\forall v\neg\varphi & \Leftrightarrow & \exists v\varphi & \overset{subcontrar.}{\leftrightarrow} & \exists v\neg\varphi & \Leftrightarrow & \neg\forall v\varphi
\end{array}
$$

For stronger logics of 'generally', we have some more information. For instance, in ultrafilter logic, ∇ commutes with the propositional connectives [16], and then we have an hexagon [17] as follows

$$
\begin{array}{ccccccc}
 & & \text{Affirmative} & & \text{Negative} & & \\
\neg\exists v\neg\varphi & \Leftrightarrow & \forall v\varphi & \overset{contrar.}{\leftrightarrow} & \forall v\neg\varphi & \Leftrightarrow & \neg\exists v\varphi \\
\Downarrow & & \Downarrow & & \Downarrow & & \Downarrow \\
\neg\nabla v\neg\varphi & \Leftrightarrow & \nabla v\varphi & \overset{contradict.}{\leftrightarrow} & \nabla v\neg\varphi & \Leftrightarrow & \neg\nabla v\varphi \\
\Downarrow & & \Downarrow & & \Downarrow & & \Downarrow \\
\neg\forall v\neg\varphi & \Leftrightarrow & \exists v\varphi & \overset{subcontrar.}{\leftrightarrow} & \exists v\neg\varphi & \Leftrightarrow & \neg\forall v\varphi
\end{array}
$$

4 Modality-Like Diagrams

In this section we examine some ideas underlying modality-like diagrams.

A *modality diagram* $\mathcal{D}$ has the following features.

- Its initial node (indicated by $\rightsquigarrow$) is labeled by the atomic letter p.
- Each node has an outgoing arrow for each modality.

We will introduce the ideas of normal form and equivalence through a simple example, the modality diagram of S5 ([6], p. 149). Explanations will follow.

The modality diagram of S5, with the modalities $\neg$, $\Box$ and $\Diamond$, is as follows

$$
\begin{array}{ccc}
\Box,\Diamond \circlearrowright \left\{\begin{array}{c}\Box p\\ \Diamond\Box p\end{array}\right\} & \overset{\neg}{\longleftrightarrow} & \left\{\begin{array}{c}\neg\Box p\\ \Diamond\neg p\end{array}\right\} \circlearrowleft \Box,\Diamond \\
\Box\ \uparrow & & \uparrow\ \Diamond \\
\rightsquigarrow \{p\} & \overset{\neg}{\longleftrightarrow} & \{\neg p\} \\
\Diamond\ \downarrow & & \downarrow\ \Box \\
\Box,\Diamond \circlearrowright \left\{\begin{array}{c}\Diamond p\\ \Box\Diamond p\end{array}\right\} & \overset{\neg}{\longleftrightarrow} & \left\{\begin{array}{c}\Box\neg p\\ \neg\Diamond p\end{array}\right\} \circlearrowleft \Box,\Diamond
\end{array}
$$

One may also choose representatives for nodes (e.g. $\Box p$ for itself and $\Box\neg\Diamond\neg p$). We then obtain a simpler diagram as follows

$$\begin{array}{ccc}
\Box,\Diamond \circlearrowright \Box p & \overset{\neg}{\longleftrightarrow} & \neg\Box p \circlearrowleft \Box,\Diamond \\
\Box \uparrow & & \uparrow \Diamond \\
\rightsquigarrow p & \overset{\neg}{\longleftrightarrow} & \neg p \\
\Diamond \downarrow & & \downarrow \Box \\
\Box,\Diamond \circlearrowright \Diamond p & \overset{\neg}{\longleftrightarrow} & \Box\neg p \circlearrowleft \Box,\Diamond
\end{array}$$

In this diagram, one can evaluate modality paths and detect their equivalence.

($\dashrightarrow$) A modality path evaluates to its end node; for instance $\rightsquigarrow p \overset{\Box\neg\Diamond\neg}{\dashrightarrow} \Box p$.
($\equiv$) Equivalent modality paths are those going to the same end node; for instance $\neg\Diamond\, p \;\equiv\; \Box\neg\, p$, because the paths labeled by $\neg\,\Diamond$ and $\Box\,\neg$ lead from the initial node $\rightsquigarrow p$ to the same end note, namely $\Box\neg p$.

In general, each modality term t, being a finite sequence of modalities, labels a path from a start node q to an end node $[t\, q]^{\mathcal{D}}$.

We will be more interested in *reachable* diagrams, where each node can be reached from the initial node by some modality term: each node n has a modality term t with $\rightsquigarrow p \overset{t}{\dashrightarrow} n$. Now, a structure $\mathfrak{M}$ gives meanings to the initial node and to the modalities. So, it evaluates reachable nodes: $[t\, p]^{\mathfrak{M}} := t^{\mathfrak{M}}(p^{\mathfrak{M}})$.

A diagram will be correct (sound and complete) for a given structure when their term equalities are the same.

We call diagram $\mathcal{D}$
sound for structure $\mathfrak{M}$ iff $[s\, p]^{\mathfrak{M}} = [t\, p]^{\mathfrak{M}}$, whenever $[s\, p]^{\mathcal{D}} = [t\, p]^{\mathcal{D}}$;
complete for structure $\mathfrak{M}$ iff $[s\, p]^{\mathcal{D}} = [t\, p]^{\mathcal{D}}$, whenever $[s\, p]^{\mathfrak{M}} = [t\, p]^{\mathfrak{M}}$.

In the case of a logic L, the structure is its *Lindenbaum-Tarski algebra*, consisting of the equivalence classes of formulas under $sp \equiv_{\mathsf{L}} tp$ iff $\vdash_{\mathsf{L}} sp \leftrightarrow tp$.

5 Constructing Modality-Like Diagrams

We now take a closer look at the construction of modality-like diagrams. The construction hinges on the stepwise addition of modalities. We will begin with a partial diagram, which can evaluate only some modality terms, and extend it to a diagram so that it can evaluate all modality terms.

We will introduce the ideas by indicating how one can construct the modality diagram for K45, with the modalities $\neg$, $\Box$ and $\Diamond$ cf. [6], p. 132).

1. Let us begin with diagram $\mathcal{P}_1$ for negation: $\rightsquigarrow p \overset{\neg}{\longleftrightarrow} \neg p$.
This diagram $\mathcal{P}_1$ is partial: it can only evaluate terms in the set $E_1 = \{\neg\}^*$. We wish to evaluate the modality $\Box p$. There is no outgoing arrow from p labeled $\Box$. Also, we have no node to represent $\Box p$ (since $\nvdash \Box p \leftrightarrow p$ and $\nvdash \Box p \leftrightarrow \neg p$). So, we must add a new node for $\Box p$.

2. We thus add a new node $\Box p$, which leads to diagram $\mathcal{P}_2$ (see Figure 1(a)). Diagram $\mathcal{P}_2$ is partial: it can only evaluate terms in the set $E_2 = \{\neg\}^* \cup \{\Box\}$. We wish to evaluate the modality $\Box\Box p$. We know that $\vdash \Box\Box p \leftrightarrow \Box p$. So we do not need a new node.

3. We thus coalesce $\Box\Box p$ to $\Box p$, which leads to diagram $\mathcal{P}_3$ (see Figure 1(b)). Diagram $\mathcal{P}_3$ is partial: it only evaluates terms in the set $E_3 = \{\neg\}^* \cup \{\Box\}^*$.

$\star$.Proceeding in this manner, we arrive at a ten-node diagram $\mathcal{P}_\star$ (cf. [6], p. 153). Diagram $\mathcal{P}_\star$ is full: it evaluates all terms in the set $E_\star = \{\neg, \Box, \Diamond\}^*$.

$$\begin{array}{lcl} \rightsquigarrow p & \overset{\neg}{\longleftrightarrow} & \neg p \\ \Box \downarrow & & \\ \Box p & & \end{array} \qquad\qquad \begin{array}{lcl} & \rightsquigarrow p & \overset{\neg}{\longleftrightarrow} \neg p \\ & \Box \downarrow & \\ \Box \circlearrowright & \Box p & \end{array}$$

(a) Diagram $\mathcal{P}_2$ (b) Diagram $\mathcal{P}_3$

Fig. 1. Intermediate diagrams for K45

Some remarks about these ideas are in order.

1) We could have started from the trivial diagram $\mathcal{P}_0$: $\rightsquigarrow p$, evaluating no modality term: $E_0 = \emptyset$.

2) Each partial diagram is correct (sound and complete) with respect to the modality terms it can evaluate.

3) The approach is somewhat robust. Imagine that one inadvertently adds an unnecessary new node. Then, the partial diagram is still sound (though not complete). To regain completeness, it suffices to coalesce the unnecessary node to its equivalent node and redirect arrows accordingly.

6 Construction Method

We now present our method for the construction of logical diagrams. The construction method hinges on extending partial diagrams.

(0) We start from a partial diagram $\mathcal{P}_0$.

(n) At stage n, we have a partial diagram $\mathcal{P}_n$, with set E_n of modality terms. We expand frontier situations, if any, obtaining diagram $\mathcal{P}_{n+1}$.

By a *frontier situation* in a partial diagram we mean a pair (n, f), such that n is a node in the diagram without outgoing arrow labeled by a modality f. A diagram without frontier situations will be full. So, we will eventually obtain a full diagram (perhaps in the limit). Expansion aims at eliminating frontier situations while preserving relative correctness (soundness and completeness). In the expanded diagram, we will have $n \xrightarrow{f} n^f$, where

$$n^f := \begin{cases} \text{old node if already have some equivalent modality} \\ \text{new node if no equivalent modality present} \end{cases}$$

The *expansion* of a partial diagram $\mathcal{P}$, with respect to a given structure $\mathfrak{M}$, produces a diagram $\mathcal{P}'$ as follows.
Given the frontier situation (n, f), with $n = [s\,p]^{\mathcal{P}}$, one has two cases to consider.

(Y) In case there is already a modality equivalent, in the sense that we have a reachable node $q = [t\,p]^{\mathcal{P}}$ in diagram $\mathcal{P}$, such that $f^{\mathfrak{M}}(n^{\mathfrak{M}}) = q^{\mathfrak{M}}$. In this case, we coalesce $n^f := q$. Note that $[f\,n]^{\mathcal{P}'} = [t\,p]^{\mathcal{P}} \in N$.

(N) In case there is no equivalent modality present, we add a new node $n^f \notin N$. Note that $[f\,n]^{\mathcal{P}'} = n^f \notin N$.

Notice that, in either case, we have an embedding $\mathcal{P} \subseteq \mathcal{P}'$.

$$\begin{array}{ccc} & \overset{s}{\dashrightarrow} & n \\ \rightsquigarrow p & & \downarrow f \\ & \underset{t}{\dashrightarrow} & q \end{array} \qquad\qquad \rightsquigarrow p \overset{s}{\dashrightarrow} n \overset{f}{\rightarrow} n^f$$

(Y) : Coalesce (N) : Add

Fig. 2. Expansion of partial diagram

Now, let us formulate more precisely our method of successive expansions. We wish to obtain a diagram that is correct with respect to a given structure $\mathfrak{M}$. The *construction method* is as follows.

$(-)$ Start with a partial diagram $\mathcal{P}_0$, with its set E_0 of modality terms.
(F) While partial diagram $\mathcal{P}_n$ has some frontier situation, expand $\mathcal{P}_n$ (with respect to structure $\mathfrak{M}$) obtaining $\mathcal{P}_{n+1}$.

The result of this process is the diagram $\mathcal{P}_\star := \bigcup_{n \in \mathbb{N}} \mathcal{P}_n$

We can now see that our construction method produces a full diagram.

Lemma 1. *Given a partial diagram $\mathcal{P}_0$, consider the diagram $\mathcal{P}_\star$ obtained by the above method of successive expansions. Then, diagram $\mathcal{P}_\star$ is full.*

Proof. The assertion follows from the preceding remarks.

We now wish to argue that our method produces a correct diagram. A partial diagram $\mathcal{P}$ can evaluate modalities in its set E: for $s \in E$, $\rightsquigarrow p \overset{s}{\dashrightarrow} q = s^{\mathcal{P}}$.

We now extend correctness to partial diagrams. We relativize our previous concepts of correctness to the modality terms a partial diagram can evaluate. A partial diagram is relatively correct for a given structure $\mathfrak{M}$ when the term equalities that can be evaluated are the same. We call a partial diagram $\mathcal{P}$

- *relatively sound* over set E for structure $\mathfrak{M}$ iff, for all $[s\,p], [t\,p] \in E$, if $[s\,p]^{\mathcal{D}} = [t\,p]^{\mathcal{D}}$, then $[s\,p]^{\mathfrak{M}} = [t\,p]^{\mathfrak{M}}$;
- *relatively complete* over set E for structure $\mathfrak{M}$ iff, for all $[s\,p], [t\,p] \in E$, if $[s\,p]^{\mathfrak{M}} = [t\,p]^{\mathfrak{M}}$, then $[s\,p]^{\mathcal{D}} = [t\,p]^{\mathcal{D}}$.

With these concepts, we can see that expansion preserves relative correctness (soundness and completeness).

Proposition 1. *Given a partial diagram $\mathcal{P}$ with its set E of modality terms, and a structure $\mathfrak{M}$, consider the corresponding expansion: diagram $\mathcal{P}'$, with set E' of modality terms. If the diagram $\mathcal{P}$ is relatively sound (respectively relatively complete) over set E for $\mathfrak{M}$, then diagram $\mathcal{P}'$ is relatively sound (respectively relatively complete) over set E' for $\mathfrak{M}$.*

Proof. *The assertion follows from the expansion construction.*

We can now see that our method eventually produces a full correct diagram.

Theorem 1. *Given a partial diagram $\mathcal{P}_0$ with its set E_0 of modality terms, consider the diagram $\mathcal{P}_\star$ obtained by the above method of successive expansions.*

- *The resulting diagram $\mathcal{P}_\star$ is full.*
- *Given structure $\mathfrak{M}$, if the starting diagram $\mathcal{P}_0$ is sound and complete over set E_0 for $\mathfrak{M}$, then the resulting diagram $\mathcal{P}_\star$ is sound and complete for $\mathfrak{M}$.*

Proof. *By Lemma 1 and Proposition 1.*

7 Variations and Extensions

In this section we present some variations and extensions of our method.

Our construction method successively expands an initial partial diagram (cf. Sections 5 and 6). A variant starts with some partial diagrams (or equivalently tables) and proceeds by merging them and expanding the results.

We will illustrate how this idea works by constructing the modality diagram for S5, now with the two modalities $\neg$ and $\Box$.

1. We start with two partial diagrams $\mathcal{P}_\neg$, for $\neg$, and $\mathcal{P}_\Box$, for $\Box$, as follows

$$\mathcal{P}_\neg : \rightsquigarrow p \overset{\neg}{\longleftrightarrow} \neg p \quad E_\neg = \{\neg\}^* \qquad\qquad \mathcal{P}_\Box : \begin{array}{c} \rightsquigarrow p \\ \Box\downarrow \\ \Box \circlearrowleft \Box p \end{array} \quad E_\Box = \{\Box\}^*$$

   These two partial diagrams can also be presented by partial tables, as follows

$\mathcal{P}_\neg$:

n	p	$\neg p$
$\neg n$	$\neg p$	p

$\mathcal{P}_\Box$:

n	p	$\Box p$
$\Box n$	$\Box p$	$\Box p$

2. We now merge two partial diagrams, obtaining $\mathcal{P}_{\neg,\Box} := \mathcal{P}_\neg \amalg \mathcal{P}_\Box$ as follows

$$\mathcal{P}_{\neg,\Box} : \begin{array}{c} \rightsquigarrow p \overset{\neg}{\longleftrightarrow} \neg p \\ \Box\downarrow \qquad\quad \\ \Box \circlearrowleft \Box p \qquad\quad \end{array} \qquad E_{\neg,\Box} = \{\neg\}^* \cup \{\Box\}^*$$

   The table presentation for $\mathcal{P}_{\neg,\Box}$ is as follows

n	p	$\neg p$	$\Box p$
$\neg n$	$\neg p$	p	–
$\Box n$	$\Box p$	–	$\Box p$

3. This diagram $\mathcal{P}_{\neg,\Box}$ is still partial. We can apply the preceding method to expand it to a full diagram $\mathcal{D}_{\neg,\Box}$ evaluating all the terms in the set $\{\neg, \Box\}^*$. The table presentation for $\mathcal{D}_{\neg,\Box}$ is as follows

n	p	$\neg p$	$\Box p$	$\neg\Box p$	$\Box\neg p$	$\neg\Box\neg p$
$\neg n$	$\neg p$	p	$\neg\Box p$	$\Box p$	$\neg\Box\neg p$	$\Box\neg p$
$\Box n$	$\Box p$	$\Box\neg p$	$\Box p$	$\neg\Box p$	$\Box\neg p$	$\neg\Box\neg p$

We now examine how one can extend our methods to other arities. The basic idea is that we are constructing a free term algebra [5]. For an n-ary operation we will have n initial nodes. It is more convenient to work with partial tables rather than diagrams.

We will illustrate this idea with a rather simple example. We will construct the diagram for classical conjunction: the binary operation $\wedge$ (see Figure 3).

1. We begin with the trivial table $\wedge_0$.
2. We can fill in the diagonals of this table $\wedge_0$, by relying on idempotence ($\vdash \varphi \wedge \varphi \leftrightarrow \varphi$). We obtain table $\wedge_1$.
3. By eliminating the frontier situation $p \wedge q$ in $\wedge_1$, we extend it to table $\wedge_2$.
4. We can now fill in some entries in table $\wedge_2$, by relying on commutativity ($\vdash \psi \wedge \theta \leftrightarrow \theta \wedge \psi$). We obtain table $\wedge_3$.
5. We can similarly fill in the remaining entries in table $\wedge_3$, by relying on properties of conjunction. We obtain the full table $\wedge_4$.

$\wedge_0$	p	q
p	—	—
q	—	—

$\wedge_1$	p	q
p	p	—
q	—	q

$\wedge_2$	p	q	$p \wedge q$
p	p	$p \wedge q$	—
q	—	q	—
$p \wedge q$	—	—	—

$\wedge_3$	p	q	$p \wedge q$
p	p	$p \wedge q$	—
q	$p \wedge q$	q	—
$p \wedge q$	—	—	—

$\wedge_4$	p	q	$p \wedge q$
p	p	$p \wedge q$	$p \wedge q$
q	$p \wedge q$	q	$p \wedge q$
$p \wedge q$	$p \wedge q$	$p \wedge q$	$p \wedge q$

Fig. 3. Partial tables for conjunction

8 Final Comments

In this section we present some closing remarks about our ideas.

We have introduced a method for the construction of symbolic diagrams, displaying term-like objects and connections between them. They are employed in many branches of science as useful tools for many purposes: visualizing connections, reasoning and experimentation.

We have presented a stepwise approach for the construction of symbolic diagrams. This approach is modular: its starts from diagrams with few operations and combines them or adds one operation at a time. The method joins diagrams and adapts the result by coalescing existing nodes and adding new nodes. We have explained and justified the method. The key idea amounts to relativization: a partial diagram evaluating a set of terms. We have introduced the ideas for the case of unary operations and indicated the extension to the general case.

Thus, we have provided a method for constructing tools with a widespread range of possible uses. One can use similar ideas for specification purposes [13].

References

1. Areces, C., Nivelle, H., de Rijke, M.: Resolution in Modal, Description and Hybrid Logic. J. Logic and Computation 11(5), 717–736 (2001)
2. Baader, F., Calvanese, D., McGuiness, D.L., Nardi, D., Pattel-Schneider, P.F.: The Description Logic Handbook: Theory, Implementations, Applications. Cambridge Univ. Press, Cambridge (2003)
3. Benevides, M.R.F., Delgado, C.A.D., de Freitas, R.P., Veloso, P.A.S., Veloso, S.R.M.: On Modalities for Vague Notions. In: Bazzan, A.L.C., Labidi, S. (eds.) SBIA 2004. LNCS (LNAI), vol. 3171, pp. 1–10. Springer, Heidelberg (2004)
4. Béziau, J.-Y.: From Consequence Operator to Universal Logic: a Survey of General Abstract Logic. In: Béziau, J.-Y. (ed.) Logica Universalis, vol. 1, pp. 3–17 (2005)
5. Burris, S., Sankappanavar, G.: A Course in Universal Algebra. Springer, New York (1980)
6. Chellas, B.: Modal Logic: an Introduction. Cambridge Univ. Press, Cambridge (1980)
7. Fagin, R., Halpern, J.Y., Moses, Y., Vardi, M.Y.: Reasoning about Knowledge. MIT Press, Cambridge (1995)
8. Kozen, D., Tuiryn, J.: Logics of Programs. In: van Leeuwen, J. (ed.) Handbook of Theoretical Computer Science, vol. B, pp. 789–840. Elsevier, Amsterdam (1980)
9. Meyer, J.J.C., van der Hoek, W.: Epistemic Logic for Computer Science and Artificial Intelligence. Cambridge Univ. Press, Cambridge (1995)
10. Moore, R.C.: Logic and Representation. Cambridge Univ. Press, Cambridge (1995)
11. Moretti, A.: Geometry for Modalities? Yes: through n-opposition Theory. In: Béziau, J.-Y., Costa-Leite, A., Facchini, A. (eds.) Aspects of Universal Logic, pp. 102–145. Univ. Neuchâtel, Neuchâtel (2004)
12. van Dalen, D.: Logic and Structure, 4th edn. Springer, Berlin (2004)
13. Pequeno, T.H.C., Veloso, P.A.S.: Do not Write more Axioms than you Have to. In: Proc. Int. Computing Symp., Taipei, pp. 487–498 (1978)
14. Rounds, W.C.: Feature Logics. In: van Benthem, J., ter Meulen, A. (eds.) Handbook of Logic and Language. Elsevier, Amsterdam (1997)
15. Veloso, P.A.S., Carnielli, W.A.: Logics for Qualitative Reasoning. In: Gabbay, D., Rahman, S., Symons, J., van Bendegem, J.P. (eds.) Logic, Epistemology and the Unity of Science, pp. 487–526. Kluwer, Dordretch (2004)
16. Veloso, S.R.M., Veloso, P.A.S.: On Special Functions and Theorem Proving in Logics for Generally. In: Bittencourt, G., Ramalho, G.L. (eds.) SBIA 2002. LNCS (LNAI), vol. 2507, pp. 1–10. Springer, Heidelberg (2002)
17. Veloso, P.A.S., Veloso, S.R.M.: On Modulated Logics for Generally: Some Metamathematical Issues. In: Béziau, J.-Y., Costa-Leite, A., Facchini, A. (eds.) Aspects of Universal Logic, pp. 146–168. Univ. Neuchâtel, Neuchâtel (2004)

Bi-objective Memetic Evolutionary Algorithm for Automated Microwave Filter Synthesis

Maria J.P. Dantas[1], Leonardo da C. Brito[2], Paulo C.M. Machado[2], and Paulo H.P. de Carvalho[3]

[1] Catholic University of Goiás, Goiânia, GO
[2] Federal University of Goiás, Goiânia, GO
[3] University of Brasília, Brasília, DF, Brazil
mjpd@cultura.com.br, brito@eeec.ufg.br, paulo@ene.unb.br

Abstract. This paper presents a Memetic Evolutionary Algorithm for automated compact filter design using two-port building-blocks. In general, complex circuits are designed with suitably arranged two-port elements. The proposed representation scheme uses a proposed Positional Matrix, associating two-port building-blocks, their possible connections (cascade, serial, parallel or hybrid), and their parameters. The candidate designs are processed through suitable bidimensional evolutionary operators. The solutions are evolved using a bi-objective classification process, taking into account the performance evaluations, based on scattering parameters and size of the circuit. Promising candidate solutions are fine-tuned by a local search method. Results demonstrated that two-port compact filters are achieved, using simple microstrip building-blocks for describing the general passive microwave circuits.

Keywords: automated synthesis, bidimensional representation, bi-objective classification, two-port building-blocks, microwave circuit.

1 Introduction

Recent works (see [1]-[6], for example) have presented a renewed interest in filter synthesis. In modern applications – wireless communication systems, for example – the rigorous filter specifications demand new effective methods to aid the designers to find new filter topologies [1]. Using evolution-based paradigms, such as Genetic Algorithms and Genetic Programming, researchers (see [7]-[10], for example) have developed methods capable to evolve both circuit topologies and the values of their parameters, without providing any prior specific design input to the algorithm. That is, they do not require expert knowledge regarding the circuit topology. However, they commonly generate extremely unconventional and unstructured circuit topologies, which can be physically unpractical. On the other hand, in contrast with arbitrary topology search methods, the arbitrary but hardly constrained topology approaches are more efficient and attractive in some design cases, reducing the circuit complexity [11]-[14], but it is not able to find innovative topologies.

G. Zaverucha and A. Loureiro da Costa (Eds.): SBIA 2008, LNAI 5249, pp. 197–206, 2008.

In this paper, we present an automated two-port filter synthesis method. The proposed algorithm applies: (1) expert knowledge to define the two-port representation scheme and appropriate evolutionary operators for imposing a set of moderate constraints on the structure of the candidate solutions, in order to reduce the search space and to avoid the occurrence of anomalous circuits, but with enough flexibility to allow the generation of novel topologies; and (2) a Memetic Evolutionary Algorithm that balance the topology search (performed by an Evolutionary Algorithm) and the parameters tuning process (performed by the Simulated Annealing method). The algorithm is focused on two-port elements, since most of the basic components of passive and active circuits are two-port building blocks, which are arranged together to produce a more complex passive or active circuit (passive microwave circuits, for example).

2 Description of the Proposed Method

Fig. 1 shows the flowchart of the Memetic Evolutionary Algorithm. In the following sections, the features of the method will be described.

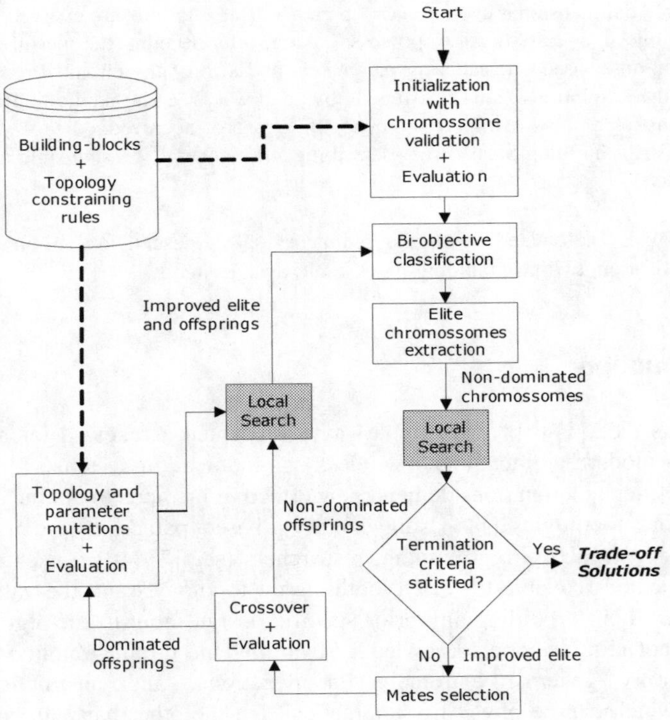

Fig. 1. The Bi-objective Memetic Evolutionary Algorithm

2.1 Two-Port Circuit Representation – Positional Matrix

The template of the evolvable circuit is shown in Fig. 2. The circuit is encoded into the proposed Positional Matrix. The initial size of this evolvable matrix is given by the number of elements of the initial circuit, which is composed by a set of elements cascaded from the source towards the load. The creation process of the initial Positional Matrix is as follows:

> (1) Randomly define the size n of the Positional Matrix;
> (2) Randomly select n circuit elements from a database and assemble the initial Positional Matrix by cascading them into its main diagonal, from the source to the load;
> (3) Repeat m times
>> Randomly select a row i and a column j of the Positional Matrix, subject to $i \leq j$. Randomly select a circuit building-block and its respective connection type based on constraining rules and encode it in the (i, j) position of the Positional Matrix (e.g., as depicted in Fig. 4).

Along the evolution process, other building blocks are placed into or removed from the variable-size Positional Matrix, as described in the following sections. The possible types of connection are: cascade, serial, parallel, or hybrid, as shown in Fig. 3. Besides the building-blocks, topology constraining rules also feeds the algorithm, as depicted by the database in Fig. 1.

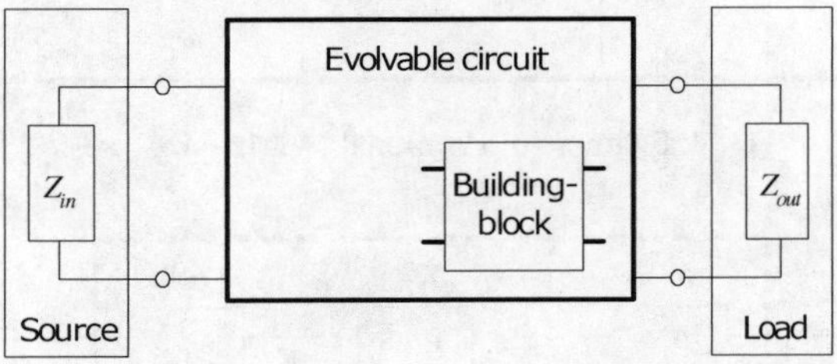

Fig. 2. Template of the evolvable circuit

Figs. 4 and 5 illustrate the proposed representation. In the Positional Matrix of Fig. 4, the circuit elements (building-blocks) $b_{11}^{(1)}$, $b_{22}^{(1)}$, and $b_{33}^{(1)}$ are connected in cascade; $b_{11}^{(2)}$ and $b_{11}^{(1)}$, are connected in parallel (p); $b_{13}^{(1)}$ is in series (s) with the input port of $b_{11}^{(1)}$ and in parallel (p) with the output port of $b_{33}^{(1)}$ (hybrid connection); $b_{23}^{(1)}$ is in parallel (p) with the input port of $b_{22}^{(1)}$ e and in series with the output port of $b_{23}^{(1)}$; (hybrid connection). Fig. 5 shows its respective topological representation.

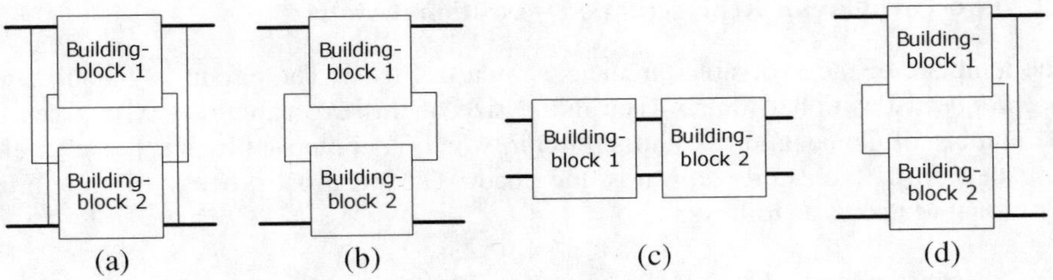

Fig. 3. The possible connections types between two-port building blocks: (a) parallel, (b) serial, (c) cascade, and (d) hybrid (serial connection in the input port and parallel in the output port)

Source	(1,1) $b_{11}^{(2)} = \begin{bmatrix} 1 & 3 \\ 2 & 4 \end{bmatrix}$ (p) $b_{11}^{(1)} = \begin{bmatrix} 1 & 3 \\ 2 & 4 \end{bmatrix}$ (c)	(1,2)	(1,3) $b_{13}^{(1)} = \begin{bmatrix} 2 & 7 \\ 10 & 8 \end{bmatrix}$ (h) - s/p	Load
$\begin{bmatrix} 1 \\ 10 \end{bmatrix}$	(2,1)	(2,2) $b_{22}^{(1)} = \begin{bmatrix} 3 & 5 \\ 4 & 6 \end{bmatrix}$ (c)	(2,3) $b_{23}^{(1)} = \begin{bmatrix} 3 & 8 \\ 4 & 9 \end{bmatrix}$ (h) - p/s	$\begin{bmatrix} 7 \\ 9 \end{bmatrix}$
	(3,1)	(3,2)	(3,3) $b_{33}^{(1)} = \begin{bmatrix} 5 & 7 \\ 6 & 8 \end{bmatrix}$ (c)	

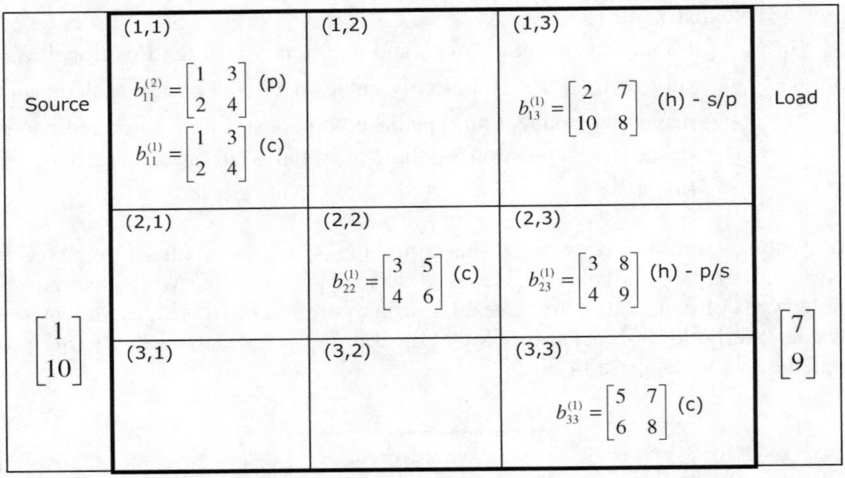

Fig. 4. Example of Positional Matrix (size 3×3)

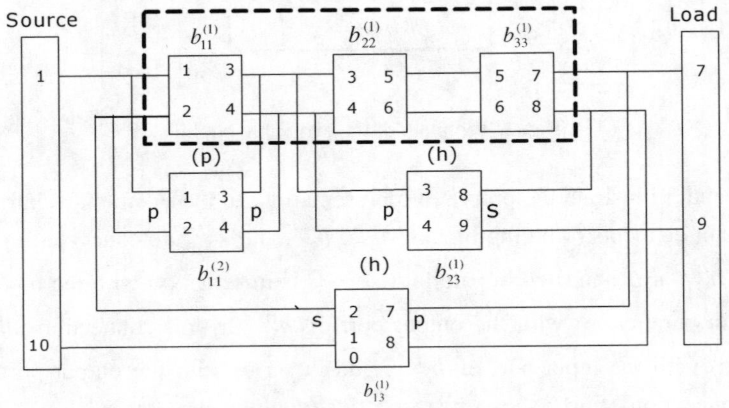

Fig. 5. Topological Representation

2.2 Evaluation Functions

Two objective-functions are defined in order to allow a trade-off relation: (1) the circuit performance, which is evaluated using a frequency-domain circuit simulator; and (2) the circuit size, given by the number of two-port building-blocks. The circuit simulator computes the frequency responses (the scattering parameters) over a set of frequencies uniformly distributed in the range defined by the user. Then, the algorithm calculates the sum of the squared deviations between the computed aggregate responses and the desired responses (sum squared error), as in (1). The desired response is provided as scattering parameters masks for the absolute values of the transmission coefficient $|S_{21}|$ and reflection coefficient $|S_{11}|$, given in dB.

$$SSE = \sum_{j=1}^{k} \left[\left(|S_{21}(f_j)| - |S_{21}^{*}(f_j)| \right)^2 + \left(|S_{11}(f_j)| - |S_{11}^{*}(f_j)| \right)^2 \right] \quad (1)$$

In (1), k is the number of evaluation frequencies, $|S_{21}^{*}(f_j)|$ is boundary value of the response mask of the respective scattering parameter at f_j. The difference $|S_{21}(f_j)| - |S_{21}^{*}(f_j)|$ in (1) is set to zero if the mask is not violated by the value $|S_{21}(f_j)|$. The same criterion is applied to $|S_{11}|$.

2.3 Evolution Schemes

The population is randomly initialized with circuits (individuals or chromosomes) that use the template circuit shown in Fig. 2, which is composed by two-port circuit elements (genes) randomly selected from the database. In order to generate high-performance small circuits, a bi-objective selection approach – the crowded-comparison operator, extracted from the NSGA-II [15] – is applied in this method at two points: to extract the elite (non-dominated chromosomes) of the current population, as well as the elite of the offspring. The two evaluation functions (objective-functions) previously defined are taken into account. The elite individuals of the population, i.e. the Pareto front, are found by applying this classification method. The selection scheme used in this work is the well-known binary tournament method.

The proposed approach provides the balance between performance and size of the solutions, and, consequently, makes it possible to naturally reduce the tendency of the process for producing larger circuits as the population evolves. Additionally, it allows the extraction of new building-blocks (as desired concerning the building-block hypothesis [16]) derived by the evolution process, which can be used in the next stage to produce the competitive circuits with some degree of structural redundancy.

A local search process assists the Evolutionary Algorithm for fitness improvement of candidate circuits, refining their parameters in order to avoid good topologies with non-optimized parameter values to be prematurely discarded. The evaluation criterion to accept new parameters for a given topology is mono-objective, based on the performance function (1). This process takes place in two points of the evolution cycle. After the classification process, the local search method is applied to each elite individual. Also, the local search procedure is carried out after the crossover/mutation procedure. Doing so, the topology space is explored and, subsequently, the parameters of the new topologies are improved. As a result, offspring solutions will be able to fairly compete with the current elite set for composing the elite of the next generation. The Simulated Annealing technique [17] with few iterations and predefined temperature values was adopted. Only a low computational effort is necessary for each local search.

2.4 Bidimensional Topology Crossover Operator

Only one crossover operator is proposed. Fig. 6 sketches this operator. Each crossover operation generates only one offspring. The crossover occurs as follows. Given two reduced matrices, a cut point in parent matrix 1 is randomly chosen, such that four regions are defined, as shown in Fig 6(a). After that, a square sub-matrix in parent matrix 2 is arbitrarily defined, as shown in Fig. 6(b) and Fig. 6(c). Fig. 6(e) e Fig. 6(f) illustrates the offspring composition. The blocks R_1, R_2, R_3 and R_4 in the offspring matrix are from the parent matrix 1, the block R_5 is from the parent matrix 2, and the block R_6, is randomly selected from the corresponding block in parent matrix 1 or from 2. Two types of offspring composition are possible and likely to occur: cut-splice, as in Fig. 6(c); cut-overlap, as in Fig. 6(d). Then, the proposed crossover operator can explore the containing knowledge in the parents, and can also promote the diversity of structures. Since selection of the cut point is independent for each parent matrix, it is obvious that the produced offspring matrix length can vary during the evolution process. Then, circuits with different sizes and complexities evolve together by exchanging their genetic material.

2.5 Bidimensional Topology and Value Mutation Operators

Four types of likely topology mutation were defined. The circuit mutation is performed via one of the following operations: (1) adding a randomly selected building-block, without position restriction in the Positional Matrix; (2) deleting a randomly selected building-block, given that the circuit remains connected; (3) deleting a randomly selected building-block from the diagonal matrix, by removing a row/column, given that the circuit remains connected; (4) inserting an arbitrary building-block in cascade into the diagonal of the Positional Matrix, by adding a row/column.

All the parameters of the two-port circuit elements of the circuit may suffer mutation. If a parameter will be mutated, a new parameter value is randomly generated through a uniform distribution bounded by the predefined range of possible values.

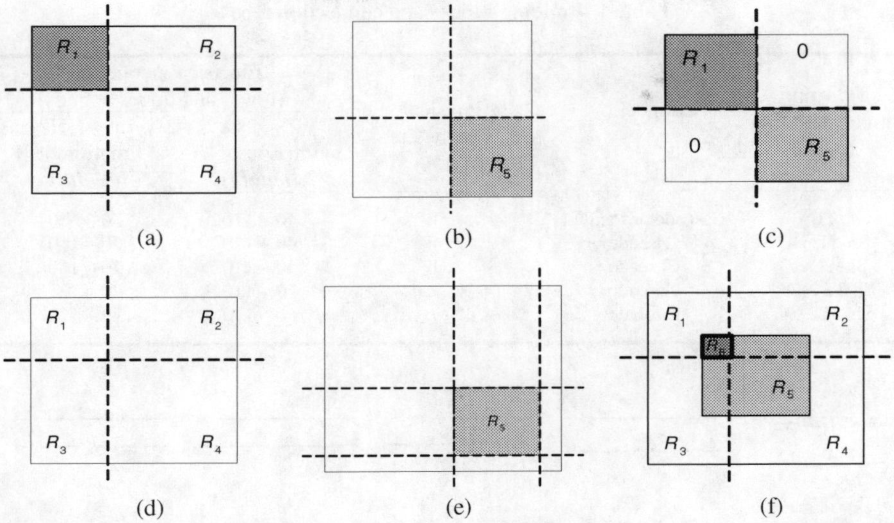

Fig. 6. Crossover operator: (a) parent matrix 1, (b) parent matrix 2, (c) cut-splice, (d) parent matrix 1, (e) parent matrix 2, and (f) cut-overlap

3 Experiment and Results

Several different two-port filters were synthesized, presenting several complexity levels. The proposed method successfully produced filters that complied with the rigorous specifications. The number of circuit evaluations for the entire synthesis process was not large. A type of dual band-pass filter, which offers a considerable difficulty degree, illustrates the application of the proposed method in this section. In recent years, dual-band filters have become extremely important components for wireless communication devices at microwave frequencies [6],[19]. Its design is a hard problem, as reported in previous works (see [12], and [18]-[19], for example).

The crossover probability was set to 100%, the topology mutation probability was set to 20%, and the parameters mutation probability was set to 5%. The circuit performance was analyzed at 80 discrete frequencies. In all 10 runs, the proposed algorithm achieved results that accomplished the specifications.

Example of experiment. In this experiment, a filter for dual-band systems was synthesized. The same filter was synthesized in [11] and [12] with the following specifications: the return losses (reflection coefficient inside the pass-bands) within 3.4—3.6 and 5.4—5.6 GHz > 10 dB, and the rejections (transmission coefficient outside the pass-bands) within 2.0—3.0.5, 4.0—5.0, and 6.0—7.0GHz > 20 dB.

Table I and Fig. 7 present the building-blocks and connection types. Besides that, during the evolution process, as a topology constraining rule, only common junctions in microstrip circuits were allowed (step-, tee-, and cross junctions) [12].

Table 1. Building-blocks and connection types

Two-port circuit building blocks	Connection types	Physical parameters	Electrical parameters (lower and upper bounds)	
			Z_{01}, as a function of L *(length)*	Θ_1 (at 4 GHz), as a function of W *(width)*
TL	Cascade or Parallel	L, W	40 – 110	30 – 100
Sh-TL-OC	Cascade	L, W	40 – 110	20 – 160
Sh-TL-SC	Cascade	L, W	40 – 110	20 – 160
Sh-TL2-OC	Cascade	L_1, W_1, L_2, W_2	40 – 110	20 – 120
Sh-TL2-SC	Cascade	L_1, W_1, L_2, W_2	40 – 110	20 – 120

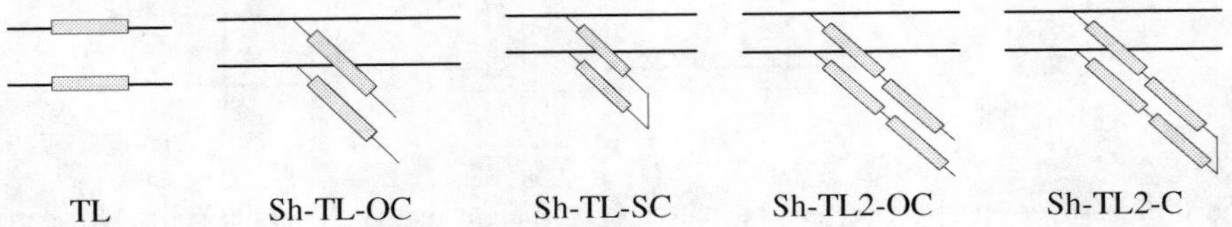

Fig. 7. Building-blocks of the experiment

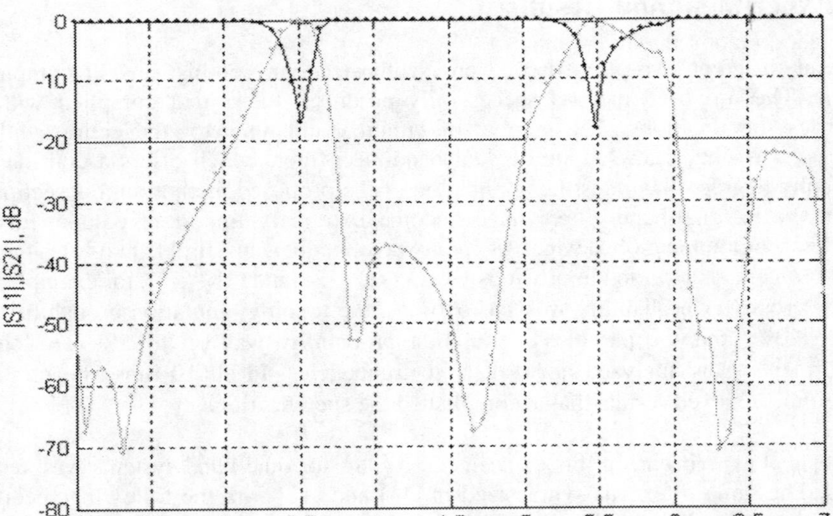

Fig. 8. The frequency responses of one Pareto solution. The thick black line represents the user-defined mask. The gray line is the transmission response $|S_{21}|$. The black line is the reflection response $|S_{11}|$.

Results. The best topology obtained with the proposed method has 11 two-port circuit elements, achieved after 20,285 circuit evaluations. It is a very compact topology and matches the specifications, as shown in Fig. 8. Besides this high-quality solution, the designer has a set of trade-off solutions available into the elite population (Pareto front). For instance, another good solution in the Pareto front has 8 circuit elements, although the frequency response was slightly worse. It can be compared with the result presented in [12], which uses a mono-objective hybrid encoded Genetic Algorithm. In [12], the best solution was composed by 10 circuit elements, and was achieved after 300 generation, with a population size of 200, or 60,000 circuit evaluations. Results as good as the ones in [12] were achieved with a lower number of circuit evaluations. On the other hand, the proposed method produced filters smaller than those presented in [11].

4 Conclusions

In this work, a Bi-objective Memetic Evolutionary Algorithm applied to design of microwave filters has been proposed. This method can be applied to the synthesis of any passive filter composed by two-port building-blocks. The advantage of making available for the designer more than a solution in Pareto front is provided. The preliminary results have demonstrated that the proposed method produces compact filters. In future works, we intend to work with multiple objective functions, using the classifying process of the NSGA-II method, but making a preference-based articulation along the evolutionary process. In the synthesis of a multi-band filter, for example, each band may correspond to an objective-function and to each of them could be assigned a preference level.

References

1. Uhm, M., Lee, J., Yom, I., Kim, J.: General Coupling Matrix Synthesis Method for Microwave Resonator Filters of Arbitrary Topology. ETRI Journal 28(2), 223–226 (2006)
2. Cameron, R.J., Faugere, J.C., Rouillier, F., Seyfert, F.: Exhaustive Approach to the Coupling Matrix Synthesis Problem And Application to the Design of High Degree Asymmetric Filters: Research Articles. International Journal of RF and Microwave Computer-Aided Engineering. Special Issue on RF and Microwave Filters, Modeling and Design Pierre Jarry and Hussein Baher Guest Editor 17, 4–12 (2007)
3. Lenoir, P., Bila, S., Seyfert, F., Baillarget, D., Verdeyme, S.: Synthesis and Design of Asymmetrical Dual-band Bandpass Filters Based on Network Simplification. IEEE Transactions on Microwave Theory and Techniques 54(7), 3090–3097 (2006)
4. Jun, D.S., Lee, H.Y., Kim, D.-Y., Lee, S.S., Nam, E.S.: A Narrow Bandwidth Microstrip Band-Pass Filter with Symmetrical Frequency Characteristics. ETRI Journal 27(5), 643–646 (2005)
5. Wang, H., Zhang, A.-J., Fang, D.-G.: The Design of Cross-Coupled Microstrip Filter Exploiting Aggressive Space Mapping Technique. In: Microwave Conference – APMC 2005 Proceedings, vol. 5, pp. 1–3 (2005)
6. Chen, C.-Y., Hsu, C.-Y.: Design of a UWB Low Insertion Loss Bandpass Filter with Spurious Response Suppression. Microwave Journal, 112–116 (February 2006)

7. Hu, J., Goodman, E.D., Rosenberg, R.: Robust and Efficient Genetic Algorithms with Hierarchical Niching and a Sustainable Evolutionary Computation Model. In: Deb, K., et al. (eds.) GECCO 2004, Part I. LNCS, vol. 3102, pp. 1220–1232. Springer, Heidelberg (2004)
8. Koza, J.R., Bennett III, Andre, F.H., Keane, M.A.: Automated Design for Both Topology and Components Values of Electrical Circuits using Genetic Programming. In: Genetic Programming 1996: Proceedings of the First Annual Conference, July 1996, pp. 123–131. Stanford University, Cambridge (1996)
9. Grimbleby, J.B.: Automatic Analogue Circuit Synthesis using Genetic Algorithms. IEEE Proceedings: Circuits, devices and systems 147(6), 319–323 (2000)
10. Zebulum, R.S., Pacheco, M.A., Vellasco, M.: Comparison of Different Evolutionary Methodologies Applied to Electronic Filter Design. In: Proceedings of the IEEE International World Congress on Computational Intelligence, ICEC 1998, Alaska, pp. 434–439 (May 1998)
11. Nishino, T., Itoh, T.: Evolutionary Generation of Microwave Line-Segment Circuit by genetic Algorithms. IEEE transactions on Microwave Theory and Techniques 50(9), 2048–2055 (2002)
12. Lai, Ming-Iu, Jeng, Shyh-Kang: Compact Microstrip Dual-Band Bandpass Filters, Design Using Genetic-Algorithm Techniques. IEEE Transactions on Microwave Theory and techniques 54(1), 160–168 (2006)
13. Dastidar, T.R., Chakrabarti, P.P., Ray, P.: A Synthesis System for Analog Circuit Based on Evolutionary Search and Topological Reuse. IEEE Transactions on Evolutionary Computation 9(2), 211–224 (2005)
14. Hou, H.-S., Chang, S.-J., Su, Y.-K.: Practical Passive Filter Synthesis using Genetic Programming. IEICE Trans. Electronic E88(6), 1180–1185 (2005)
15. Deb, K., Pratap, A., Agarwl, S., Meyarivan, T.: A Fast and Elitist Multiobjective Genetic Algorithm: NSGA-II. IEEE Transactions on Evolutionary Computation 6(2), 182–197 (2002)
16. Goldberg: Genetic Algorithms in Search, Optimization, and Machine Learning, 1st edn. Addison-Wesley Professional, Reading (1989)
17. Michalewicz, Z., Fogel, D.B.: How to Solve It: Modern Heuristics, 2nd edn. Springer, Heidelberg (2004)
18. Tsai, L.-C., Hsue, C.-W.: Dual-Band Bandpass Filters using Equal-Length Coupled-Serial-Shunted Lines and Z-Transform Technique. IEEE Transactions on Microwave Theory and Techniques 52(4), 1111–1117 (2004)
19. Zhang, H.: Compact, Reconfigurable, and dual-band microwave circuits. Doctorate Thesis, Dep. of Electronic and Computer Engineering, Hong Kong, pp. 1–187 (January 2007)

Fitting a Least Absolute Deviation Regression Model on Interval-Valued Data

André Luis Santiago Maia and Francisco de A.T. de Carvalho

Centro de Informática - CIn/UFPE, Av. Prof. Luiz Freire, s/n
Cidade Universitária, CEP: 50740-540 - Recife - PE - Brasil
{alsm3,fatc}@cin.ufpe.br

Abstract. This paper introduces a least absolute deviation (LAD) regression model suitable for manage interval-valued data. Each example of the data set is described by a feature vector where each feature value is an interval. In the approach, it is fitted two LAD regressions, respectively, on the mid-point and range of the interval values assumed by the variables. The prediction of the lower and upper bound of the interval value of the dependent variable is accomplished from its mid-point and range which are estimated from the fitted LAD regression models applied to the mid-point and range of each interval values of the independent variables. The evaluation of the proposed prediction method is based on the estimation of the average behaviour of root mean squared error and of the correlation coefficient in the framework of a Monte Carlo experience in comparison with the method proposed in [5].

Keywords: Interval-Valued Data, LAD Regression, Symbolic Data Analysis.

1 Introduction

The classical model of regression for usual data is used to predict the behaviour of a dependent variable Y as a function of other independent variables that are responsible for the variability of variable Y. To fit this model to the data, it is necessary the estimation of a vector $\boldsymbol{\beta}$ of parameters from the data vector $\mathbf{Y}$ and the model matrix $\mathbf{X}$, supposed with complete rank p. The estimation using the *method of least square* does not require any probabilistic hypothesis on the variable Y. This method consists of minimizing the sum of the square of residuals. However, the least squares linear regression estimator is well-known to be highly sensitive to unusual observations (outliers) in the data, and as a result many more robust estimators have been proposed as alternatives. One of the earliest proposals was regression performed through minimization of the ℓ_1 norm of the residuals, $\sum_{i=1}^{n} |y_i - \mathbf{x}_i\hat{\beta}|$, also called least-sum of absolute deviations (LAD) regression, where the regression coefficients are estimated through minimization of the sum of the absolute values of the residuals. There is a good deal of empirical evidence going back more than 30 years that LAD regression is more robust than least squares in the presence of fat-tailed errors (see, e.g., Sharpe, [6]).

G. Zaverucha and A. Loureiro da Costa (Eds.): SBIA 2008, LNAI 5249, pp. 207–216, 2008.

In classical data analysis, the items are usually represented as a vector of quantitative or qualitative measurements where each column represents a variable. In particular, each individual takes just one single value for each variable. In practice, however, this model is too restrictive to represent complex data since to take into account variability and/or uncertainty inherent to the data, variables must assume sets of categories or intervals, possibly even with frequencies or weights. The aim of *Symbolic Data Analysis* (SDA) is to extend classical data analysis techniques (clustering, factorial techniques, decision trees, etc.) to these kinds of data (sets of categories, intervals, or weight (probability) distributions) called symbolic data [1]. SDA is a domain in the area of knowledge discovery and data management related to multivariate analysis, pattern recognition and artificial intelligence.

In the framework of *Symbolic Data Analysis*, Billard and Diday [3] presented for the first time an approach to fitting a linear regression model to an interval-valued data-set. Their approach consists of fitting a linear regression model to the mid-point of the interval values assumed by the variables in the learning set and applies this model to the lower and upper bounds of the interval values of the independent variables to predict, respectively, the lower and upper bounds of the interval value of the dependent variable.

In [5], it is presented a new approach based on two linear regression models, the first regression model over the mid-points of the intervals and the second one over the ranges, which reconstruct the bounds of the interval-values of the dependent variable in a more efficient way when compared with the Billard and Diday method.

This paper introduces a new approach to fit a linear regression model for interval-valued data which, in the presence of outliers, is more robust than the approach proposed in [5]. In the proposed approach, analogous to [5], it is fitted two least absolute deviation (LAD) regression models, respectively, on the mid-point and range of the interval values assumed by the variables on the learning set. The prediction of the lower and upper bound of the interval value of the dependent variable is accomplished from its mid-point and range which are estimated from the fitted LAD regression models applied to the mid-point and range of each interval values of the independent variables.

Section 2 presents the approaches considered to fit a linear regression model to interval-valued data. Section 3 describes the framework of the Monte Carlo simulations and presents experiments with synthetic and real interval-valued data sets. Finally, section 4 gives the concluding remarks.

2 Linear Regression Models for Interval-Valued Data

In this section, we present the method introduced in [5] (named *center and range method* - CRM) and the method proposed in this paper (here named *center and range least absolute deviation method - CRMLAD*).

2.1 The Center and Range Method (CRM)

CRM method [5] fits two linear regression models, respectively, on the mid-point and range of the interval values assumed by the variables on the learning set.

Let $E = \{e_1, \ldots, e_n\}$ be a set of examples that are described by $p+1$ interval-valued variables Y, $X_1, \ldots, X_p$. Each example $e_i \in E$ $(i = 1, \ldots, n)$ is represented as an interval quantitative feature vector $\mathbf{z}_i = (\mathbf{x}_i, y_i)$, $\mathbf{x}_i = (x_{i1}, \ldots, x_{ip})$, where $x_{ij} = [a_{ij}, b_{ij}] \in \Im = \{[a, b] : a, b \in \Re, a \leq b\}$ $(j = 1, \ldots, p)$ and $y_i = [y_{Li}, y_{Ui}] \in \Im$ are, respectively, the observed values of X_j and Y.

Let Y^c and X_j^c $(j = 1, 2, \ldots, p)$ be, respectively, quantitative variables that assume as their value the midpoint of the interval assumed by the interval-valued variables Y and X_j $(j = 1, 2, \ldots, p)$. Also let Y^r and X_j^r $(j = 1, 2, \ldots, p)$ be, respectively, quantitative variables that assume as value the half range of the interval assumed by the interval-valued variables Y and X_j $(j = 1, 2, \ldots, p)$. This means that each example $e_i \in E$ $(i = 1, \ldots, n)$ is represented by two vectors $\mathbf{w}_i = (\mathbf{x}_i^c, y_i^c)$ and $\mathbf{r}_i = (\mathbf{x}_i^r, y_i^r)$, with $\mathbf{x}_i^c = (x_{i1}^c, \ldots, x_{ip}^c)$ and $\mathbf{x}_i^r = (x_{i1}^r, \ldots, x_{ip}^r)$, where $x_{ij}^c = (a_{ij} + b_{ij})/2$, $x_{ij}^r = (b_{ij} - a_{ij})/2$, $y_i^c = (y_{Li} + y_{Ui})/2$ and $y_i^r = (y_{Ui} - y_{Li})/2$ are, respectively, the observed values of X_j^c, X_j^r, Y^c and Y^r.

Consider Y^c and Y^r as dependent variables and X_j^c and X_j^r $(j = 1, 2, \ldots, p)$ as independent predictor variables that are related according to the following linear regression relationship:

$$y_i^c = \beta_0^c + \beta_1^c x_{i1}^c + \ldots + \beta_p^c x_{ip}^c + \epsilon_i^c, \tag{1}$$
$$y_i^r = \beta_0^r + \beta_1^r x_{i1}^r + \ldots + \beta_p^r x_{ip}^r + \epsilon_i^r. \tag{2}$$

Thus, in CRM, the *sum of squares of deviations* is given by

$$S(\boldsymbol{\beta}^c) = \sum_{i=1}^{n} (\epsilon_i^c)^2 = \sum_{i=1}^{n} (y_i^c - \beta_0^c - \beta_1^c x_{i1}^c - \ldots - \beta_p^c x_{i1}^c)^2, \tag{3}$$
$$S(\boldsymbol{\beta}^r) = \sum_{i=1}^{n} (\epsilon_i^r)^2 = \sum_{i=1}^{n} (y_i^r - \beta_0^r - \beta_1^r x_{i1}^r - \ldots - \beta_p^r x_{ip}^r)^2. \tag{4}$$

It is possible to find the values of β_0^c, $\beta_1^c, \ldots, \beta_p^c$ and β_0^r, $\beta_1^r, \ldots, \beta_p^r$ that minimize the above expressions, differentiating equations (3) and (4) with respect to the parameters and setting the results equal to zero.

The prediction of the lower and upper bound of the interval value of the dependent variable is accomplished from its mid-point and range which are estimated from the fitted linear regression models applied to the mid-point and range of each interval value of the independent variables.

Thus, given a new example e, described by $\mathbf{z} = (\mathbf{x}, y)$, $\mathbf{w} = (\mathbf{x}^c, y^c)$ and $\mathbf{r} = (\mathbf{x}^r, y^r)$, where $\mathbf{x} = (x_1, \ldots, x_p)$ with $x_j = [a_j, b_j]$, $\mathbf{x}^c = (x_1^c, \ldots, x_p^c)$ with $x_j^c = (a_j + b_j)/2$ and $\mathbf{x}^r = (x_1^r, \ldots, x_p^r)$ with $x_j^r = (b_j - a_j)/2$ $(j = 1, \ldots, p)$, the value $y = [y_L, y_U]$ of Y is predicted from the predicted values $\hat{y}^c$ of Y^c and $\hat{y}^r$ of Y^r, as follows:

$$\hat{y}_L = \hat{y}^c - (1/2)\hat{y}^r \text{ and } \hat{y}_U = \hat{y}^c + (1/2)\hat{y}^r, \tag{5}$$

where, $\hat{y}^c = (\tilde{\mathbf{x}}^c)^\top \hat{\boldsymbol{\beta}}^c$, $\hat{y}^r = (\tilde{\mathbf{x}}^r)^\top \hat{\boldsymbol{\beta}}^r$, $(\tilde{\mathbf{x}}^c)^\top = (1, x_1^c, \ldots, x_p^c)$, $(\tilde{\mathbf{x}}^r)^\top = (1, x_1^r, \ldots, x_p^r)$, $\hat{\boldsymbol{\beta}}^c = (\hat{\beta}_0^c, \hat{\beta}_1^c, \ldots, \hat{\beta}_p^c)^\top$ and $\hat{\boldsymbol{\beta}}^r = (\hat{\beta}_0^r, \hat{\beta}_1^r, \ldots, \hat{\beta}_p^r)^\top$.

2.2 The Center and Range Least Absolute Deviation Method (CRMLAD)

The CRM method minimizes the sum of squared errors. More robust parameter estimates can be obtained by minimizing the sum of absolute values of errors. This approach gives less weight to outliers. CRMLAD model fits also two linear regression models, respectively, on the mid-point and range of the interval values assumed by the variables on the learning set.

Consider again Y^c and Y^r as dependent variables and X_j^c and X_j^r $(j = 1, 2, \ldots, p)$ as independent predictor variables that are related according to the following linear regression relationship:

$$y_i^c = \beta_0^c + \beta_1^c x_{i1}^c + \ldots + \beta_p^c x_{ip}^c + \epsilon_i^c, \tag{6}$$

$$y_i^r = \beta_0^r + \beta_1^r x_{i1}^r + \ldots + \beta_p^r x_{ip}^r + \epsilon_i^r. \tag{7}$$

Thus, in CRMLAD, the *sum absolute deviation* is given by

$$T(\boldsymbol{\beta}^c) = \sum_{i=1}^{n} |\epsilon_i^c| = \sum_{i=1}^{n} |y_i - \beta_0^c - \beta_1^c x_{i1}^c - \ldots + \beta_p^c x_{ip}^c|, \tag{8}$$

$$T(\boldsymbol{\beta}^r) = \sum_{i=1}^{n} |\epsilon_i^r| = \sum_{i=1}^{n} |y_i - \beta_0^r - \beta_1^r x_{i1}^r - \ldots + \beta_p^r x_{ip}^r|. \tag{9}$$

It is possible to find the values of β_0^c, β_1^c,..., β_p^c and β_0^r, β_1^r,..., β_p^r that minimize the above expressions using a simplex based algorithm for ℓ_1-regression developped by Barrodale and Roberts [2]. This algorithm is a modification of the simplex method of linear programming applied to the primal formulation of the ℓ_1 problem.

According to Buchinsky [4], the representation of the LAD regression as a linear program has some important implications. Two of these implications are the guarantee of that the regression coefficients are estimated in a finite number of simplex iterations and the robustness of the vector of coefficients estimated in the presence of outliers in the dependent variable.

Thus, given a new example e, described by $\mathbf{z} = (x_1, \ldots, x_p, y)$, $\mathbf{z}^c = (x_1^c, \ldots, x_p^c, y^c)$ and $\mathbf{z}^r = (x_1^r, \ldots, x_p^r, y^r)$, the value $y = [y_L, y_U]$ of Y will be predicted from the predicted values y^c of Y^c and y^r of Y^r as follow:

$$\hat{y}_L = \hat{y}^c - (1/2)\hat{y}^r \text{ and } \hat{y}_U = \hat{y}^c + (1/2)\hat{y}^r,$$

where $\hat{y}^c = (\tilde{\mathbf{x}}^c)^\top \hat{\boldsymbol{\beta}}^c$, $\hat{y}^r = (\tilde{\mathbf{x}}^r)^\top \hat{\boldsymbol{\beta}}^r$, $(\tilde{\mathbf{x}}^c)^\top = (1, x_1^c, \ldots, x_p^c)$, $(\tilde{\mathbf{x}}^r)^\top = (1, x_1^r, \ldots, x_p^r)$, $\hat{\boldsymbol{\beta}}^c = (\hat{\beta}_0^c, \hat{\beta}_1^c, \ldots, \hat{\beta}_p^c)^\top$ and $\hat{\boldsymbol{\beta}}^r = (\hat{\beta}_0^r, \hat{\beta}_1^r, \ldots, \hat{\beta}_p^r)^\top$.

Thus, CRMLAD method differs from CRM method in that the sum of the absolute, not squared, deviations of the fit from the observed values is minimized to obtain estimates. CRMLAD method estimates the conditional median (0.5 regression quantile of $\mathbf{y}^c$ and $\mathbf{y}^r$) of the dependent variable given independent variables.

3 The Monte Carlo Experiences

Experiments with synthetic interval-valued data sets with different degrees of difficulty to fit a linear regression model are considered in this section as well as a cardiological interval-valued data set. The results of the proposed approach are compared with the method presented in [5].

3.1 Simulated Interval-Valued Data Sets

Initially, it is considered standard continuous quantitative data sets in $\Re^2$ and in $\Re^4$. Each data set (in $\Re^2$ or in $\Re^4$) has 375 points partitioned in a learning set (250 points) and a test set (125 points). Each data point, in $\Re^2$ or in $\Re^4$, belonging to a standard data set is a seed for an interval data set (a rectangle in $\Re^2$ or hypercube in $\Re^4$) and in this way, from these standard data sets it is obtained the interval data sets.

The construction of the standard data sets and of the corresponding interval-valued data sets is accomplished in the following steps:
s1): Let us suppose that each random variables X_j^c ($j = 1$ if the data is in $\Re^2$ or $j = 1, 2, 3$ if the data is in $\Re^4$), which assume as value the mid-point of the interval value assumed by the interval-valued variables X_j ($j = 1, 2, 3$), is uniformly distributed in the interval $[a, b]$; at each iteration it is randomly selected 375 values of each variable X_j^c, which are the mid-points of these intervals;
s2): The random variable Y^c, which assume as value the mid-point of the interval value assumed by the interval-valued variable Y, is supposed to be related to variables X_j^c according to $Y^c = (\mathbf{X}^c)^\top \beta + \epsilon$, where $(\mathbf{X}^c)^\top = (1, X_1^c)$ (if the data is in $\Re^2$) or $(\mathbf{X}^c)^\top = (1, X_1^c, X_2^c, X_3^c)$ (if the data is in $\Re^4$), $\beta = (\beta_0 = 2, \beta_1 = 4)^\top$ (if the data is in $\Re^2$) or $\beta = (\beta_0 = 1, \beta_1 = 2, \beta_2 = 3, \beta_3 = 4)^\top$ (if the data is in $\Re^4$) and $\epsilon = U[c, d]$; the mid-points of these 375 intervals are calculated according this linear relation;
s3): Once obtained the mid-points of the intervals, let us consider now the range of each interval. Let us suppose that each random variable Y^r, X_j^r ($j = 1, 2, 3$), which assume as value, respectively, the range of the interval assumed by the interval-valued variables Y and X_j ($j = 1, 2, 3$), is uniformly distributed, respectively, in the intervals $[e, f]$ and $[g, h]$; at each iteration it is randomly selected 375 values of each variable Y^r, X_j^r, which are the range of these intervals;
s4): At each iteration, the interval-valued data set is partitioned in a learning (250 observations) and test (125 observations) set.

Table 1 shows four different configurations for the interval data sets which are used to compare de performance of the CRM and CRMLAD methods in different situations.

Moreover, each one of these data configurations was carried out with a different number of outliers n_{out} in the variable Y^c. In this paper, the number of outliers has been arbitrarily selected among 0, 1 and 25. Then, n_{out} observations of Y^c have been randomly selected and substituted in the data set by four times its value.

Table 1. Data set configurations

C_1	$X_j^c \sim U[20, 40]$	$X_j^r \sim U[20, 40]$	$Y^r \sim U[20, 40]$	$\epsilon \sim U[-20, 20]$
C_2	$X_j^c \sim U[20, 40]$	$X_j^r \sim U[20, 40]$	$Y^r \sim U[20, 40]$	$\epsilon \sim U[-5, 5]$
C_3	$X_j^c \sim U[20, 40]$	$X_j^r \sim U[1, 5]$	$Y^r \sim U[1, 5]$	$\epsilon \sim U[-20, 20]$
C_4	$X_j^c \sim U[20, 40]$	$X_j^r \sim U[1, 5]$	$Y^r \sim U[1, 5]$	$\epsilon \sim U[-5, 5]$

These configurations take into account the combination of two factors (range and error on the mid-points) with two degrees of variability (low and high): low variability range ($U[1, 5]$), high variability range ($U[20, 40]$), low variability error ($U[-5, 5]$) and high variability error ($U[-20, 20]$).

The configuration C_1, for example, represents observations with a high variability range and with a poor linear relationship between Y and X_1, X_2 and X_3 due the high variability error on the mid-points. Figure 1 shows the configuration C_1 when the data is in $\Re^2$. In the other hand, the configuration C_4 represent observations with a low variability range and with a rich linear relationship between Y and X_1, X_2 and X_3 due the low variability error on the mid-points.

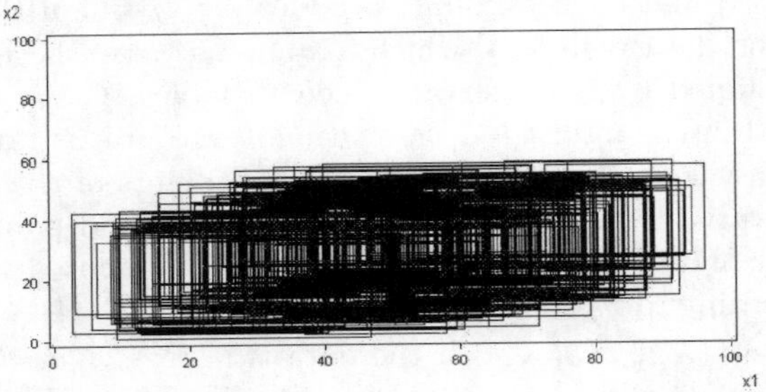

Fig. 1. Configuration C_1 showing a poor linear relationship between Y and X_1

3.2 Experimental Evaluation

The evaluation of the performance of the linear regression models (CRM and CRMLAD approaches) is based on the following measures: the *lower bound root mean-squared error* ($RMSE_L$), the *upper bound root mean-squared error* ($RMSE_U$), the *lower bound correlation coefficient* (r_L) and the *upper bound correlation coefficient* (r_U).

These measures are estimated for the CRM and CRMLAD methods in the framework of a Monte Carlo simulation with 100 replications for each independent test interval-valued data set, for each of the four fixed configurations, as well as for different numbers of independent variables in the model matrix $\mathbf{X}$ and different numbers of outliers in the dependent variables Y^c. At each replication, a linear regression model is fitted on the training interval-valued data set for each method and these models are used to predict the interval values of the dependent variable Y on the test interval-valued data set and these measures are calculated. Finally, it is calculated, for each measure, its average and standard deviation.

Table 2. Comparision between CRM and CRMLAD methods - Average and standard deviation of each measure; no outliers in Y^c ($n_{out} = 0$)

Conf.	p	Stat.	$RMSE_L$		$RMSE_U$		r_L (%)		r_U (%)	
			CRM	CRM LAD	CRM	CRM LAD	CRM	CRM LAD	CRM	CRM LAD
C_1	1	$\overline{x}$	12.01	12.09	12.01	12.09	88.76	88.76	88.78	88.77
		S	0.55	0.56	0.51	0.53	1.45	1.45	1.29	1.29
	3	$\overline{x}$	11.82	11.99	11.83	12.01	93.48	93.38	93.44	93.34
		S	0.47	0.50	0.47	0.53	0.99	1.02	0.95	0.98
C_2	1	$\overline{x}$	4.08	4.11	4.11	4.14	98.49	98.49	98.47	98.47
		S	0.22	0.23	0.21	0.22	0.19	0.19	0.21	0.21
	3	$\overline{x}$	4.10	4.16	4.13	4.19	99.14	99.12	99.13	99.11
		S	0.23	0.24	0.20	0.21	0.14	0.14	0.13	0.13
C_3	1	$\overline{x}$	11.53	11.63	11.52	11.62	89.56	89.56	89.56	89.56
		S	0.43	0.46	0.42	0.45	1.22	1.22	1.23	1.23
	3	$\overline{x}$	11.63	11.79	11.62	11.77	93.60	93.52	93.63	93.54
		S	0.49	0.50	0.50	0.51	0.98	1.00	0.97	0.98
C_4	1	$\overline{x}$	2.95	2.97	2.97	2.99	99.19	99.19	99.18	99.18
		S	0.13	0.14	0.13	0.14	0.09	0.09	0.09	0.09
	3	$\overline{x}$	2.94	2.98	2.92	2.96	99.55	99.54	99.56	99.55
		S	0.13	0.13	0.13	0.14	0.06	0.07	0.07	0.07

Tables 2, 3 and 4 presents the results of this Monte Carlo experience for a specific vector of parameters $\boldsymbol{\beta}$. When there is no outliers in the data set (see Table 2), the methods CRM and CRMLAD present almost the same performance according to the selected measures.

However, when there are outliers in the data set (Tables 3 and 4), CRMLAD method clearly outperforms CRM method concerning the *mean-squared error* and *correlation coefficient*.

These tables show also that CRMLAD is robust concerning the number of outiliers regardless the number of independent variables in the model. Unfortunatly, this is not the case concerning CRM model. In conclusion, both methods

Table 3. Comparision between CRM and CRMLAD methods - Average and standard deviation of each measure; one outlier in Y^c ($n_{out} = 1$)

Conf.	p	Stat.	$RMSE_L$		$RMSE_U$		r_L (%)		r_U (%)	
			CRM	CRM LAD	CRM	CRM LAD	CRM	CRM LAD	CRM	CRM LAD
C_1	1	$\overline{x}$	12.12	12.02	12.16	12.08	88.91	88.90	88.89	88.88
		S	0.62	0.59	0.63	0.60	1.40	1.39	1.52	1.53
	3	$\overline{x}$	13.67	12.23	13.67	12.23	92.19	93.01	92.14	92.99
		S	1.12	0.60	1.07	0.61	1.37	0.94	1.35	0.95
C_2	1	$\overline{x}$	4.56	4.12	4.58	4.10	98.48	98.48	98.49	98.49
		S	0.48	0.23	0.46	0.22	0.23	0.23	0.19	0.19
	3	$\overline{x}$	7.47	4.14	7.47	4.15	98.15	99.14	98.18	99.13
		S	1.46	0.23	1.42	0.21	0.88	0.13	0.86	0.14
C_3	1	$\overline{x}$	11.87	11.70	11.87	11.70	89.35	89.35	89.35	89.35
		S	0.55	0.46	0.57	0.48	1.27	1.27	1.29	1.29
	3	$\overline{x}$	13.45	11.77	13.45	11.76	92.74	93.72	92.74	93.73
		S	1.13	0.51	1.12	0.51	1.41	0.98	1.41	0.99
C_4	1	$\overline{x}$	3.51	2.95	3.51	2.96	99.21	99.21	99.21	99.21
		S	0.42	0.13	0.41	0.14	0.09	0.09	0.09	0.09
	3	$\overline{x}$	7.14	2.99	7.14	2.99	98.44	99.54	98.44	99.54
		S	1.52	0.12	1.51	0.12	0.87	0.07	0.87	0.07

Table 4. Comparision between CRM and CRMLAD methods - Average and standard deviation of each measure; twenty five outliers in Y^c ($n_{out} = 25$)

Conf.	p	Stat.	$RMSE_L$		$RMSE_U$		r_L (%)		r_U (%)	
			CRM	CRM LAD	CRM	CRM LAD	CRM	CRM LAD	CRM	CRM LAD
C_1	1	$\overline{x}$	39.19	12.16	39.29	12.25	88.85	88.84	88.77	88.76
		S	2.82	0.68	2.73	0.56	1.47	1.47	1.45	1.45
	3	$\overline{x}$	86.64	12.44	86.70	12.48	75.00	93.14	75.02	93.07
		S	4.73	0.77	4.70	0.62	23.69	0.85	23.41	0.79
C_2	1	$\overline{x}$	37.99	4.17	37.96	4.16	98.45	98.44	98.46	98.46
		S	3.02	0.24	2.95	0.24	0.22	0.22	0.20	0.20
	3	$\overline{x}$	85.47	4.26	85.46	4.22	83.83	99.09	83.85	99.10
		S	5.11	0.24	5.18	0.25	15.33	0.13	15.41	0.14
C_3	1	$\overline{x}$	39.72	11.87	39.71	11.87	89.56	89.56	89.58	89.58
		S	3.07	0.58	3.07	0.58	1.28	1.28	1.29	1.29
	3	$\overline{x}$	86.25	12.12	86.20	12.12	78.37	93.47	78.38	93.46
		S	4.88	0.66	4.89	0.66	13.78	0.97	13.76	0.98
C_4	1	$\overline{x}$	38.24	3.04	38.26	3.05	99.19	99.19	99.18	99.18
		S	3.17	0.16	3.18	0.17	0.09	0.08	0.10	0.10
	3	$\overline{x}$	84.85	3.05	84.86	3.07	85.83	99.54	85.84	99.53
		S	4.72	0.15	4.70	0.17	12.40	0.07	12.42	0.07

presented almost the same performance when there is no outliers in the dependent interval-valued variable. However, CRMLAD clearly outperform CRM method when outliers are present and, moreover, it is robust concerning the number of outliers.

3.3 Cardiological Interval-Valued Data Set

This data set (Table 5) concerns the record of the pulse rate Y, systolic blood pressure X_1 and diastolic blood pressure X_2 for each of eleven patients [3]. The aim is to predict the interval values y of Y (the dependent variable) from x_j ($j = 1, 2$) through a linear regression model.

The fitted linear regression models to the CRM and CRMLAD methods in this symbolic interval data set are presented below:
CRM: $\hat{y}^c = 21.17 + 0.33x_1^c + 0.17x_2^c$ and $\hat{y}^r = 20.21 - 0.15x_1^r + 0.35x_2^r$;
CRMLAD: $\hat{y}^c = 21.35 + 0.35x_1^c + 0.13x_2^c$ and $\hat{y}^r = 14.00 - 0.20x_1^r + 0.60x_2^r$.

Table 5. Cardiological interval data set

u	Pulse rate	Systolic blood pressure	Diastolic blood pressure
1	[44 - 68]	[90 - 100]	[50 - 70]
2	[60 - 72]	[90 - 130]	[70 - 90]
3	[56 - 90]	[140 - 180]	[90 - 100]
4	[70 - 112]	[110 - 142]	[80 - 108]
5	[54 - 72]	[90 - 100]	[50 - 70]
6	[70 - 100]	[130 - 160]	[80 - 110]
7	[72 - 100]	[130 - 160]	[76 - 90]
8	[76 - 98]	[110 - 190]	[70 - 110]
9	[86 - 96]	[138 - 180]	[90 - 110]
10	[86 - 100]	[110 - 150]	[78 - 100]
11	[63 - 75]	[60 - 100]	[140 - 150]

Table 6. Predicted interval values of the dependent variable *pulse rate*, according to CRM and CRMLAD methods

Interval data set with no outlier		Interval data set with 1 outlier	
CRM	CRMLAD	CRM	CRMLAD
[50 - 75]	[51 - 75]	[117 - 142]	[54 - 78]
[60 - 82]	[62 - 80]	[97 - 118]	[64 - 82]
[81 - 99]	[85 - 97]	[69 - 86]	[85 - 97]
[66 - 91]	[66 - 91]	[78 - 103]	[67 - 92]
[50 - 75]	[51 - 75]	[117 - 142]	[54 - 78]
[72 - 98]	[72 - 98]	[70 - 96]	[73 - 99]
[73 - 93]	[76 - 92]	[82 - 103]	[76 - 93]
[75 - 97]	[75 - 97]	[74 - 96]	[76 - 98]
[80 - 101]	[82 - 100]	[63 - 84]	[82 - 100]
[68 - 90]	[70 - 89]	[82 - 104]	[71 - 90]
[63 - 81]	[63 - 75]	[56 - 74]	[63 - 75]

Table 7. Performance of the methods in the cardiological interval data set with 1 outlier and with no outlier

Interval data set without outliers				
Method	$RMSE_L$	$RMSE_U$	r_L (%)	r_U (%)
CRM	9.81	8.94	64.45	79.59
CRMLAD	10.44	8.66	61.43	81.17
Inteval data set with 1 outlier				
Method	$RMSE_L$	$RMSE_U$	r_L (%)	r_U (%)
CRM	32.35	34.28	-59.48	-49.34
CRMLAD	10.55	8.80	60.44	80.66

In order to show the interest of the least absolute deviation approach, we introduced an outlier in the cardiological interval-valued data set and we applied CRM and CRMLAD models on this modified cardiological interval-valued data set. The outlier has been introduced in the description of the first individual for the dependent variable where we change the value $y_1 = [44, 68]$ by value $\tilde{y}_1 = [212, 236]$.

Table 6 show the predicted interval values of the dependent variable (*pulse rate*) for CRM and CRMLAD models applied on the cardiologial interval-valued data set and for these models applied on this interval data set with one outlier. Notice that the interval values predicted by CRM repression model are strongly affected by the introduced outlier. This is not the case for the CRMLAD model.

The performance of the CRM and CRMLAD methods is evaluated also through the calculation of $RMSE_L$, $RMSE_U$, r_L and r_U measures on the cardiological interval data set with 1 outlier and without outlier. Table 7 shows the results. Note that these mesures are only slightly affected in the case of CRMLAD regression model. This is not the case for the CRM regression model.

4 Concluding Remarks

In this paper, we presented a new method to fit a linear regression model on interval-valued data. The proposed approach fits two least absolute deviation regression models, respectively, on the mid-point and range of the interval values assumed by the variables on the learning set. The prediction of the lower and upper bound of the interval value of the dependent variable is accomplished from its mid-point and range which are estimated from the fitted least absolute deviation regression models applied to the mid-point and range of each interval values of the independent variables. Thus, with these two fitted regression models it was possible the reconstruction of the intervals bounds in a suitable way.

Monte Carlo simulations with synthetic data and an application with a cardiological interval-valued data set showed the superiority of the method introduced in this paper, measured by the average behavior of the *root mean-squared error* and of the *correlation coefficient*, when compared with the method proposed by [5] in the case of presence of outliers in the dependent variable.

Acknowledgments. The authors would like to thank CNPq, CAPES and FACEPE (Brazilian Agencies) for their financial support.

References

1. Bock, H.H., Diday, E.: Analysis of Symbolic Data, Exploratory methods for extracting statistical information from complex data. Springer, Heidelberg (2000)
2. Barrodale, I., Roberts, F.D.K.: Solution of an overdetermined system of equation in the ℓ_1 norm. Communications of the Association for Computing Machinery 17, 319–320 (1974)
3. Billard, L., Diday, E.: Regression Analysis for Interval-Valued Data. In: Kiers, H.A.L., et al. (eds.) Data Analysis, Classification and Related Methods: Proceedings of the IFCS-2000, Namur, Belgium, vol. 1, pp. 369–374. Springer, Heidelberg (2000)
4. Buchinsky, M.: Recent advances in quantile regression: a pratical guideline for empirical research. Journal of Human Resources 33, 88–126 (1997)
5. Lima Neto, E.A., De Carvalho, F.A.T.: Centre and Range method for fitting a linear regression model to symbolic interval data. Computational Statistics and Data Analysis 52, 1500–1515 (2008)
6. Sharpe, W.F.: Mean-absolute-deviation characteristic lines for securities and portfolios. Management Science 18, 1–13 (1971)

Missing Value Imputation Using a Semi-supervised Rank Aggregation Approach

Edson T. Matsubara, Ronaldo C. Prati,
Gustavo E.A.P.A. Batista, and Maria C. Monard

Institute of Mathematics and Computer Science at University of São Paulo
P. O. Box 668, ZIP Code 13560-970, São Carlos, SP, Brazil
{edsontm,prati,gbatista,mcmonard}@icmc.usp.br

Abstract. One relevant problem in data quality is the presence of missing data. In cases where missing data are abundant, effective ways to deal with these absences could improve the performance of machine learning algorithms. Missing data can be treated using imputation. Imputation methods replace the missing data by values estimated from the available data. This paper presents CORAI, an imputation algorithm which is an adaption of CO-TRAINING, a multi-view semi-supervised learning algorithm. The comparison of CORAI with other imputation methods found in the literature in three data sets from UCI with different levels of missingness inserted into up to three attributes, shows that CORAI tends to perform well in data sets at greater percentages of missingness and number of attributes with missing values.

1 Introduction

Machine learning (ML) algorithms usually take a set of cases as input (also known as examples or instances) to generate a model. These cases are generally represented by a vector, where each vector position represents the value of an attribute (feature) of a given case. However, in many applications of ML algorithms in real world data sets, some attribute values might be missing. For example, patient data may contain unknown information due to tests which were not taken, patients' refusal to answer certain questions, and so on.

In cases where missing data are abundant, effective ways to deal with these absences could improve the performance of ML algorithms. One of the most common approaches of dealing with missing values is imputation [1]. The main idea of imputation methods is that, based on the values present in the data set, missing values can be guessed and replaced by some plausible values. One advantage of this approach is that the missing data treatment is independent of the learning algorithm used, enabling the user to select the most suitable imputation method for each situation before the application of the learning algorithm.

A closely related research topic that has emerged as exciting research in ML over the last years is Semi-Supervised Learning (SSL) [2]. To generate models, SSL aims to use both labeled (*i.e.*, cases where the values of the class attribute, that is a special attribute we are interested in predicting based on the others

G. Zaverucha and A. Loureiro da Costa (Eds.): SBIA 2008, LNAI 5249, pp. 217–226, 2008.

attributes, are known in advance when generating the model) and unlabeled data (*i.e.*, cases where the values of the class attribute are not known when generating the model). To accomplish this task, some SSL algorithms attempt to infer the label of the unlabeled cases based on few labeled cases. Therefore, in a broad sense, these SSL algorithms might be seen as "imputation methods for the class attribute", which is the view considered in this work.

In this work we propose an algorithm named CORAI (CO-TRAINING Ranking Aggregation Imputation), which is an adaptation of the SSL algorithm CO-TRAINING, that can be used to deal with missing data. The comparison of CORAI with other imputation methods found in the literature in three data sets from UCI with different levels of missingness inserted into up to three attributes, assessed in terms of imputation error rate, shows that CORAI tends to perform well in data sets at greater percentages of missingness and number of attributes with missing values.

The outline of this paper is as follows: Section 2 presents related work. Section 3 describes CORAI. Section 4 presents the experimental results and Section 5 concludes this paper.

2 Related Work

According to the dependencies among the values of the attributes and the missingness, missing values can be divided into three groups [1]: (1) missing completely at random (**MCAR**) is the highest level of randomness and occurs where missingness of attribute values is independent of the values (observed or not); (2) missing at random (**MAR**) occurs when the probability of a case having a missing value may depend on the known values, but not on the value of the missing data itself; (3) not missing at random (**NMAR**) occurs when the probability of a case having a missing value for an attribute could depend on the value of that attribute.

A straightforward way to deal with missing values is to completely discard the cases and/or the attributes where missing values occur. Removing the cases is the most standard approach although, in case the missing values are concentrated into a few attributes, it may be interesting to remove them instead of removing the cases. Case and attribute removal with missing data should be applied only if missing data are MCAR, as not MCAR missing data have non-random elements, which can make the results biased.

Another approach is to fill in the missing data by guessing their values [3]. This method, known as imputation, can be carried out in a rather arbitrary way by imputing the same value to all missing attribute values. Imputation can also be done based on the data distribution inferred from known values, such as the "cold-deck/hot-deck" approach [4], or by constructing a predictive model based on the other attributes. An important argument in favor of this latter approach is that attributes usually have correlations among themselves. Therefore, these correlations could be used to create a predictive model for attributes with missing data. An important drawback of this approach is that the model estimated values are usually more well-behaved than the true values would be. In other words, since the missing values are predicted from a set of attributes, the predicted values are likely to be more consistent with this set of attributes than the true (not known) values are. A second drawback is the requirement for correlation among the attributes. If there are no relationships among other attributes in the data set and the attribute with missing data, then the model will not be appropriate for estimating missing values.

As already mentioned, some SSL algorithms can be viewed as a way of "guessing the class" of a set of unlabeled cases. SSL algorithms have recently attracted considerable attention from the ML community, and numerous SSL approaches have been proposed (see [5] for an up-to-date review on the subject). In this paper, we are interested in investigating whether SSL approaches might be used to deal with the missing data problem. To the best of our knowledge, the only algorithm that is used to handle both missing data and SSL problems is Expectation Maximization [6]. In this paper, however, we are interested in a special family of SSL that can take advantage of alternative predictive patterns in the training data, such as multi-view SSL algorithms [7,8,9]. Our research hypothesis is that, by exploiting these alternative predictive patterns, missing data can be imputed in a better way than other methods.

3 Proposed Method

Let $X = A_1 \times ... \times A_M$ be the instance space over the set of attributes $\{A_1, ..., A_M\}$, and let $y \in Y = \{y_1, ..., y_Z\}$ be the class attribute. Assume that instances $(\mathbf{x}, y)$, where $\mathbf{x} \in X$ and $y \in Y$, are drawn from an unknown underlying distribution D. The supervised learning problem is to find $h : X \to Y$ from a training set of labeled examples $L = \{(\mathbf{x}_l, y_l) : l = 1, ..., n\}$ that are drawn from D. In semi-supervised learning, we also have unlabeled data $U = \{\mathbf{x}_u : u = n + 1, ..., N\}$ in the training set that are drawn from D without their corresponding class y_u. In our problem, some examples may have attributes with missing values and those attributes are denoted as A_i^*, where $dom(A_i^*) = dom(A_i) \cup \{\text{"?"}\}$ and "?" denotes a missing value. An instance space which contains A_i^* is represented as X^*. The imputation method to deal with missing values is to find $h^* : (X^*, Y) \to X$ which can be used to map all A_i^* back to A_i.

Numerous approaches can be used to construct h^*. Among them are predictive models, which can be used to induce relationships between the available attribute values and the missing ones. In this paper, we propose to adapt SSL algorithms to deal with missing data by imputation by considering each attribute A_i^* that has missing values as the class attribute into a SSL algorithm[1]. Therefore, examples which do not have missing values in A_i^* are treated as "labeled" attributes and examples with missing values are treated as "unlabelled."

Numerous SSL algorithms have been proposed in recent years. In this work, we have selected CO-TRAINING [7], a well known SSL algorithm, which was the first

[1] As we are dealing with missing value imputation as a semi-supervised classification problem, in this work we restrict the domain of A_i^* to be qualitative.

Algorithm 1. CORAI

Input: L, U
Output: L
Build U' ;
$U = U - U'$;
while *stop criteria do not met* **do**
  Induce h_1 from L;
  Induce h_2 from L;
  $R'_1 = h_1(U')$;
  $R'_2 = h_2(U')$;
  $R = bestExamples(R'_1, R'_2)$;
  $L = L \cup R$;
  if $U_1 = \emptyset$ **then return**(L) **else**
    Randomly select examples from U to replenish U';
  end
end
return(L_1);

to introduce the notion of multi-view learning in this area. Multi-view learning is applied in domains which can naturally be described using different views. For instance, in web-page classification, one view might be the text appearing on the page itself and a second view might be the words appearing on hyperlinks pointing to this page from other pages on the web. CO-TRAINING assumes compatible views (examples in each view have the same class label) and each different view has to be in itself sufficient for classification. Although it is not always possible to find different views on data sets which meet these assumptions, we can only use one view and different learning algorithms to compose the views. This is an idea proposed in [8] which extends CO-TRAINING for problems restricted to one view data sets.

The main differences between CORAI and CO-TRAINING are the use of a different strategy to select the best examples to be labeled on each iteration and the use of two learning algorithms rather than two views. Our method can be described as follows: initially, a small pool of examples U' withdrawn from U are created, and the main loop of Algorithm 1 starts. First, the set of labeled examples L are used to induce two classifiers h_1 and h_2 using two different learning algorithms (in our case NAÏVE BAYES and C4.5). Next, the subset of unlabeled examples U' is labeled using h_1 and inserted into R'_1, and U' is used again but now it is labeled using h_2 and inserted into R'_2. Both sets of labeled examples are given to the function *bestExamples* which is responsible for ranking good examples according to some criterion and inserting them into L. After that, the process is repeated until a stop criteria is met.

We also modify the *bestExamples* function as proposed in the CO-TRAINING method. Originally, this function first filters examples which disagrees with their classification, *i.e*, $h_1(\mathbf{x}) \neq h_2(\mathbf{x})$. However, attributes with missing values may assume many different values, and when examples are filtered, $h_1(x) \neq h_2(x)$ for almost all examples. This occurs because it is less likely that classifiers agree with their classification in multi-class problems rather than binary problems. To deal with this problem, we proposed the use of ranking aggregation to select the best examples.

From now on, assume that the missing value problem has been mapped to a semi-supervised learning problem by swapping the class attribute with an attribute A_i^*. Thus, the reader should be aware that Y is actually referring to A_i^*.

First, we need to define classification in terms of *scoring classifiers*. Scoring classifiers maps $s : X \rightarrow \mathbb{R}^{|Y|}$, assigning a numerical score $s(\mathbf{x})$ to each instance $\mathbf{x} \in X$ for each class. NAÏVE BAYES actually computes a sort of scoring classifier where the classification is given by the class with the largest score. Decision trees can also be adapted to output scores by counting the distribution of examples for each possible classification in their leaves.

A rank aggregation combines the output of scoring classifiers on the same set of examples in order to obtain a "better" ordering. In this paper, rank aggregation uses two scoring classifiers obtained from two sets of examples R_1' and R_2' scored by h_1 and h_2, respectively (Algorithm 1). Let y_{1z} be the scores given by h_1 and y_{2z} be the scores given by h_2 for the class y_z ($z = 1...Z$). The method which implements best examples orders examples according to scores for each class and compute $rpos_{1z}$ which is the rank position of an instance with regards to y_z on R_1'. For instance, to compute $rpos_{11}$, initially the instances according to y_{11} are ordered and then the rank position from this ordering is stored in $rpos_{11}$. This is done for all classes $y_{11}, ..., y_{1Z}$ to obtain $rpos_{11}, ..., rpos_{1Z}$. In the same way, the method uses R_2' to compute $rpos_{2z}$. Finally, the rank position obtained from R_1' and R_2' is given by $rpos_z = rpos_{1z} + rpos_{2z}$ for each class y_z, and the selected instances are the ones with low $rpos_z$. Taking the instances with low $rpos_z$ means that these examples have a good (low) rank position on average, which is similar to selecting the examples with high confidence in a ranking perspective. In our implementation, we preserve the class distribution observed in the initial labeled data by selecting the number of examples proportional to this distribution.

4 Experimental Analysis

4.1 Experimental Setup

The experiments were carried out using three data sets from UCI Machine Learning Repository [10]. Originally, all data sets had no missing values and missing data were artificially implanted into the data sets. The artificial insertion of missing data allows a more controlled experimental setup. First of all, we can control the pattern of missing data. In this work, missing data were inserted in the MCAR pattern. Secondly, as the values replaced by missing data are known, imputation errors can be measured. Finally, this experimental setup allows the missing data to be inserted using different rates and attributes.

Table 1 summarizes the data sets used in this study. It shows, for each data set, the number of examples (#Examples), number of attributes (#Attributes),

Table 1. Data sets summary description

Data set	# Examples	#Attributes (quanti., quali.)	#Classes	Majority error
CMC	1473	9 (2,7)	3	57.29%
German	1000	20 (7,13)	2	30.00%
Heart	270	13 (7,6)	2	44.44%

together with the number of quantitative and qualitative attributes, number of classes (#Classes), and the majority class error.

Ten-fold cross-validation was used to partition the data sets into training and test sets. Finally, missing values were inserted into the training sets. Missing values were inserted in 20%, 40%, 60% and 80% of the total number of examples of the data set. In addition, missing values were inserted in one, two or three attributes. In order to choose in which attributes to implant missing data, we conducted some experiments to select attributes that are relevant to predict the class attribute. Observe that it is important to insert missing values into relevant attributes, otherwise the analysis might be hindered by dealing with non-representative attributes which will not be incorporated into the classifier by a learning algorithm such as a decision tree. Since finding the most representative attributes of a data set is not a trivial task, three feature subset algorithms, available in Weka software [11] were used: Chi-squared ranking filter; Gain ratio feature evaluator and ReliefF ranking filter. All three feature selection algorithms generate a ranking of the most representative attributes. For each data set the three rankings were composed into an average ranking, and the three top ranked qualitative attributes were chosen. Table 2 shows the selected attributes for each data set, as well as the number of values (#Values) of each attribute.

As mentioned before, missing values were inserted in 20%, 40%, 60% and 80% of the total number of examples for one (the attribute selected in first place), two (the attributes selected in first and second places) and the three selected attributes (the attributes in first, second and third places). Inserting missing values into more than one attribute is performed independently for each attribute. For instance, 20% of the missing values inserted into two attributes means that, for each attribute, two independent sets with 20% of examples each were sampled. In other words, the first set's values were altered to missing for

Table 2. Selected attributes for data set

Data set	Selected Attributes (Position) Name	#Values
CMC	1st (1) wife education	4
	2nd (2) husband education	4
	3th (7) standard of living	4
German	1st (0) status	4
	2nd (2) credit history	5
	3th (5) savings account	5
Heart	1st (12) thal	3
	2nd (2) chest pain	4
	3th (8) angina	2

the first attribute, similarly, the second set's values were altered to missing for the second attribute. As the two sets are independently sampled, some examples may have missing values in one, two or none of the selected attributes. A similar procedure was performed when missing data was inserted into three attributes.

Our experimental analysis involves the following methods to deal with missing data: CORAI, the proposed method; 9NNI [3], an imputation method based on k-nearest neighbor; and mode imputation, an imputation method that substitutes all the missing data by the attributes' most frequent value. In order to deal with missing values in multiple attributes, CORAI is executed independently for each attribute. In each execution, one different attribute with missing data is considered as class, and all other attributes are left in the data set to build the classification model.

4.2 Experimental Results

The main purpose of our experimental analysis is to evaluate the imputation error of the proposed method compared with other methods in the literature. As stated before, missing data were artificially implanted and this procedure allows us to compare the imputed values with the true values. Imputation error rate was calculated for each possible attribute value.

In order to analyze whether there is a difference among the methods, we ran the Friedman test[2]. Due to lack of space, only results of these tests are reported here[3]. Friedman test was ran with four different null-hypotheses: (1) that the performance of all methods are comparable considering all results; (2) that the performance of all methods are comparable for each percentage of missing data; (3) that the performance of all methods are comparable for each amount of attributes with missing data; (4) that the performance of all methods are comparable for each percentage of missing data and amount of attributes with missing data. When the null-hypothesis is rejected by the Friedman test, at 95% of confidence level, we can proceed with a post-hoc test to detect which differences among the methods are significant. We ran the Bonferroni-Dunn multiple comparison with a control test, using CORAI as a control. Therefore, the Bonferroni-Dunn test points-out whether there is a significant difference among CORAI and the other methods involved in the experimental evaluation.

Figure 1 shows the results of the Bonferroni-Dunn test with our first null-hypothesis: that the performance of all methods are comparable considering all results. This test does not make any distinction among percentage of missing data or amount of attributes with missing data. As seen in Figure 1, CORAI performs best, followed by 9NNI and MODE. The Bonferroni-Dunn test points out that CORAI outperforms MODE, but there is no significant difference between CORAI and 9NNI.

---

[2] The Friedman test is the nonparametric equivalent of the repeated-measures ANOVA. See [12] for a thorough discussion regarding statistical tests in machine learning research.

[3] All tabulated results can be found in `http://www.icmc.usp.br/~gbatista/corai/`

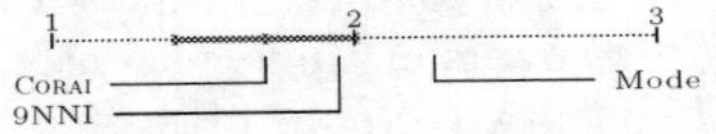

Fig. 1. Results of the Bonferroni-Dunn test on all imputation errors. The thick line marks the interval of one critical difference, at 95% confidence level.

The second null-hypothesis is that all methods perform comparably for each percentage of missing data. The objective is to analyze whether some methods perform better than others when the percentage of missing values varies, or in a critical situation, when the percentage of missing values is high. Figure 2 shows the results of the Bonferroni-Dunn test with our second null-hypothesis. CORAI frequently outperforms the other methods.

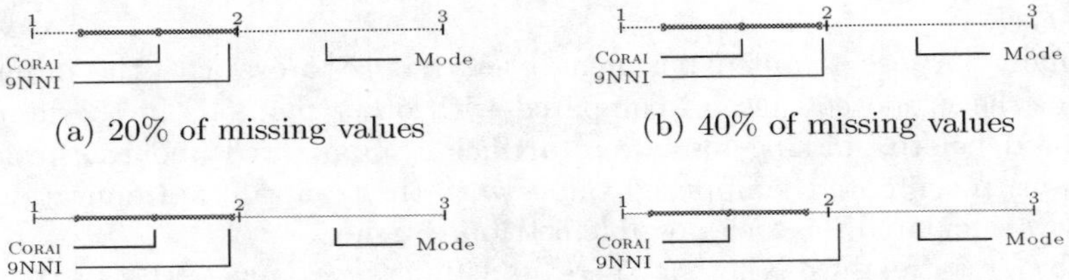

(a) 20% of missing values

(b) 40% of missing values

(c) 60% of missing values

(d) 80% of missing values

Fig. 2. Results of the Bonferroni-Dunn test considering the percentage of missing values. The thick line marks the interval of one critical difference, at 95% confidence level.

Our third null-hypothesis is that all methods perform comparably for different amounts of attributes with missing data. The objective is to analyze whether some methods perform better than others when the number of attributes with missing values increases. Figure 3 shows the results of the Bonferroni-Dunn test. CORAI obtained the lowest imputation error followed by 9NNI and MODE. Furthermore, CORAI outperformed all other methods, at 95% confidence level, when missing values were present in two or three attributes. When missing values were implanted in only one attribute CORAI performs better than MODE, but do not outperform 9NNI.

Table 3 shows the results of the Friedman and Bonferroni-Dunn tests for the fourth hypothesis. With this hypothesis we analyze whether all methods perform comparably given different combinations of percentage of missing data and number of attributes with missing data. In this table, column "%Missing" represents the percentage of missing data inserted into the attributes; "#Attributes" stands for the number of attributes with missing values; columns "CORAI", "9NNI" and "MODE" show the results of the Friedman test for the respective method; and finally, column "CD" presents the critical different produced by the Bonferroni-Dunn test. In addition, results obtained by CORAI that outperform 9NNI and MODE at 95% confidence level are colored with dark gray, and results of CORAI

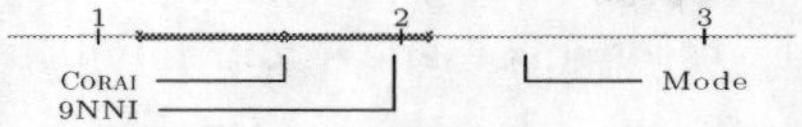

(a) 1 attribute with missing values

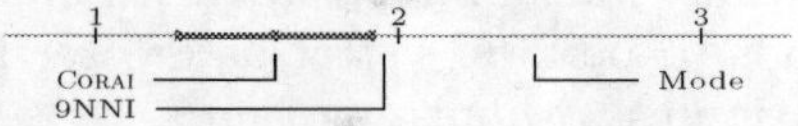

(b) 2 attributes with missing values

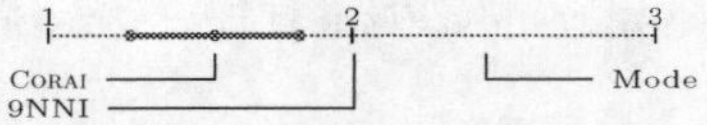

(c) 3 attributes with missing values

Fig. 3. Results of the Bonferroni-Dunn test considering the number of attributes with missing values. The thick line marks the interval of one critical difference, at 95% confidence level.

Table 3. Results of the Friedman and Bonferroni-Dunn tests considering the amount of missing data and number of attributes with missing values. The thick line marks the interval of one critical difference.

%Missing	#Attributes	Imputation Methods			CD
		CORAI	9NNI	Mode	
m20	#At=1	1.727	1.864	2.409	0.96
	#At=2	1.625	1.958	2.417	0.65
	#At=3	1.571	1.986	2.443	0.54
m40	#At=1	1.545	2.045	2.409	0.96
	#At=2	1.625	1.917	2.458	0.65
	#At=3	1.571	2.000	2.429	0.54
m60	#At=1	1.636	1.955	2.409	0.96
	#At=2	1.583	1.938	2.479	0.65
	#At=3	1.571	1.957	2.471	0.54
m80	#At=1	1.545	2.045	2.409	0.96
	#At=2	1.542	2.000	2.458	0.65
	#At=3	1.486	2.086	2.429	0.54

that outperform MODE only are colored with light gray. As can be observed, CORAI always perform better than the other imputation methods.

As a final analysis we ran the C4.5 and NAÏVE BAYES learning algorithms on the imputed data sets and measured the misclassification error on the test sets. Due to the lack of space, these results are not reported here. We also ran the Friedman test in order to verify whether there is a significant difference among the classifiers. Following the Friedman test there was no significant difference at 95% confidence level.

5 Conclusion

This paper presented CORAI, an algorithm for missing values imputation inspired in the multi-view SSL algorithm CO-TRAINING. Imputation using CORAI was compared with MODE and 9NNI, two imputation methods found in the literature, with four percentage of missingness (20%, 40%, 60% and 80%) artificially introduced in one, two and three attributes in three data sets from UCI machine learning repository. Results in these three data set show that CORAI tends to perform better at greater percentages of missingness and number of attributes with missing values.

One limitation of CORAI is that it only handles qualitative attributes. We plan to extend CORAI to quantitative attribute as a future work. Another important research direction is to evaluate how CORAI performs in patterns of missingness other than MCAR.

Acknowledgements. We wish to thank the anonymous reviewers for their helpful comments. This research was partially supported by the Brazilian Research Councils CAPES, FAPESP, CNPq and FPTI-BR.

References

1. Little, R.J.A., Rubin, D.B.: Statistical analysis with missing data. John Wiley & Sons, Inc., New York (1986)
2. Chapelle, O., Schölkopf, B., Zien, A. (eds.): Semi-Supervised Learning. MIT Press, Cambridge (2006)
3. Batista, G.E.A.P.A., Monard, M.C.: An analysis of four missing data treatment methods for supervised learning. Applied Art. Intell. 17(5-6), 519–533 (2003)
4. Levy, P.: Missing data estimation, 'hot deck' and 'cold deck'. In: Encyclopedia of Biostatistics. Wiley, Chichester (1998)
5. Zhu, X.: Semi-supervised learning literature survey. Computer Sciences TR 1530, University of Wisconsin Madison (2007), `http://pages.cs.wisc.edu/~jerryzhu/research/ssl/semireview.html`
6. Dempster, A.P., Laird, N.M., Rubin, D.B.: Maximum likelihood from incomplete data via the em algorithm. Journal of Royal Stat. Soc. B39, 1–38 (1977)
7. Blum, A., Mitchell, T.M.: Combining labeled and unlabeled sata with co-training. In: Conference on Learning Theory, pp. 92–100 (1998)
8. Goldman, S., Zhou, Y.: Enhancing supervised learning with unlabeled data. In: International Conference on Machine Learning, pp. 327–334 (2000)
9. Zhou, Z.H., Li, M.: Tri-training: Exploiting unlabeled data using three classifiers. IEEE Transactions on Knowledge and Data Engineering 17(11), 1529–1541 (2005)
10. Asuncion, A., Newman, D.: UCI machine learning repository (2007)
11. Witten, I.H., Frank, E.: Data Mining: Practical Machine Learning Tools and Techniques with Java Implementations, 2nd edn. Morgan Kaufmann, San Francisco (2005)
12. Demšar, J.: Statistical comparisons of classifiers over multiple data sets. Journal of Machine Learning Research 7, 1–30 (2006)

A Learning Function for Parameter Reduction in Spiking Neural Networks with Radial Basis Function

Alexandre da Silva Simões[1] and Anna Helena Reali Costa[2]

[1] Automation and Integrated Systems Group (GASI)
São Paulo State University (UNESP)
Av. Três de Março, 511, Alto da Boa Vista, 18.087-180, Sorocaba, SP, Brazil
[2] Intelligent Techniques Laboratory (LTI)
São Paulo University (USP)
Av. Prof. Luciano Gualberto, trav. 3, 158, 05508-900, São Paulo, SP, Brazil
assimoes@sorocaba.unesp.br, anna.reali@poli.usp.br

Abstract. Spiking neural networks – networks that encode information in the timing of spikes – are arising as a new approach in the artificial neural networks paradigm, emergent from cognitive science. One of these new models is the pulsed neural network with radial basis function, a network able to store information in the axonal propagation delay of neurons. Learning algorithms have been proposed to this model looking for mapping input pulses into output pulses. Recently, a new method was proposed to encode constant data into a temporal sequence of spikes, stimulating deeper studies in order to establish abilities and frontiers of this new approach. However, a well known problem of this kind of network is the high number of free parameters – more that 15 – to be properly configured or tuned in order to allow network convergence. This work presents for the first time a new learning function for this network training that allow the automatic configuration of one of the key network parameters: the synaptic weight decreasing factor.

1 Introduction

Spiking neural networks (SNNs) have been proposed as a new approach in the artificial neural network (ANN) paradigm. One of these new biologically inspired models is the pulsed neural network with radial basis function (RBF) [1], which is able to store information in the time delay of neuron axons. Learning algorithms have been proposed to one of these neurons [2], as well as supervised [3] and unsupervised [4] methods has been proposed to a whole network of these neurons, allowing this model to map a sequence of input pulses into a sequence of output pulses. However, practical applications of this network in real computational problems were only possible since the recent proposition of a codification system using Gaussian receptive fields [3], which present a suitable codification system to encode continuous data into a pulsed form.

G. Zaverucha and A. Loureiro da Costa (Eds.): SBIA 2008, LNAI 5249, pp. 227–236, 2008.

Time representation of information, the one used by SNNs, is richer than the average firing rate, that has been the code used in ANN models in the last decades [5]. So, a general hypothesis considered here is that new abilities can be possible when working in time domain, hypothesis that itself justifies a deeper investigation on this class of neural networks. The first results on the application of this model in color classification and clustering domains can be found in [3]. However, some of the opened questions with this approach are: *i)* what are the temporal codes proper to deal with problems not solved by traditional neural networks? *ii)* what are the proper neuron models to process these temporal signals? *iii)* what are the learning methods suitable for this networks?

Considering the last opened question, an additional problem is the high number of free parameters – more than 15 – to be configured or adapted in this approach. In practical situations, the automatic configuration of the synaptic weight decreasing factor has shown itself in particular a very difficult task [4] [6]. In a general way, even the reproduction of results showed for the first time in [3] has shown itself difficulty, since the values of this great number of network parameters are not informed.

This work presents a review of the SNN using the spike response model as radial basis function neurons and with information encoded using Gaussian receptive fields and presents a new methodology to automatically configure the synaptic weight decreasing factor parameter. This automatic configuration is allowed by a change in the network unsupervised learning function. The *proben1* methodology [7] was adopted in the tests documentation, allowing further comparisons and experiments repetition.

This paper is organized as follows: section 2 describes the spiking neuron model. Section 3 presents a network model using these neurons. The learning method is discussed in section 4. Section 5 describes the synaptic weight sum problem. Section 6 shows the proposed method to solve the problem. Section 7 present materials and methods for tests. Section 8 shows results and discussion. Section 9 presents this work conclusions.

2 Neuron Model

This section presents the pulsed neuron model called spiking response model (SRM) and the conditions to establish its equivalence with a RBF neuron.

2.1 Spiking Neuron Model

In biological neurons, the presence or absence of pulses looks to be highly relevant, whereas their form and intensity do not [5]. Analogous concept is used in the SNN. The potential μ_j of neuron j is modeled as the sum of the contributions coming from all its synapses, considering a possible refraction of neuron j (not considered in present work). When a pulse is received in the synapse between a neuron i and neuron j, this neuron is excited by a function ε_{ij}. The neuron is said to fire if its potential crosses the threshold ϑ_j. This process is illustrated in figure 1a.

Mathematically:

$$u_j(t) = \sum_{t_j^{(f)} \in T_i} \gamma_j(t - t_j^{(f)}) + \sum_{i \in \Gamma_j} \sum_{t_i^{(f)}} w_{ij}.\epsilon_{ij}(t - t_i^{(f)}), \tag{1}$$

where: t is the time; $t_j(f)$ is the time of the fth spike produced by neuron j; T_j is the set of spikes produced by neuron j; γ_j is a function that models the refraction; Γ_j is the set of neurons pre-synaptics to j; w_{ij} is the weight associated with the synapse between neurons $i \in \Gamma_j$ and j; ε_{ij} is a function that models the increasing of the potential of neuron j when it receives a spike. A function usually adopted to model the increasing of the potential of j is:

$$\epsilon_{ij}(s) = \left[exp\left(-\frac{s - \delta_{ij}}{\tau_m} \right) - exp\left(-\frac{s - \delta_{ij}}{\tau_s} \right) \right] .\mathcal{H}(s - \delta_{ij}), \tag{2}$$

where: δ_{ij} is the axonal propagation delay between i and j; τ_s and τ_m are respectively the rising and decreasing constants, with $0 < \tau_s < \tau_m$; $H(s)$ is the Heaviside function.

2.2 SRM as a RBF Neuron

Consider the following situation: *i)* threshold ϑ_j is high and can only be reached if all pre-synaptic spikes arise simultaneously; *ii)* values of the axonal propagation delay δ_{ij} of each synapse are similar to the time difference between the input pulses and a reference. In this case, the SRM neuron implements a RBF neuron. This situation is shown in figure 1b, where the application of the pulses $t_{i1}, \cdots, t_{in}$ in the synapses of neuron j changes its potential μ_j to reach the threshold ϑ_j if and only if:

$$X_i = \{x_{i_1}, x_{i_2}, ..., x_{i_n}\} \approx \Delta_j = \{\delta_{j1}, \delta_{j2}, ..., \delta_{jn}\}, \tag{3}$$

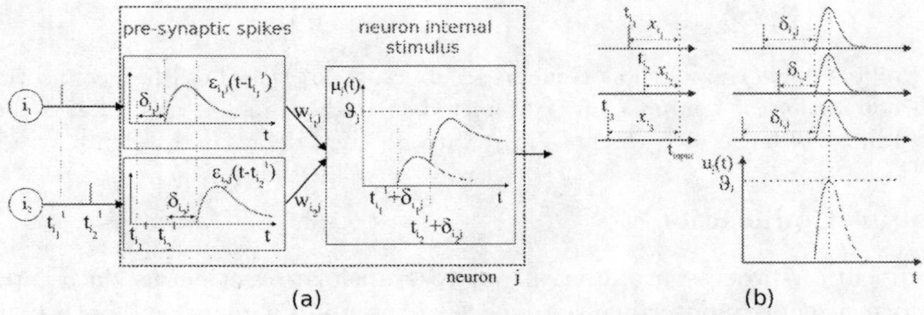

Fig. 1. Neuron model: a) Potential $\mu_j(t)$ of neuron j receiving contributions from pre-synaptic neurons. Neuron j does not produce a spike because its potential does not reach threshold ϑ_j; b) A SRM neuron as a typical RBF neuron.

3 SNN Model

3.1 SNN General Model

As initially proposed by [1], this network has two layers – namely input (I) and output (J) layers – of neurons fully connected, as shown in figure 2a. Neurons of the input layer do not have processing function and mainly generate pulses to the output layer, which executes the processing. The output layer is composed of m SRM neurons modeled as RBF neurons, as previously discussed. The main task of the learning algorithm is to properly tune the free parameters δ in each synapse.

3.2 SNN with Synaptic Spread Model

An alternative architecture to the SNN was proposed by [4]. In this approach, each synapse between each neuron $i \in I$ and $j \in J$ is splited into a set of k connections. In each connection, different propagation delays and synaptic weights are assigned. The propagation delays are usually adopted as an increasing sequence of delays $0, 1, ..., k$. The task of properly tuning a propagation delay δ between i and j can now be seen as the task of choosing only one of the k connections between i and j. This choice is done saturating only one weight w_{ij} in w_{max} and resetting all remaining weights, as shown in figure 2b.

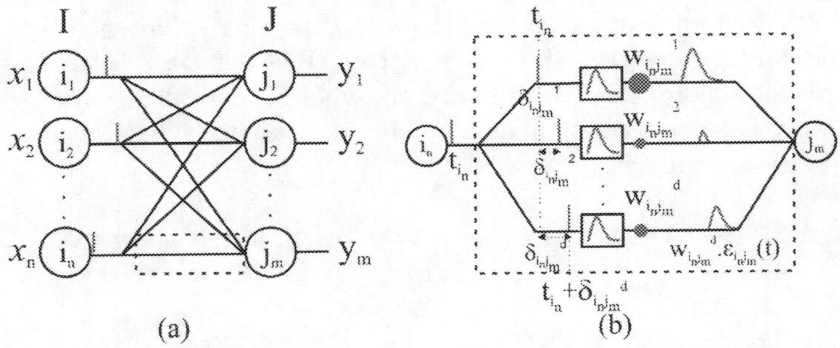

Fig. 2. Spiking network model: a) General architecture; b) Detail of the k connections for each connection of neuron i and j depicted in dashed lines if figure 2a: a pulse generated in i will be received k times in j, with different delays and weights

3.3 Input Codification

The codification process proposed in [3] to transform continuous data into a pulsed form is shortly described here: each of the g input neurons have a Gaussian receptive field covering the whole extension of the input dimension. The Gaussian centers – points of maximum excitation of each neuron – of the g neurons belonging to the same input dimension are progressively distributed over the whole range of the input data. The same distribution is associated to each one of the n input dimensions. The input neuron will produce a spike depending on the stimulus of its Gaussian receptive field: the more excited is the neuron, quicker is the spike. Each input neuron is allowed to produce only one discharge.

3.4 Typical Processing

For simplification, let us consider a network with synaptic spread with the following conditions: *i)* all weights of a neuron are small and equal; *ii)* the decreasing time τ_m is much larger that the rise time τ_s; and *iii)* the number of connections k between neurons i and j is high. The network processing in these conditions is shown in figure 3. After the arrival of the first spike in neuron j (in this case in t_{i2}), a new stimulus is assigned to the potential μ_j in each time step due to the synaptic spread. The potential is linearly increased. When the second pulse arrives (in this case in t_{i1}), the potential is hardly increased, and so on. After the arrival of the last spike, the increment of the potential in one time step can be estimated as $n.g$, since there are $n.g$ input neurons.

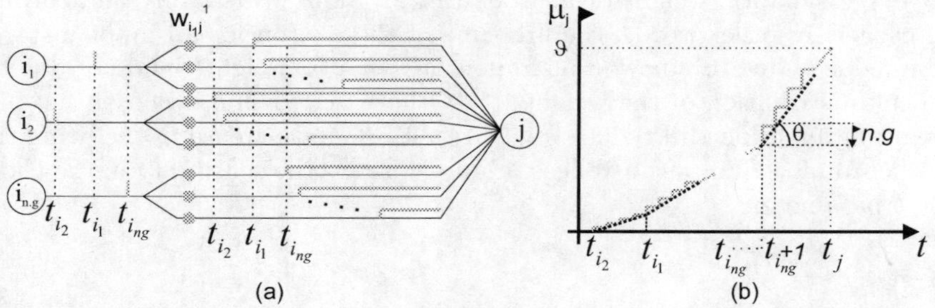

Fig. 3. a) a neuron in the output layer receiving pulses from the $n.g$ pre-synaptic neurons. Each pulse will contribute with a successive sequence of increasing in the neuron potential; b) potential changing its inclination when each new pulse is received. After the last pulse ting the potential inclination will be approximately constant.

4 Learning

The goal of learning in this model is to store in the propagation delays of each RBF neuron the center of one cluster, or, in other words, to incite the fire of the neuron j which the center is closer to the input instance in $\Re^n$, where n is the number of input dimensions.

According to the unsupervised learning algorithm proposed by [4], the synaptic weighs should be randomly initialized. When an input pattern is presented to the network, neurons are expected to fire, due to its typical processing discussed above. The first neuron to fire is called the winner of the competitive process. Only the weights of the winner neuron are updated using a Hebbian learning rule L(Δ_t). This function increases the weights of the connections that received spikes immediately before the fire of j and decrease remaining weights. A usual learning function is [4]:

$$\Delta w_{ij}^{k} = \eta . L(\Delta t) = \eta . \left((1-b).e^{-\frac{(\Delta t - c)^2}{\beta^2}} + b \right), \tag{4}$$

where: Δw_{ijk} is the increase of the kth connection between neurons i and j; Δ is the learning rate; L(.) is the learning function; b is the decreasing of the weights; Δ_t is the difference between the arriving of the spike and the fire of neuron j; c fix the peak of the learning function; β is the generalization parameter. This learning function is shown in figure 5a.

5 The Synaptic Weight Sum Stability Problem

Consider the situation where the value of the learning parameter b is chosen too small. When the learning function is applied over the synaptic weights of a neuron, this will cause an increase in a synaptic weight, a smaller increase in this weight neighbor connections (the neighborhood will depend on β value), and a small decrease in remaining connections. This process will significantly increase the average synaptic weight of neurons (w_{med}). This will result the neuron to spike earlier. In other words, the fire time of the neuron for all input instances will be reduced due to a new scale in neuron synaptic weights, and not due to a new peak distribution of the weights [4]. Otherwise, when b is chosen too high, the learning function will reduce w_{med}, resulting in a increase in the firing time of the neuron for all input instances. Figure 4 shows the sensibility in the chosen of the b parameter.

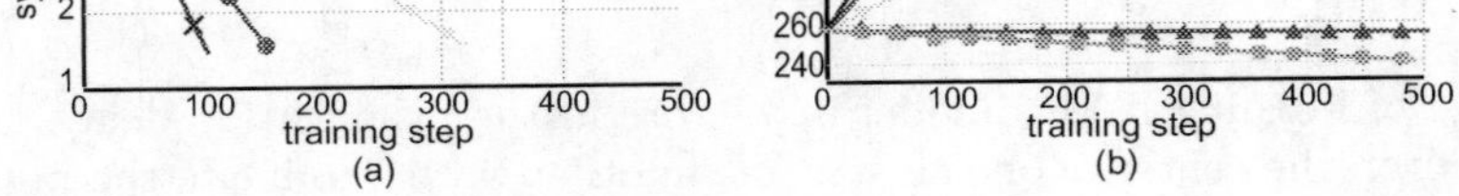

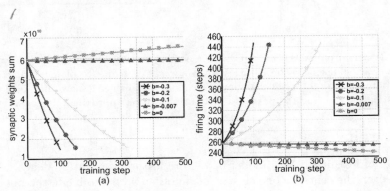

Fig. 4. A SNN trained with the learning rule in eq. 4 for 500 interactions with a single training example with n=3, g=3, m=3, c=-2.85, d=450, h=200, β=1.67, ϑ=3600, $w_{initial}$=20. a) Evolution of the synaptic weight sum of the output layers neurons by training step; b) Evolution of the winner neurons firing time by training step. Networks trained with decreasing synaptic weights sum have significant increasing in fire time. When the synaptic weight sum is stable (b=-0.007), the firing time tends to evolute only according to the competitive learning process. It is important to remark that system is extremely sensible to the b parameter, since a range from 0 to -0.3 leads to completely different dynamics in the learning process.

In order to keep stable the synaptic weight sum, [4] proposed to chose a b parameter that keep the integral of the learning function over time equal to zero. However, authors could not apply this relation to find a suitable b since the learning function proposed by them uses a term e^{x^2}, which is well known to do not have integral.

$$\int_{-\infty}^{\infty} L(\Delta t).dt \approx 0 \tag{5}$$

6 Proposed Approach

In order to allow a quick tune of the b parameter of the network – which is extremely difficult in traditional approach – this work proposes to change the learning function in eq. 4 by:

$$\Delta w_{ij}^{k} = \eta.L(\Delta t) = \eta.\left[(b-1).tanh^2\left(\frac{\Delta t - c}{\xi}\right) + 1\right], \tag{6}$$

where: Δw_{ij}^{k} is the change in the synaptic weight of the kth synapse between neurons i and j; η is the learning rate; $L(.)$ is the learning function; t is the time of competition between neurons in each training step; b is the decrement of the synaptic weights parameter; c is the back in time to reward synapses; ξ is the generalization rate.

Figure 5 presents booth (eq. 4 and 6) learning functions. Althought they are very similar, the learning function in eq. 6 allow algebraic integration. In other words it is possible to handle the expression:

$$\int_{-d}^{0}\left[(b-a).tanh^2\left(\frac{\Delta_t - c}{\xi}\right) + a\right] d\Delta_t = 0 \tag{7}$$

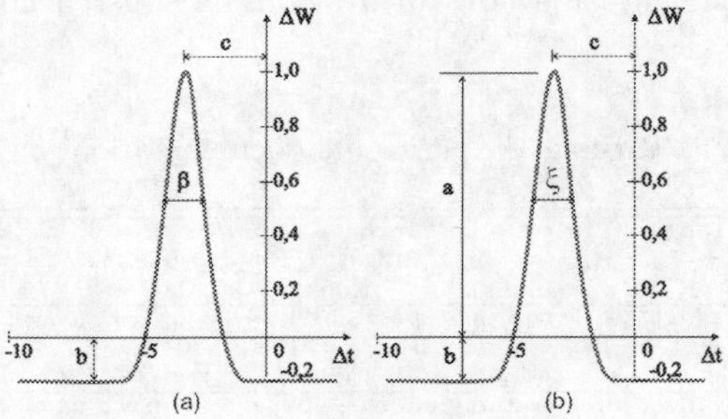

Fig. 5. Learning functions: a) learning function proposed by [4]; b) Tanh learning function. Booth functions have a=1, b=-0.3, c=-2.85, $\beta = \xi$=1.67.

It is possible to estimate the synaptic weight sum stability point showing [6] that the integral in eq. 7 is equal to zero when:

$$b \cong \kappa_e.a.\frac{-tanh(-\frac{c}{\xi}) - 1}{\frac{d}{\xi} - tanh(-\frac{c}{\xi}) - 1} \tag{8}$$

where κ_e is a stability parameter (integral is null for κ_e=1, decreases for $\kappa_e > 1$ and increases for $\kappa_e < 1$). Working with κ_e slightly above 1 has shown itself capable of stimulating competitions between neurons [6].

Equation 8 has shown itself sufficient to completely automate the b network parameter, which traditionally has a very difficult tune.

7 Materials and Methods

In order to test the convergence capability of the SNN-RBF with the *tanh* learning function, a clustering task was proposed. Three artificially generated data sets in $\Re^3$ were submited to the network for classification: *i) EXP-01*: three linearly separable clusters with equal size; *ii) EXP-02*: three linearly separable clusters with distinct size; and *iii) EXP-03*: two linearly inseparable clusters, which centers are positioned outside of the cluster. Experiments data are available in [6].

A SNN as shown in previous sections was implemented in a DevC++ environment. Convergence tests were conducted with both learning functions (eq. 4 and eq. 6). In the second case we adopted κ_e slightly above 1. The input data were presented to the network – point by point – and the displacement of the center of the winner RBF neurons were estimated in each training step. The training was suspended when weights reached w_{max} or w_{min} values. After convergence, the mean square error (MSE) was observed and compared with those obtained with the well known *k-means* algorithm in different initial conditions. It is important to remark that there are no previously documented tests in SNN-RBF that allow comparison with this work. Table 1 presents tests documentation for future comparisons, according to *proben1* methodology [7].

Table 1. Experiments documentation

Exp-01	Input range: [0,1]; training epochs: 430; $n = 3, m = 3, k = 250, g = 9, h = 100, \sigma = 1, w_{initial} = 40 \pm 10\%, a = 1, b = -0,023255, c = -2, w_{max} = 1.497, \eta = 10, \vartheta = 45.000, \tau_m = 1.000, \tau_s = 0,1, \beta = 1,67$
Exp-02	Input range: [0,1]; training epochs: 399; $n = 3, m = 3, k = 250, g = 9, h = 100, \sigma = 1, w_{initial} = 40 \pm 10\%, a = 1, b = -0,0258395, c = -2, w_{max} = 1.100, \eta = 10, \vartheta = 39.600, \tau_m = 1.000, \tau_s = 0,1, \beta = 1,67$
Exp-03	Input range: [0,1]; training epochs: 402; $n = 3, m = 3, k = 250, g = 9, h = 100, \sigma = 1, w_{initial} = 58 \pm 10\%, a = 1, b = -0,0258395, c = -2, w_{max} = 1.581, \eta = 10, \vartheta = 56.925, \tau_m = 1.000, \tau_s = 0,1, \beta = 1,67$

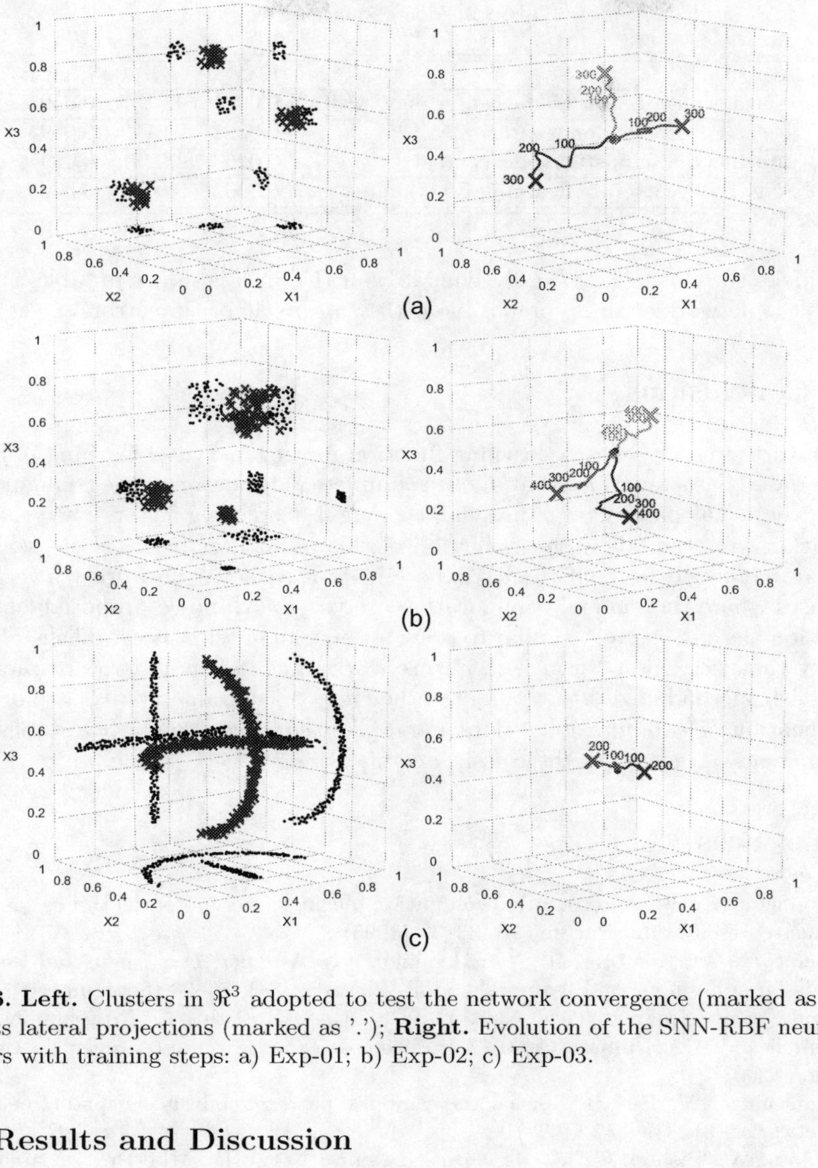

Fig. 6. Left. Clusters in $\Re^3$ adopted to test the network convergence (marked as 'x') and its lateral projections (marked as '.'); **Right.** Evolution of the SNN-RBF neurons centers with training steps: a) Exp-01; b) Exp-02; c) Exp-03.

8 Results and Discussion

Tests did not show network convergence using eq. 4 since b parameter has proven to be extremely difficult to manually configure. Using equation 6, synaptic weights sum remained close to stability allowing convergence, as expected. Figure 6 presents the SNN convergence process. In a similar approach to *k-means*, after an initial phase of competition in first training steps, a convergence phase arises and all neuron centers converged to the center of a different cluster, even if it is

Table 2. Mean square error

Algorithm	EXP-01	EXP-02	EXP-03
k-means with center in the center of $\Re^3 \pm 10$ %	0,0021	0,0137	0,0595
k-means with random centers	0,0021	0,0192	0,0595
k-means taking instances as centers	0,0363	0,0231	0,0595
SNN-RBF with *tanh* learning function	0,0062	0,0056	0,0608

outside of the cluster (*EXP-03*). The MSE for the SNN is shown in table 2. In all cases the network error is compatible with *k-means* algorithm error.

9 Conclusions

This work presented a new learning function for unsupervised learning of SNNs with radial basis functions that allow a complete automation of the configuration of one of its key parameters: the synaptic weight decreasing factor. The proposed equations, in true, allow the reduction of this parameter from the network set of parameters to be configured. Experiments has shown network convergence in cases where the manual configuration are very hard. Since proposed learning function has a shape very similar to previous one, all previously established learning systematic remain intouched. Future works must follow in two directions: *i)* establish methodologies to configure other network parameters; *ii)* explore the codification of computational data into spike sequences trying to establish its possible advantages over the classical codification.

References

1. Hopfield, J.: Pattern recognition computation using action potential timing for stimulus representation. Nature 376, 33–36 (1995)
2. Gerstner, W., Kempter, R., Van Hemmen, J.L., Wagner, H.: A neuronal learning rule for sub-millisecond temporal coding. Nature (384), 76–78 (September 1996)
3. Bohte, S.M.: Spiking Neural Networks. PhD thesis, Institute for Programming Research and Algorithmics. Centre for Mathematics and Computer Science, Amsterdam (2003)
4. Natschläger, T., Ruf, B.: Spatial and temporal pattern analysis via spiking neurons. Network 9(3), 319–332 (1998)
5. Maass, W., Bishop, C.M. (eds.): Pulsed Neural Networks. MIT Press, Cambridge (1999)
6. da Simöes, A.S.: Unsupervised learning in spiking neural networks with radial basis functions. PhD thesis, Escola Politécnica da Universidade de Säo Paulo (2006)
7. Prechelt, L.: PROBEN1 – A set of neural network benchmark problems and benchmarking rules. Technical Report 21/94, Technical Report 21/94. University of Karlsruhe, Germany (September 1994)

A Robust Methodology for Comparing Performances of Clustering Validity Criteria

Lucas Vendramin, Ricardo J.G.B. Campello, and Eduardo R. Hruschka

Department of Computer Sciences – University of São Paulo at São Carlos
SCC/ICMC/USP, C.P. 668, São Carlos, SP, 13560-970, Brazil
vendra@grad.icmc.usp.br, campello@icmc.usp.br, erh@icmc.usp.br

Abstract. Many different clustering validity measures exist that are very useful in practice as quantitative criteria for evaluating the quality of data partitions. However, it is a hard task for the user to choose a specific measure when he or she faces such a variety of possibilities. The present paper introduces an alternative, robust methodology for comparing clustering validity measures that has been especially designed to get around some conceptual flaws of the comparison paradigm traditionally adopted in the literature. An illustrative example involving the comparison of the performances of four well-known validity measures over a collection of 7776 data partitions of 324 different data sets is presented.

1 Introduction

Clustering is a data mining task in which one aims at determining a finite set of categories to describe a data set according to similarities among its objects [1,2]. Clustering techniques can be broadly divided into three main types [3]: overlapping (so-called non-exclusive), partitional, and hierarchical. The last two are related to each other in that a hierarchical clustering is a nested sequence of hard partitional clusterings, each of which represents a partition of the data set into a different number of mutually disjoint subsets. A hard partition of a data set $\mathbf{X} = \{ \mathbf{x}(1), \cdots, \mathbf{x}(N) \}$, composed of n-dimensional feature or attribute vectors $\mathbf{x}(j)$, is a collection $\mathbf{S} = \{\mathbf{S}_1, \cdots, \mathbf{S}_k\}$ of k non-overlapping data subsets $\mathbf{S}_i$ (clusters) such that $\mathbf{S}_1 \cup \mathbf{S}_2 \cup \cdots \cup \mathbf{S}_k = \mathbf{X}$, $\mathbf{S}_i \neq \oslash$ and $\mathbf{S}_i \cap \mathbf{S}_l = \oslash$ for $i \neq l$. Overlapping techniques search for soft or fuzzy partitions by somehow relaxing the mutually disjointness constraints $\mathbf{S}_i \cap \mathbf{S}_l = \oslash$.

Many different clustering validity measures exist that are very useful in practice as quantitative criteria for evaluating the quality of data partitions – e.g. see [4,5] and references therein. These measures are important, for instance, to help determine the number of clusters contained in a given data set [3,2]. Some of the most well-known validity measures, also referred to as *relative* validity (or quality) criteria[1], are possibly the Davies-Bouldin Index [6,3], the Variance Ratio Criterion – VRC (so-called Calinski-Harabasz Index) [7,2], Dunn's Index

[1] The terms *measure* and *criterion* will be freely interchanged throughout the paper.

G. Zaverucha and A. Loureiro da Costa (Eds.): SBIA 2008, LNAI 5249, pp. 237–247, 2008.

[8,4], and the Silhouette Width Criterion of Kaufman and Rousseeuw [1,2], just to mention a few. It is a hard task for the user, however, to choose a specific measure when he or she faces such a variety of possibilities. To make things even worse, new measures have still been proposed from time to time. For this reason, a problem that has been of interest for more than two decades consists of comparing the performances of existing clustering validity measures and, eventually, that of a new measure to be proposed. A cornerstone in this area is the work by Milligan and Cooper [5], who compared 30 different measures through an extensive set of experiments involving several labeled data sets. Twenty three years later, that seminal work is still used and cited by many authors who deal with clustering validity criteria. In spite of this, there are some conceptual problems with the comparison methodology adopted by Milligan and Cooper. First, their methodology relies on the assumption that the accuracy of a criterion can be quantified by the number of times it indicates as the best partition (among a set of candidates) a partition with the *right* number of clusters for a specific data set, over many different data sets for which such a number is known in advance – e.g. by visual inspection of 2D data or by labeling synthetically generated data. This assumption has also been implicitly made by other authors who worked on more recent papers involving the comparison of clustering validity criteria (e.g. see [9,4,10]). Note, however, that there may exist various partitions of a data set into the *right number* of clusters that are very *unnatural* with respect to the spatial distribution of the data. On the other hand, there may exist numerous partitions of a data set into the *wrong number* of clusters, but clusters that exhibit a high degree of compatibility with the spatial distribution of the data.

Another problem with Milligan and Cooper's methodology is that the assessment of each validity criterion relies solely on the correctness of the partition elected as the best one according to that criterion. The accuracy of the criterion when evaluating all the other candidate partitions is ignored. Accordingly, the capability of the criterion to properly distinguish among a set of partitions that are not good in general is not taken into account. This capability indicates a particular kind of robustness of the criterion that is very important in many real-world application scenarios for which no clustering algorithm can provide precise solutions due to noise contamination, cluster overlapping, high dimensionality, and other possible complicative factors.

Before proceeding with a proposal to get around the drawbacks described above, it is important to remark that some very particular criteria are only able to estimate the number of clusters in data – by suggesting when a given iterative (e.g. hierarchical) clustering procedure should stop increasing this number. Such criteria are referred to as stopping rules [5]. Stopping rules should not be seen as clustering validity measures in a broad sense, since they are not able to quantitatively measure the quality of data partitions. In other words, they are not optimization-like validity measures[2]. When comparing the efficacy of stopping rules, there is not much to do except using Milligan and Cooper's methodology. However, when dealing with optimization-like validity measures, which is the

[2] Such measures can be used as stopping rules [5], but the reverse is not true.

kind of criterion subsumed in the present paper, a more rigorous comparison methodology may be desired. This is precisely the subject of the present work.

The remaining of this paper is organized as follows. In Section 2, the problem of comparing relative validity measures is discussed in details. Specifically, Section 2.1 describes the traditional methodology, whereas the proposed methodology is introduced in Section 2.2. Then, an illustrative example is reported in Section 3. Finally, the conclusions and perspectives are addressed in Section 4.

2 Comparing Relative Clustering Validity Criteria

2.1 Traditional Methodology

As previously mentioned in Section 1, the traditional methodology for comparing relative clustering validity criteria consists essentially of counting the number of times each criterion indicates as the best partition (among a set of candidates) a partition with the *right* number of clusters, over many different data sets for which such a number is known in advance – e.g. by visual inspection of 2D data or by labeling synthetically generated data. An example for the Ruspini data set (Figure 1) is shown in Figure 2. Figure 2 displays the values of four relative validity criteria for a set of partitions of the Ruspini data obtained by running the classic k-means algorithm multiple times with the number of clusters ranging from 2 to 10. In this figure, maximum peaks are evident for the Silhouette, Calinski-Harabasz (VRC), and Dunn's indexes when $k = 4$. Similarly, a distinguishable minimum is also verified for the Davies-Bouldin index when $k = 4$. Unlike Dunn, Calinski-Harabasz, and Silhouette, which are maximization criteria, Davies-Bouldin is a minimization criterion. In summary, all these measures point out that the optimal number of clusters is 4. This is in accordance with common sense evaluation of the data by visual inspection, even though 5 clusters is also quite acceptable in this case [1]. However, real data sets are not that simple. Furthermore, as discussed in Section 1, there are many pitfalls along the way if one considers the accuracy in estimating the right number of clusters as the only indicative of the performance of a relative clustering validity criterion. For this reason, a complementary, alternative methodology for comparing such criteria is described in the next section.

2.2 Proposed Methodology

The proposed methodology for comparing relative clustering validity criteria can be better explained by means of a pedagogical example. For the sake of simplicity and without any loss of generality, let us assume in this example that our comparison involves only the performances of two validity criteria in the evaluation of a small set of partitions of a single labeled data set. The hypothetic results are displayed in Table 1, which shows the values of the relative criteria under investigation as well as the values of a given external (absolute rather than relative) criterion for each of the five data partitions available. One of the most well-known and efficient external criteria is the Jaccard coefficient [4,3].

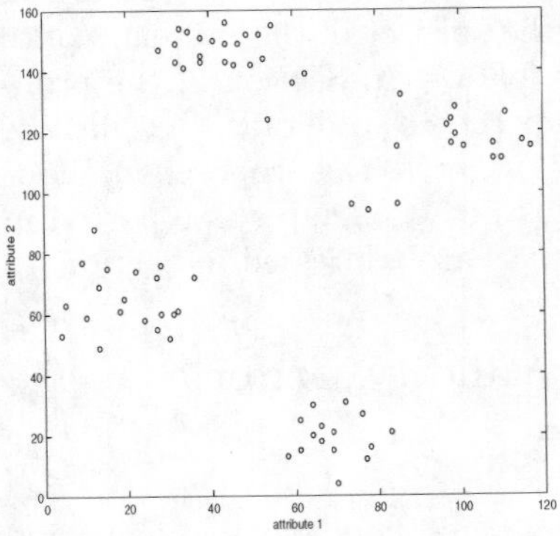

Fig. 1. Ruspini data set

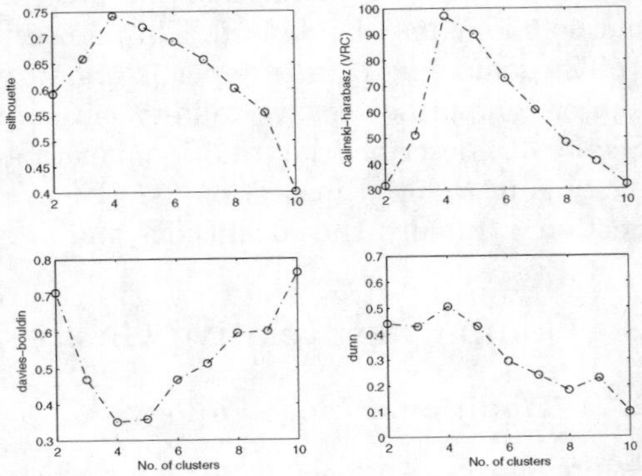

Fig. 2. Relative criteria for the Ruspini data

Table 1. Example of evaluation results for five partitions of a given data set

Partition	Relative Criterion 1	Relative Criterion 2	External Criterion
1	0.75	0.92	0.82
2	0.55	0.22	0.49
3	0.20	0.56	0.31
4	0.95	0.63	0.89
5	0.60	0.25	0.67

Assuming that the evaluations performed by the external criterion are trustable measures of the quality of the partitions[3], it is expected that the better the relative criterion the greater its capability of evaluating the partitions according to an ordering and magnitude proportions that are similar to those established by the external criterion. Such a degree of similarity can be straightforwardly computed using a sequence correlation index, such as the well-known Pearson product-moment correlation coefficient [11]. Clearly, the larger the correlation value the higher the capability of a relative measure to unsupervisedly mirror the behavior of the external index and properly distinguish between better and worse partitions. In the example of Table 1, the Pearson correlation between the first relative criterion and the external criterion (columns 2 and 4, respectively) is 0.9627. This high value reflects the fact that the first relative criterion ranks the partitions in the same order that the external criterion does. Unitary correlation is not reached only because there are some differences in the relative importance given to the partitions. Contrariwise, the Pearson correlation between the second relative criterion and the external criterion scores only 0.4453. This is clearly in accordance with the strong differences that can be visually observed between the evaluations in columns 3 and 4 of Table 1.

[3] This is based on the fact that such a sort of criterion relies on supervised information about the underlying structure in the data (known referential clusters) [3,2].

In a practical comparison procedure, there should be multiple partitions of varied qualities for a given data set. Moreover, a practical comparison procedure should involve a representative collection of different data sets that fall within a given class of interest (e.g. mixtures of Gaussians). This way, if there are N_D labeled data sets available, then there will be N_D correlation values associated with each relative validity criterion, each of which represents the agreement level between the evaluations of the partitions of one specific data set when performed by that relative criterion and by an external criterion. The mean of such N_D correlations is a robust measure of resemblance between those particular (relative and external) validity criteria, at least in what regards that specific collection of data sets. Despite this, besides just comparing such means for different relative criteria in order to rank them, one should also apply to the results an appropriate statistical test to check the hypothesis that there are (or there are not) significant differences among those means.

In summary, the proposed methodology is the following:

1. Take or generate N_D different data sets with known clusters.
2. For each data set, get a collection of N_p data partitions of varied qualities and numbers of clusters. For instance, such partitions can be obtained from a single run of a hierarchical algorithm or from multiple runs of a partitional algorithm with different numbers and initial positions of prototypes.
3. For each data set, compute the values of the relative and external validity criteria for each one of the N_p partitions available. Then, for each relative criterion, compute the correlation between the corresponding vector with N_p relative validity values and the vector with N_p external validity values.
4. For each relative validity criterion, compute the mean of its N_D correlation values (one per data set); then, rank all the relative criteria according to their means. To check whether the difference between the means of two specific criteria is significant or not from a statistical perspective, apply an appropriate statistical test to the results.

Remark 1. Since the external validity criteria are always maximization criteria, minimization relative criteria, such as Davies-Bouldin, must be converted into maximization ones (flipped around their means) before comparing them.

Remark 2. The methodology proposed above is a generalization of the one proposed in a less recognized paper by Milligan, published in 1981 [12], in which the author stated that "Logically, if a given criterion is succeeding in indicating the degree of correct cluster recovery, the index should exhibit a close association with an external criterion which reflects the actual degree of recovery of structure in the proposed partition solution". Milligan, however, conjectured that the above statement would possibly not be justified for comparisons involving partitions with different numbers of clusters, due to a kind of monotonic trend that some external indexes may exhibit as a function of this quantity. For this reason, the analyses in [12] were limited to partitions obtained with the right number (k^*) of clusters known to be present in the data. This limitation itself causes two major impacts on the reliability of the comparison procedure. First, the robustness of the criteria under investigation in what regards their ability to properly distinguish among partitions that are not good in general is not taken into account. Second, using only k^* implies that there will be a single pair of relative and external criteria associated with a given data set, which means that a single correlation value will result from the values of relative and external criteria computed over the whole collection of available data sets. Obviously, a single sample does not allow any statistical evaluation of the results.

The more general methodology proposed here rules out those negative impacts resulting from Milligan's conjecture mentioned above. But what about the conjecture? Such a conjecture is possibly one of the reasons that made Milligan and Cooper not to adopt the same idea in their 1985 paper [5], which was focused on procedures for determining the number of clusters in data. In our opinion, such a conjecture was not properly elaborated because, while it is true that some external indexes (e.g. Rand Index [2]) do exhibit a monotonic trend as a function of the number of clusters, such a trend is observed when the number of clusters of both the referential and evaluated partitions increase [13] – or in the specific situations in which there is no structure in the data and in the corresponding referential partition [14]. Neither of these is the case, however, when one takes a well-defined referential partition of a structured data set with a given (fixed) number of clusters and compares it against partitions of the same data produced by some clustering algorithm with variable number of clusters, as verified by Milligan and Cooper themselves in a later paper [14].

3 Illustrative Example

In this section, an illustrative example is presented that involves the comparison of those well-known relative validity criteria already mentioned in previous sections, namely, Dunn's Index [8,4], the Calinski-Harabasz Index [7,2], the Davies-Bouldin Index [6,3], and the Silhouette Width Criterion of Kaufman and Rousseeuw [1,2]. The data sets adopted here are reproductions of the artificial data sets used in the classic study of Milligan and Cooper [5]. The data generator was developed following strictly the descriptions in that reference. In brief, the data sets consist of a total of $N = 50$ objects each, embedded in either an $n = 4$, 6, or 8 dimensional Euclidean space. Each data set contains either $k^* = 2$, 3, 4, or 5 distinct clusters for which overlap of cluster boundaries is not permitted on the first dimension of the variables space. The actual distribution of the objects within clusters follows a (mildly truncated) multivariate normal distribution, in such a way that the resulting structure could be considered to consist of natural clusters that exhibit the properties of external isolation and internal cohesion.

Following Milligan and Cooper's procedure, the design factors corresponding to the number of clusters and to the number of dimensions were crossed with each other and both were crossed with a third factor that determines the number of objects within the clusters. Provided that the number of objects in each data set is fixed, this third factor directly affects not only the cluster densities, but the overall data balance as well. This factor consists of three levels, where one level corresponds to an equal number of objects in each cluster (or as close to equality as possible), the second level requires that one cluster must always contain 10% of the data objects, whereas the third level requires that one cluster must contain 60% of the objects. The remaining objects were distributed as equally as possible across the other clusters present in the data. Overall, there were 36 cells in the design (4 numbers of clusters × 3 dimensions × 3 balances). Milligan an Cooper generated three sampling replications for each cell, thus producing a total of 108 data sets. For greater statistical confidence, we generate here nine sampling replications for each cell, thus producing a total of 324 data sets.

With the data sets in hand, the average linkage version of the popular AGNES (AGlomerative NESting) hierarchical clustering algorithm [1] was systematically applied to each data set. For each data set, the algorithm produced a hierarchy of data partitions with the number of clusters ranging from $k = 2$ through $k = N = 50$. Every such a partition can then be evaluated by the relative clustering validity criteria under investigation. The evaluation results corresponding to a typical data set are displayed in Figure 3.

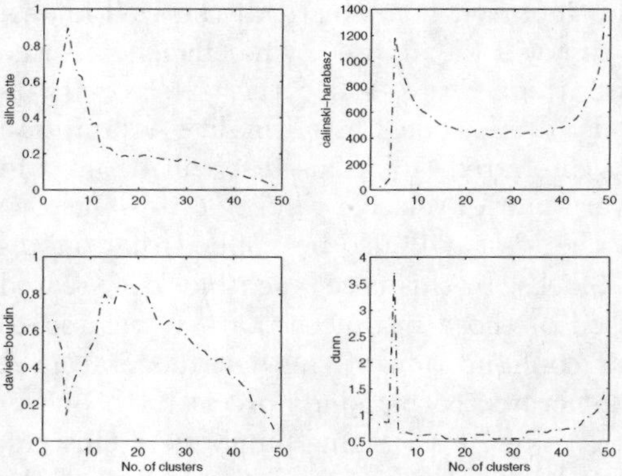

Fig. 3. Relative validity values for a typical data set: partitions for $k = 2, \cdots, 50$ produced by the AGNES hierarchical algorithm

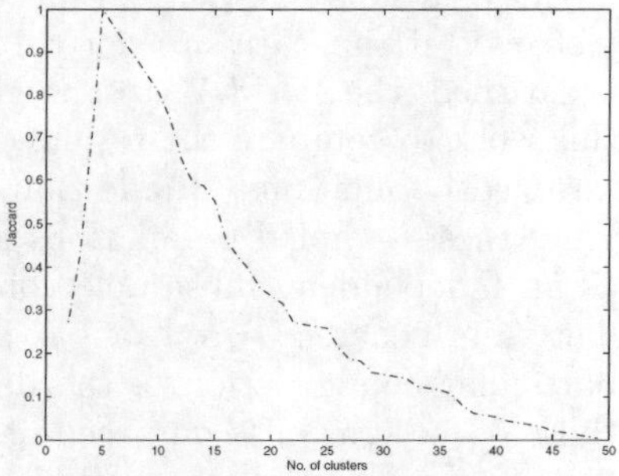

Fig. 4. External validity values (Jaccard coefficient) for the same data partitions of Figure 3

Note in Figure 3 that VRC and Dunn's indexes exhibit a secondary peak when evaluating partitions with the number of clusters k approaching the number of objects $N = 50$. Such a behavior is also observed in other relative clustering validity criteria not comprised in the present study and must be carefully dealt with when assessing their performances. The reason is that this eventual secondary peak, which may be less or more intense depending on the criterion and on the data set as well, clearly affects the correlation between the relative and external validity values (please, visually compare Figures 3 and 4). It is worth remarking that, in most applications, partitions with k approaching N are completely out of scope for any practical clustering analysis. In these cases, it is strongly recommended to perform the methodology proposed in Section 2.2 over a set of partitions with the number of clusters limited within an acceptable range. Such an acceptable range depends on the application domain, but there are some general rules that can be useful to guide the less experienced user in practical situations. A commonly used rule of thumb is to set the upper bound of this interval to $\sqrt{N}$. Another possibility, which can be particular useful if the number of objects is small, is to set this upper bound to $N/2$. The rationale behind such a more conservative number is that, conceptually speaking, a cluster is expected to be composed of at least two objects, otherwise being an outlier rather than a group of objects. Both these intervals will be adopted in the present study, i.e. $k \in \{2, \cdots, 8\}$ and $k \in \{2, \cdots, 25\}$. This means that only a subset of the partitions produced by the AGNES algorithm for each data set will be used in the experiments. Having such selected partitions in hand ($324 \times 7 = 2268$ for the 1st interval and $324 \times 24 = 7776$ for the 2nd one), the relative and external validity criteria for all of them can be computed and the comparison procedure described in Section 2.2 can be concluded.

Histograms of the results strongly suggest that the observed sample distributions hardly satisfy the normality assumption. For this reason, outcomes of parametric statistical tests will not be reported here. Instead, the well-known Wilcoxon/Mann-Whitney (W/M-W) test will be adopted. The efficiency of the W/M-W test is 0.95 with respect to parametric tests like the t-test or the z-test even if the data are normal. Thus, even when the normality assumption is satisfied, the W/M-W test might be preferred [15]. This test will be used in this work to compare the results of every pair of relative validity criteria as two sampled populations. In addition, another test will also be applied that subdivides these sampled populations into blocks. In this case, each block is treated as an independent subsample composed of those instances that are related to data sets generated from a particular configuration of the data generator. A particular configuration of the data generator corresponds precisely to one of those 36 design cells composed of data sets with the same numbers of clusters, dimensions, and the same balance. This is called *two-way randomized block design*, one way referring to samples coming from different relative criteria and another way referring to samples coming from a different configuration of the data generator. A non-parametric test of this class is named Friedman test (a relative of ANOVA), which will be adopted here to reinforce the results of the W/M-W test in a conservative, duplicated manner.

The final results are summarized in Figures 5 and 6. The correlation mean between every relative criterion and the Jaccard index – computed over the whole collection of 324 correlation values – is shown in the bottom bars. The value displayed in each cell of the top tables corresponds to the difference between the correlation means of the corresponding pair of relative criteria. A shaded cell indicates that the corresponding difference is statistically significant at the $\alpha = 5\%$ level (one-sided test). Darker shaded cells indicate that significance has been observed with respect to both W/M-W and Friedman tests, whereas lighter shaded cells denote significance with respect to one of these tests only.

		A	B	C	D
silhouette	A	0.000	0.096	0.145	0.161
calinski-harabasz (VRC)	B	-0.096	0.000	0.050	0.065
dunn	C	-0.145	-0.050	0.000	0.015
davies-bouldin	D	-0.161	-0.065	-0.015	0.000
	Mean	0.814	0.718	0.668	0.653

Fig. 5. Correlation means of four relative validity criteria with respect to the Jaccard index over 324 data sets and their differences: AGNES with $k \in \{2, \cdots, 8\}$

		A	B	C	D
silhouette	A	0.000	0.092	0.260	0.832
calinski-harabasz (VRC)	B	-0.092	0.000	0.168	0.740
dunn	C	-0.260	-0.168	0.000	0.572
davies-bouldin	D	-0.832	-0.740	-0.572	0.000
	Mean	0.896	0.804	0.636	0.064

Fig. 6. Correlation means of four relative validity criteria with respect to the Jaccard index over 324 data sets and their differences: AGNES with $k \in \{2, \cdots, 25\}$

In Figures 5 and 6, the relative validity criteria have been placed in the rows and columns of the top tables according to the ordering established by their correlation means (decreasing order from top to bottom and from left to right). Note that the same order is kept in both scenarios. Aside from some particular exceptions observed either in Figure 5 or 6, the results suggest that the performances of the four relative validity criteria under investigation could be ranked as Silhouette > VRC > Dunn > Davies-Bouldin, where ">" means that there exists some statistical evidence of the corresponding difference in performances in at least one of the scenarios. A more conservative reading of the results, in which better performance is asserted if and only if there are significant differences in both scenarios and with respect to both statistical tests, would lead to the conclusion that (Silhouette, VRC) > (Dunn, Davies-Bouldin).

The separate analyses in Figures 5 and 6 reveal a very interesting fact: by comparing the bottom bars of those figures it can be seen that the performance of the Davies-Bouldin index drops drastically when partitions with much more clusters than the actual number of clusters in the data ($k^* \in \{2, 3, 4, 5\}$) are taken into account. This suggests that Davies-Bouldin is not robust to distinguish among poor quality partitions and, as such, this index may not be recommended for tight real-world application scenarios involving complicative factors like noise contamination, cluster overlapping, high dimensionality, among others. But a word of caution is needed here. It is worth remarking that the above results and conclusions hold for a particular collection of data sets. Since such a collection is reasonably representative of a particular class, namely, those data sets with volumetric clusters of approximately hyperspherical shapes, then one can believe that similar results are likely to be observed for other data sets of this class. However, nothing can be presumed about data sets that do not fall within this category, at least not before new experiments involving such data are performed.

4 Conclusions and Future Work

The present paper has introduced an alternative methodology for comparing relative clustering validity criteria. This methodology is based upon the use of labeled data sets and an external (absolute) validity criterion to guide the comparisons. It has been especially designed to get around the conceptual flaws of the comparison paradigm traditionally adopted in the literature. In particular: (i) it does not rely on the assumption that the accuracy of a validity criterion can be quantified by the relative frequency with which it indicates as the best partition (among a set of candidates) a partition with the right number of clusters; and (ii) it does not completely rely on the single partition elected as the best one according to that criterion. Getting rid of such over-simplified assumptions makes the proposed comparison methodology more robust to distinguish between better and worse partitions embedded in difficult application scenarios. As a word of caution, it is important to recall that the proposed methodology is appropriate to compare optimization-like validity criteria, which are those able to quantitatively measure the quality of data partitions. When comparing simple stopping rules, which are only able to estimate the number of clusters in data, the traditional methodology is possibly still the only choice. From this viewpoint, the proposed methodology can be seen as complementary to the traditional one. An illustrative example involving the comparison of the performances of four well-known validity measures over a collection of 7776 data partitions of 324 different data sets has been presented. This example has been intended to illustrating the proposed methodology, rather than making a broad comparison of several different existing criteria. A more extensive study involving the comparison of several clustering (and fuzzy clustering) validity criteria is on the way.

Acknowledgments

This work has been supported by CNPq and Fapesp.

References

1. Kaufman, L., Rousseeuw, P.: Finding Groups in Data. Wiley, Chichester (1990)
2. Everitt, B.S., Landau, S., Leese, M.: Cluster Analysis, 4th edn. Arnold (2001)
3. Jain, A.K., Dubes, R.C.: Algorithms for Clustering Data. Prentice-Hall, Englewood Cliffs (1988)
4. Halkidi, M., Batistakis, Y., Vazirgiannis, M.: On clustering validation techniques. Journal of Intelligent Information Systems 17, 107–145 (2001)
5. Milligan, G.W., Cooper, M.C.: An examination of procedures for determining the number of clusters in a data set. Psychometrika 50(2), 159–179 (1985)
6. Davies, D.L., Bouldin, D.W.: A cluster separation measure. IEEE Trans. on Pattern Analysis and Machine Intelligence 1, 224–227 (1979)
7. Calinski, R.B., Harabasz, J.: A dendrite method for cluster analysis. Communications in Statistics 3, 1–27 (1974)
8. Dunn, J.C.: Well separated clusters and optimal fuzzy partitions. Journal of Cybernetics 4, 95–104 (1974)
9. Bezdek, J.C., Pal, N.R.: Some new indexes of cluster validity. IEEE Trans. on Systems, Man and Cybernetics – B 28(3), 301–315 (1998)
10. Maulik, U., Bandyopadhyay, S.: Performance evaluation of some clustering algorithms and validity indices. IEEE Transactions on Pattern Analysis and Machine Intelligence 24(12), 1650–1654 (2002)

11. Casella, G., Berger, R.L.: Statistical Inference, 2nd edn. Duxbury Press (2001)
12. Milligan, G.W.: A monte carlo study of thirdy internal criterion measures for cluster analysis. Psychometrika 46(2), 187–199 (1981)
13. Fowlkes, E.B., Mallows, C.L.: A method for comparing two hierarchical clusterings. Journal of the American Statistical Association 78, 553–569 (1983)
14. Milligan, G.W., Cooper, M.C.: A study of the comparability of external criteria for hierarchical cluster analysis. Multivariate Behavioral Research 21, 441–458 (1986)
15. Triola, M.F.: Elementary Statistics. Addison Wesley Longman (1999)

An Experiment in Spanish-Portuguese Statistical Machine Translation

Wilker Ferreira Aziz[1], Thiago Alexandre Salgueiro Pardo[1], and Ivandré Paraboni[2]

[1] University of São Paulo – USP / ICMC
Av. do Trabalhador São-Carlense, 400 - São Carlos, Brazil
[2] University of São Paulo – USP / EACH
Av. Arlindo Bettio, 1000 - São Paulo, Brazil
wilker.aziz@usp.br, taspardo@icmc.usp.br, ivandre@usp.br

Abstract. Statistical approaches to machine translation have long been successfully applied to a number of 'distant' language pairs such as English-Arabic and English-Chinese. In this work we describe an experiment in statistical machine translation between two 'related' languages: European Spanish and Brazilian Portuguese. Preliminary results suggest not only that statistical approaches are comparable to a rule-based system, but also that they are more adaptive and take considerably less effort to be developed.

1 Introduction

As in many other Natural Language Processing (NLP) tasks, statistical techniques have emerged in recent years as a highly promising approach to Machine Translation (MT), and are now part of mainstream research in the field. By automatically analysing large collections of parallel text, statistical MT is often capable of outperforming existing (e.g., rule-based) systems even for so-called 'distant' language pairs such as English and Arabic. Regular international contests on MT organized by NIST (*National Institute of Standards and Technology*) have shown the advances of statistical approaches. In the 2005 contest, when statistical and rule-based MT systems competed, the best-performing statistical system – the Google Translator – fared about 376% higher than the widely known rule-based system Systran for the language pair English-Arabic. For the English-Chinese pair, Google Translator outperformed Systran by 140%.

Researchers in the area argue that, since the statistical MT techniques are able to perform so well for distant languages, with all likelihood even better results should be achieved for 'closely-related' (e.g., Romance) languages. As an attempt to shed light on this issue, in this work we describe an experiment in statistical machine translation between two 'related' languages, namely, European Spanish and Brazilian Portuguese. In doing so, rather than seeking to produce a fully developed MT system for this language pair, we limit ourselves to investigate our basic statistical MT system as compared to the shallow-transfer approach in [1]. We shall then argue that our preliminary results - although not overwhelming given the small scale of the experiment - provide remarkable evidence in favour of the statistical approach and pave the way for further research.

G. Zaverucha and A. Loureiro da Costa (Eds.): SBIA 2008, LNAI 5249, pp. 248–257, 2008.

The rest of this paper is structured as follows. Section 2 describes the statistical translation model used in our work. Section 3 describes the training data, the sentence and word alignment procedures and the alignment revision techniques involved. Section 4 presents results of a preliminary evaluation work and a comparison with the system described in [1]. Finally, Section 5 summarizes our work so far.

2 Translation Model

Generally speaking, a statistical approach to translate from, e.g., Spanish to Portuguese involves finding the Portuguese sentence *p* that maximizes the probability of *p* given a Spanish sentence *e*, that is, the probability $P(p \mid e)$ of *p* being a translation of *e*. Given the probabilities $P(p)$ and $P(e)$ obtained from Portuguese and Spanish language models, respectively, the problem can be expressed in Bayesian terms as

$$P(p \mid e) = \frac{P(p)P(e \mid p)}{P(e)}$$

Since $P(e)$ remains constant for the input sentence *e*, this amounts to maximizing

$$P(p)P(e \mid p)$$

in which $P(e \mid p)$ is the *translation model*. The *language model* $P(p)$ may be computed as the probability of the sequence of n-grams in the translation *p*.

Given a pair *<e, p>* of sentences, it is said that *e* and *p* are mutual translations if there is at least one possible lexical alignment among them, i.e., correspondences among their words and/or phrases. Assuming the set of all possible alignments among *e* and *p* to be *a*, obtaining $P(e \mid p)$ can be seen as the maximization of the sum of individual contributions of every single alignment a_i, a process called *decoding*:

$$P(p \mid e) = P(p)\sum_i P(a_i, e \mid p)$$

The expression $P(a_i, e \mid p)$ can be obtained empirically from a number of parameters intended to capture relevant aspects of the translation process. In this work we use the basic IBM 4 model [10], in which the following parameters are defined:

- *<a>* the set of alignments for *<e, p>*.
- ε the probability of a target-sentence of length *m* being the translation of a source-sentence of length *l*.
- *t* the probability of *e* being a translation of *p*.
- *d* the probability of a target word being placed at the jth position in the target sentence given that it is a translation of a source word at the ith position in the source sentence.
- p_1 the probability of obtaining a spurious word.
- p_0 (or $1 - p_1$) the probability of not obtaining a spurious word.
- φ_0 the fertility of spurious cases[1].
- φ_i fertility rate associated with the ith word.
- *n* the probability of a word p_i having fertility φ_i.

[1] Defined as the <NULL> symbol in the word alignment – see Example 2 and 3 in Section 3.

Fertility is defined as the number of target words generated from a given source word. A zero fertility value corresponds to a deletion operation. A spurious word appears so to speak "spontaneously" in the target sentence when there is no corresponding source word. This is modelled by the parameters p_1 and φ_0 and the NULL symbol in the word alignment representation (see Section 3.)

The above parameters are actually the same ones defined in model 3 described in [10]. However, the presently discussed model 4 also distinguishes among three word types (heads, non-heads and NULL-generated), allowing for more control over word order. For the interested reader, the complete model is defined as

$$P(a,e \mid p) = \binom{m-\varphi_0}{\varphi_0} p_0^{(m-2\varphi_0)} p_1^{\varphi_0} \prod_{i=1}^{l} n(\varphi_i \mid p_i) \prod_{i=1}^{l} \prod_{k=1}^{\varphi_i} t(\tau_{ik} \mid p_i) \times$$

$$\times \prod_{i=1,\varphi_i>0}^{l} d_1(\pi_{i1} - c_{\rho i} \mid class(e_{pi}), class(\tau_{i1})) \prod_{i=1}^{l} \prod_{k=2}^{\varphi_i} d_{>1}(\pi_{ik} - \pi_{i(k-1)} \mid class(\tau_{ik})) \prod_{k=1}^{\varphi_0} t(\tau_{0k} \mid NULL)$$

For more details, we refer to [10]. In the reminder of this paper we will focus on the above model only.

3 Training

In order to build a translation model as described in the previous section, we collected 645 Portuguese-Spanish text pairs from the Environment, Science, Humanities, Politics and Technology supplements of the on-line edition of the "Revista Pesquisa FAPESP", a Brazilian magazine on scientific news. The corpus consists of 17,681 sentence pairs comprising 908,533 words in total (being 65,050 distinct.) The Portuguese version consists of 430,383 words (being 32,324 distinct) and the Spanish version consists of 478,150 words (being 32,726 distinct.) We are aware that our data set is considerably smaller than standard training data used in statistical MT (standard training data in the area comprises about 200 million words, while the Google team have been using over 1 billion words in their experiments.) However, as we shall see in the next sections, the amount of training data that we used in this work turned out to be sufficient for our initial investigation.

Portuguese sentential segmentation was performed using SENTER [2]. The tool was also employed in the segmentation of the Spanish texts with a number of changes (e.g., in order to handle Spanish abbreviations.) Despite the similarities between the two languages, it is immediate to observe that word-to-word translation is not feasible, as Example 1 should make clear: besides the differences in word order, there are subtle changes in meaning (e.g., "espantosa" vs. "impresionante", analogous to "amazing" vs. "impressive"), additional words (e.g. "ubicada") and so on.

Example 1. A Portuguese text fragment (left) and corresponding Spanish translation (right)

Ao desencadear uma cascata de eventos físico-químicos poucos quilômetros acima da floresta, a espantosa concentração de aerossóis na Amazônia no auge da estação (...)	Esa impresionante concentración de aerosoles en la Amazonia, al desencadenar una cascada de eventos fisicoquímicos ubicada a algunos kilómetros arriba del bosque, en el auge de la estación (...)

For the purposes of this work, a *sentence alignment* a is taken to be an ordered set of p(a) sentences in our Portuguese corpus and an ordered set of s(a) related sentences in the Spanish corpus. Values of p(a) and s(a) can vary from zero to an arbitrary large number. For example, a Portuguese sentence may correspond to exactly one sentence in the Spanish translation, and such 1-to-1 relation is called a *replacement* alignment. On the other hand, if a Portuguese sentence is simply omitted from the Spanish translation then we have a 1-to-0 alignment or *deletion*. In our work we are interested in replacement alignments only. This will not only reduce the computational complexity of our next task – word alignment – but will also provide the required input format for MT tools that we use, such as GIZA++ [5].

For the sentence alignment task, we used an implementation of the *Translation Corpus Aligner* (TCA) method [3] called TCAalign [4]. The choice was based on the high precision rates reported for Portuguese-English (97.10%) and Portuguese-Spanish (93.01%) language pairs [4]. The set of alignments produced by TCAalign consists of m-to-n relations marked with XML tags. As our goal is to produce an aligned corpus as accurate as possible, the data was inspected semi-automatically for potential misalignments (which were in turn collapsed into appropriate 1-to-1 alignments.) As a result of the sentence alignment procedure, 10% of the alignments in our corpus were classified as unsafe, and their manual inspection revealed that 1,668 instances (9.43%) were indeed incorrect.

A major source of misalignment was due to segmentation errors, which caused two or more sentences to be regarded as a single unit. These cases were adjusted manually so that the resulting corpus contained a set of Portuguese sentences and their Spanish counterparts in 1-to-1 relationships. There were also cases in which *n* Portuguese sentences were (correctly) aligned to *n* Spanish sentences in a different order, and had to be split into individual 1-to-1 alignments. Some kinds of misalignment were introduced by the alignment tool itself, and yet others were simply due to different choices in translation leading to correct n-to-m alignments. Since punctuation will be removed in the generation of our translation models, in these cases it was possible - when there was no change in meaning - to manually split the Spanish sentence and create two individual 1-to-1 alignments.

Two versions of the corpus have been produced: one represents the aligned corpus in its original format (as the examples in the previous section), with capital letters, punctuation marks and alignment tags; the other represents the aligned corpus in GIZA++ [5] format, in which the entire text was converted to lower case, punctuation marks and tags were removed, and the correspondence between sentences is given simply by their relative position within each text (i.e., with each sentence representing one line in the text file.)

We used GIZA++ to align the second version of the corpus at word level. A word alignment contains a sentence pair identification (sentence pair id, source and target sentence length and the alignment score) and the target and source sentences, in that order. The word alignment information is represented entirely in the source sentence. Each source word is followed by a reference to the corresponding target, or an empty set { } for source words that do not have a correspondence, i.e., n-0 (deletion) relationships. The source sentence contains also a NULL set representing the target words that do not occur in the source, i.e., 0-n (inclusion) relationships.

The following examples illustrate the output of the word alignment procedure for two Spanish-Portuguese sentence pairs. In the first case, the source words "de" and "los" were not aligned with any word in the target sentence, i.e., they disappeared in the Portuguese translation. On the other hand, all Portuguese words were aligned with their Spanish correspondents as shown by the empty NULL set. In the second case, the source words "la", "de" and "el" were not aligned to any Portuguese words, but "niño" was aligned with two of them (i.e., "el niño".) In the same example the NULL set shows that the target word 4 (i.e., "do") is not aligned with any Spanish words, i.e., it was simply included in the Portuguese translation.

Example 2. Deletion of source (Spanish) words "de" and "los"

Sentence pair (1) source length 8 target length 6 alignment score : 0.00104896
os dentes do mais antigo orangotango
NULL ({ }) hallan ({ 1 }) dientes ({ 2 }) del ({ 3 }) más ({ 4 }) antiguo ({ 5 }) de ({ }) los ({ }) orangutanes ({ 6 })

Example 3. Deletion of source (Spanish) words "la", "de" and "el", inclusion of target (Portuguese) word "do" and 1-2 alignment of "niño" with "el niño".

Sentence pair (14) source length 8 target length 6 alignment score : 1.28219e-08
salinidade indica chegada do el niño
NULL ({ 4 }) la ({ }) salinidad ({ 1 }) indica ({ 2 }) la ({ }) llegada ({ 3 }) de ({ }) el ({ }) niño ({ 5 6 })

The word alignments were generated using the following GIZA++ parameters. The probability p_0 was left to be determined automatically; language model parameters (smoothing coefficient, minimum and maximum frequencies and smoothing constant for null probabilities) were left at their default values; the maximum sentence length was set to 182 words as seen in our corpus, and the maximum fertility parameter was set to 10 (i.e., word fertilities will range from 0 to 9) as suggested in [5]. As a result, 489,594 word alignments were produced according to the following distribution.

Table 1. Word alignment distribution

Alignment Type	Instances	Probability
0-1	15040	3.072 %
1-0	71543	14.613 %
1-1	398175	81.328 %
1-2	3344	0.683 %
1-3	791	0.167 %
1-4	305	0.062 %
1-5	187	0.038 %
1-6	85	0.017 %
1-7	42	0.009 %
1-8	27	0.005 %
1-9	55	0.011 %

About 82% of all alignments were of the word-to-word type. Moreover, very few mappings (701 cases or 0.14% in total) involved more than three words in the target language (i.e., alignments 1-4 to 1-9.) These results may suggest a strong similarity between Portuguese and Spanish as argued in [1] for the languages spoken in Spain.

4 Decoding and Testing

The translation model described in Section 2 and associated data (e.g., word class tables, zero fertility lists etc. - see [5] for details) were generated from a small training set of about 17,000 sentence pairs using a simple trigram-based language model produced by the CMU-Cambridge Tool Kit [8]. Both Spanish-Portuguese and Portuguese-Spanish translations were then decoded using the ISI ReWrite Decoder tool [9] as follows:

Spanish-Portuguese translator: $P(p) = P(p)P(e \mid p)$
Language Model (Portuguese): $P(p)$
Translation Model (Portuguese-Spanish): $P(e \mid p)$

Portuguese-Spanish translator: $P(e) = P(e)P(p \mid e)$
Language Model (Spanish): $P(e)$
Translation Model (Spanish-Portuguese): $P(p \mid e)$

Recall from Section 2 that given a translation from Spanish to Portuguese, the decoding process intends to find the sentence p that maximizes

$$P(p \mid e) = P(p)\sum_i P(a_i, e \mid p)$$

The decoding task was based on the IBM model 4 (described in Section 2) using 5 iterations during training procedure. The selected decoding strategy was the A* search algorithm described in [5] using the translation heuristics for maximizing $P(t \mid s) * P(s \mid t)$.

For evaluation purposes, a test set of 649 previously unseen sentence pairs was taken from recent issues of "Revista Pesquisa FAPESP" and it was used as input to our translator. In the Appendix at the end of this paper we present a number of instances of Spanish-Portuguese (Table 4) and Portuguese-Spanish (Table 5) translations obtained in this way. In each example, the first line shows a test sentence in the source language; the second line shows the reference (i.e., human-made) translation, and the third line presents the output of our statistical machine translator. Occasional target words shown in the original (source) language are due to the overly small size of out training data.

Using the same test data, we compared the BLEU scores [6] obtained by our statistical approach to those obtained by the rule-based system Apertium [1]. Briefly, BLEU is an automatic evaluation metric which computes the number of n-grams shared between an automatic translation and a human (reference) translation. The higher the BLEU score, the better the translation quality. BLEU scores have been shown to be comparable to human judgements about translated texts. Accordingly, BLEU has been extensively used in MT evaluation. Table 2 summarizes our findings.

Table 2. BLEU scores for Statistical versus rule-based translations

Approach	Portuguese-Spanish	Spanish-Portuguese
Statistical	0.5673	0.5734
Apertium	0.6282	0.5955

From the above results a number of observations can be made. Firstly, although the Apertium system slightly outperforms our statistical approach, their BLEU scores remain nevertheless remarkably close. This encouraging first trial allows us to predict that these results would most likely be the other way round had we used a more realistic amount of training data[2]. The growth potential of our approach is still vast, and it can be achieved at a fairly low cost. This seems less obvious in the case of tailor-made translation rules. Secondly, our development efforts probably took only a fraction of the time that would be required for developing a comparable rule-based system, and that would remain the case even if we limited ourselves to one-way translation. Our system naturally produces two translation models, and hence translates both ways, whereas a rule-based system would require two sets of language-specific translation rules to be developed by language experts. Finally, it should be pointed out that as language evolves and translation rules need to be revised, a rule-based system may ultimately have to be re-written from scratch, whereas a statistical translation model merely requires additional (e.g., more recent) instances of text to adapt.

It may be argued however that BLEU does not capture translation quality accurately in the sense that it does not reflect the degree of difficulty experienced by humans in the task of post-edition. For that reason, we decided to carry out a (manual) qualitative evaluation of both statistical and rule-based MT in the Spanish-Portuguese direction. More specifically, we compute the number of lexical and syntactical errors as suggested in [11] and word error rates (WER) as follows:

$$WER = \frac{I + D + R}{S}$$

In the above, 'I' stands for the number of insertion steps necessary to transform the automatic translation into the reference one, 'D' is the number of deletions, 'R' is the number of replacements and 'S' is the number of words in a source reference set.

A random selection of 20 instances of translations (482 words in total) taken from our test data was analysed at lexical and syntactical levels. The lexical level takes into account *dictionary* errors (e.g., words not translated or wrong translations of proper names, abbreviations, etc.), *homonyms* (wrong translation choices), *idioms* (incorrect or word-to-word translation of idiomatic expressions), and *connotative* errors (literal translation of expressions with no literal meaning.) At the syntactic level we looked into sentence structure, i.e., errors stemming from verb or noun agreement, the choice of determiners and prepositions, and the appropriate use of verbal complements. The results of the analysis are show in Table 3 below, in which lower WERs are better.

[2] And possibly a more powerful language model e.g., [7].

Table 3. Statistical versus rule-based Spanish-Portuguese translations

Approach	Lexical Level				Syntactic Level	WER
	Dictionary	Homonyms	Connotative	Idioms		
Statistical	19	7	0	2	32	0.3216
Apertium	40	5	0	2	24	0.2635

At first glance the results show that the statistical approach presented a higher WER, which may suggest that its output would require more effort to become identical to the reference translation if compared to Apertium. However, we notice that the WER difference was mainly due to the lack of preposition + determiner contractions as in "de" (of) + "a" (the) = "da", causing WER to compute one replacement and one deletion for each translation. Thus, given that preposition post-processing can be trivially implemented, we shall once again interpret these results as indicative of comparable translation performances under the assumption that the present difficulties could be overcome by using a larger training data set. Due to the highly subjective nature of this evaluation method, however, we believe that more work on this issue is still required.

5 Final Remarks

In this paper we have described a first experiment in Portuguese-Spanish statistical machine translation using a small parallel corpus as training data. Preliminary results seem to confirm long held claims of the statistical MT community suggesting that this approach may be indeed comparable to rule-based MT in both performance (particularly as measured by BLEU scores) and development costs.

We now intend to expand the current training data to the (much higher) levels commonly seen in modern work in the field, and possibly make use of a more sophisticated translation model. We are also aware that the translation task between such 'closely-related' languages is somewhat simpler and does not reveal the full extent of translation difficulties found in other (i.e., so-called 'distant') language pairs. For that reason, we will also investigate the translation of Portuguese texts to and from English, having as an ultimate goal the design of a robust, state-of-art statistical translation model of practical use for Brazilian Portuguese speakers.

Acknowledgments

The authors acknowledge support from FAPESP (2006/04803-7, 2006/02887-9 and 2006/03941-7) and CNPq (484015/2007 9.)

References

1. Corbí-Bellot, A.M., Forcada, M.L., Ortiz-Rojas, S., Pérez-Ortiz, J.A., Ramírez-Sánchez, G., Sánchez-Martínez, F., Alegria, I., Mayor, A., Sarasola, K.: An open-source shallow-transfer machine translation engine for the romance languages of Spain. In: 10th Annual Conference of the European Association for Machine Translation, pp. 79–86 (2005)

2. Pardo, T.A.S.: SENTER: Um Segmentador Sentencial Automático para o Português do Brasil. NILC Technical Reports Series NILC-TR-06-01. University of São Paulo, São Carlos, Brazil (2006)
3. Hofland, K., Johansson, S.: The Translation Corpus Aligner: A program for automatic alignment of parallel texts. In: Johansson, S., Oksefjell, S. (eds.) Corpora and Cross-linguistic research Theory, Method, and Case Studies. Rodopi, Amsterdam (1998)
4. Caseli, H.M.: Indução de léxicos bilíngües e regras para a tradução automática. Doctoral thesis, University of São Paulo, São Carlos, Brazil (2007)
5. Och, F.J., Ney, H.: A Systematic Comparison of Various Statistical Alignment Models. Computational Linguistics 29(1), 19–51 (2003)
6. Papineni, K., Roukos, S., Ward, T., Zhu, W.: BLEU: a Method for Automatic Evaluation of Machine Translation. In: Proceedings of he 40th Annual Meeting of the Association for Computational Linguistics, pp. 311–318 (2002)
7. Chen, S.F., Goodman, J.: An empirical study of smoothing techniques for language modeling. Computer Speech and Language 13, 359–394 (1999)
8. Clarkson, P.R., Rosenfeld, R.: Statistical Language Modeling Using the CMU-Cambridge Toolkit. In: Proceedings of ESCA Eurospeech (1997)
9. Germann, U., Jahr, M., Knight, K., Marcu, D., Yamada, K.: Fast Decoding and Optimal Decoding for Machine Translation. In: Proceedings of the 39th Annual Meeting of the Association for Computational Linguistics (2001)
10. Brown, P.E., Pietra, S.A.D., Pietra, V.J.D., Mercer, R.L.: The Mathematics of Statistical Machine Translation: Parameter Estimation. Computational Linguistics 16(2), 79–85 (1993)
11. Oliveira Jr., O.N., Marchi, A.R., Martins, M.S., Martins, R.T.: A Critical Analysis of the Performance of English-Portuguese-English MT Systems. V Encontro para o processamento computacional da Língua Portuguesa Escrita e Falada (2000)

Appendix – Examples of Human and Machine Translation

Table 4. Spanish source sentences followed by (r) Portuguese reference, (s) statistical and (a) rule-based (Apertium) machine translations

si fuera así probaremos con otro hasta lograr el resultado deseado

(r) mas aí testaremos outro até conseguirmos o resultado desejado
(s) se fora assim probaremos com outro até conseguir o resultado desejado
(a) se fosse assim provaremos com outro até conseguir o resultado desejado

en esa región viven cerca de 40 mil personas en nueve favelas y tres conjuntos habitacionales

(r) na região vivem cerca de 40 mil pessoas em nove favelas e três conjuntos habitacionais
(s) em essa região vivem cerca de 40 mil pessoas em nove favelas e três conjuntos habitacionales
(a) nessa região vivem cerca de 40 mil pessoas em nove favelas e três conjuntos habitacionais

sus textos escritos en lengua cuneiforme en forma de cuña grabada en tablas de barro son poesías en homenaje a una diosa llamada inanna adorada por la sacerdotisa

(r) seus textos escritos na linguagem cuneiforme em forma de cunha gravada em tábuas de barro são poesias em homenagem a uma deusa chamada inanna adorada pela sacerdotisa
(s) seus textos escritos em língua cuneiforme em forma de cuña gravada em tábuas de barro são poesias em homenagem à uma deusa chamada inanna adorada por a sacerdotisa
(a) seus textos escritos em língua cuneiforme em forma de cuña gravada em tabelas de varro são poesias em homenagem a uma deusa chamada inanna adorada pela sacerdotisa

Table 5. Portuguese source sentences followed by (r) Spanish reference, (s) statistical and (a) rule-based (Apertium) machine translations

mas aí testaremos outro até conseguirmos o resultado desejado

(r) si fuera así probaremos con otro hasta lograr el resultado deseado
(s) pero entonces testaremos otro hasta logremos el resultado deseado
(a) pero ahí probaremos otro hasta conseguir el resultado deseado

na região vivem cerca de 40 mil pessoas em nove favelas e três conjuntos habitacionais

(r) en esa región viven cerca de 40 mil personas en nueve favelas y tres conjuntos habitacionales
(s) en región viven alrededor de 40 mil personas en nueve favelas y tres conjuntos habitacionais
(a) en la región viven cerca de 40 mil personas en nueve favelas y tres conjuntos habitacionales

seus textos escritos na linguagem cuneiforme em forma de cunha gravada em tábuas de barro são poesias em homenagem a uma deusa chamada inanna adorada pela sacerdotisa

(r) sus textos escritos en lengua cuneiforme en forma de cuña grabada en tablas de barro son poesías en homenaje a una diosa llamada inanna adorada por la sacerdotisa
(s) sus textos escritos en lenguaje cuneiforme en forma de cunha grabada en tablas de barro son poesías en homenaje la una diosa llamada inanna adorada por sacerdotisa
(a) sus textos escritos en el lenguaje cuneiforme en forma de cunha grabada en tábuas de barro son poesías en homenaje a una diosa llamada inanna adorada por la sacerdotisa

On the Automatic Learning of Bilingual Resources: Some Relevant Factors for Machine Translation

Helena de M. Caseli, Maria das Graças V. Nunes, and Mikel L. Forcada

[1] NILC – ICMC, University of São Paulo
CP 668P – 13.560-970 – São Carlos – SP – Brazil
helename,gracan@icmc.usp.br
[2] Departament de Llenguatges i Sistemes Informàtics,
Universitat d'Alacant, E-03071 Alacant, Spain
mlf@ua.es

Abstract. In this paper we present experiments concerned with automatically learning bilingual resources for machine translation: bilingual dictionaries and transfer rules. The experiments were carried out with Brazilian Portuguese (pt), English (en) and Spanish (es) texts in two parallel corpora: pt–en and pt–es. They were designed to investigate the relevance of two factors in the induction process, namely: (1) the coverage of linguistic resources used when preprocessing the training corpora and (2) the maximum length threshold (for transfer rules) used in the induction process. From these experiments, it is possible to conclude that both factors have an influence in the automatic learning of bilingual resources.

Keywords: Machine translation, bilingual resources, automatic learning, parallel corpora.

1 Introduction

The ability to translate from one language to another has became not only a desirable but also a fundamental skill in the multilingual world every day accessible through the Internet. Due to working or social needs, there is a growing necessity of being able to get at least the gist of a piece of information in a different language. Unfortunately, this ability is not inherent.

The good news is that computers are getting more and more available to ease this task. This is one of the challenges of machine translation (MT) research and also of this paper. In particular, this paper is concerned not only with translating from one language to another but also with the automatic learning of bilingual resources useful for rule-based machine translation (RBMT) systems.

Traditionally, the bilingual resources used in RBMT systems are built by means of hard manual work. However, in the last years several methods have been proposed to automatically learn bilingual resources such as dictionaries [1,2,3,4] and transfer rules [5,6,7] from translation examples (parallel corpora).

G. Zaverucha and A. Loureiro da Costa (Eds.): SBIA 2008, LNAI 5249, pp. 258–267, 2008.

A bilingual dictionary is a bilingual list of words and multiword units (possibly accompanied by morphological information) that are mutual translations. A transfer (or translation) rule, in turn, is a generalization of structural, syntactic or lexical correspondences found in the parallel sentences (translation examples).

In line with these initiatives, this paper presents experiments carried out to investigate the relevance of certain factors when automatically learning bilingual dictionaries and transfer rules following the methodology of the ReTraTos project.[1]

Following this methodology, the bilingual dictionaries and the transfer rules are induced from automatically word-aligned (or lexically aligned) parallel corpora processed with morphological analysers and part-of-speech (PoS) taggers. The aspects under investigation in this paper are: (1) the coverage of linguistic resources used in preprocessing the training corpora and (2) the maximum length threshold used in the transfer rule induction process, that is, the number of source items that a transfer rule can contain. The evaluation of the translation quality was carried out with Brazilian Portuguese (`pt`), Spanish (`es`) and English (`en`) texts in two parallel corpora —`pt–es` and `pt–en`— by using the automatic metrics BLEU [8] and NIST [9].

Our interest in the above factors is motivated by previous results [10]. We now intend to investigate if a better coverage of the preprocessing linguistic resources can bring better MT results, and if the previous poor `pt–en` MT performance was due to the threshold length (too restrictive) used during the induction of the transfer rules. We think that the small length of source patterns was not sufficient to learn relevant syntactic divergences between those languages. To our knowledge, it is the first time that such a study is carried out for Brazilian Portuguese or other languages.

This paper is organized as follows. Section 2 presents related work on automatic induction of bilingual dictionaries and transfer rules. Section 3 describes briefly the induction methodology under study in this paper. Section 4 shows the experiments and their results, and section 5 ends this paper with some conclusions and proposals for future work.

2 Related Work

According to automatic evaluation metrics like BLEU [8] and NIST [9], the phrase-based statistical MT (SMT) systems such as [11] and [12] are the state-of-the-art in MT. Despite this fact, this paper is concerned with RBMT since the symbolic resources (dictionaries and rules) of RBMT suit better than the models of SMT to our research purpose: to know how translation is performed and what affects its performance.

This section presents briefly some of the methods proposed in the literature to automatically induce bilingual dictionaries and transfer rules. These methods can follow many different approaches, but usually they perform induction from a sentence-aligned parallel corpus.

[1] `http://www.nilc.icmc.usp.br/nilc/projects/retratos.htm`

Usually, a bilingual dictionary is obtained as a by-product of a word alignment process [13,14,15]. In [1], for example, an English–Chinese dictionary was automatically induced by means of training a variant of the statistical model described in [13]. By contrast, the method proposed in [2] uses a non-aligned parallel corpus to induce bilingual entries for nouns and proper nouns based on co-occurrence positions. Besides the alignment-based approaches, others have also been proposed in the literature such as [3] and [4]. While [3] builds a bilingual dictionary from unrelated monolingual corpora, [4] combines two existing bilingual dictionaries to build a third one using one language as a bridge.

The transfer rule induction methods also use the alignment information to help the induction process. For example, the method proposed in [6] uses shallow information to induce transfer rules in two steps: monolingual and bilingual. In the monolingual step, the method looks for sequences of items that occur at least in two sentences by processing each side (source or target) separately —these sequences are taken as monolingual patterns. In the bilingual step, the method builds bilingual patterns following a co-occurrence criterion: one source pattern and one target pattern occurring in the same pair of sentences are taken to be mutual translations. Finally, a bilingual similarity (distance) measure is used to set the alignment between source and target items that form a bilingual pattern.

The method proposed in [7], by its turn, uses more complex information to induce rules. It aligns the nodes of the source and target parse trees by looking for word correspondences in a bilingual dictionary. Then, following a best-first strategy (processing first the nodes with the best word correspondences), the method aligns the remaining nodes using a manually defined alignment grammar composed of 18 bilingual compositional rules. After finding alignments between nodes of both parse trees, these alignments are expanded using linguistic constructs (such as noun and verb phrases) as context boundaries.

The method in [16] infers hierarchical syntactic transfer rules, initially, on the basis of the constituents of both (manually) word-aligned languages. To do so, sentences from the language with more resources (English, in that case) are parsed and disambiguated. Value and agreement constraints are set from the syntactic structure, the word alignments and the source/target dictionaries. Value constraints specify which values the morphological features of source and target words should have (for instance, masculine as gender, singular as number and so on). The agreement constraints, in turn, specify whether these values should be the same.

Finally, in [17] the authors used an aligned parallel corpus to infer shallow-transfer rules based on the alignment templates approach [12]. This research makes extensive use of the information in an existing manually-built bilingual dictionary to guide rule extraction.

3 Induction and Translation in the ReTraTos Environment

The general scheme of the induction and translation in the ReTraTos environment is shown in Figure 1. The input for both induction systems (bilingual

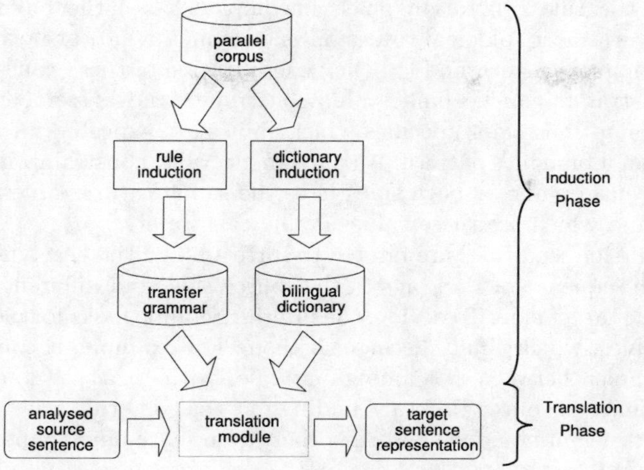

Fig. 1. Scheme of the induction and translation phases in the ReTraTos environment

dictionary and transfer rule) is a PoS-tagged and word-aligned parallel corpus. After having been induced, the resources —transfer grammar and bilingual dictionary— are used to translate source sentences into target sentences.

3.1 Induction in the ReTraTos Environment

The induction processes are briefly described in this section. A more detailed description of these processes can be found in [10] and [18].

The bilingual dictionary induction process comprises the following steps: (1) the compilation of two bilingual dictionaries, one for each translation direction (one source–target and another target–source); (2) the merging of these two dictionaries in one specifying the valid translation direction if necessary; (3) the generalization of morphological attribute values in the bilingual entries; and (4) the treatment of morphosyntactic differences related to entries in which the value of the target gender/number attribute has to be determined from information that goes beyond the scope of the bilingual entry itself.[2]

The rule induction method, in turn, induces the transfer rules following four phases: (1) pattern identification, (2) rule generation, (3) rule filtering and (4) rule ordering. Firstly, similarly to [6], the bilingual patterns are extracted in two steps: monolingual and bilingual. In the monolingual step, source patterns are identified by an algorithm based on the *sequential pattern mining* technique and the PrefixSpan algorithm [19]. In the bilingual step, the target items aligned to each source pattern are looked for (in the parallel translation example) to form the bilingual pattern.

[2] For example, the es noun *tesis* (thesis) is valid for both number (singular and plural) and it has two possible pt translations: *tese* (singular) and *teses* (plural).

Secondly, the rule generation phase encompasses: (a) the building of constraints between morphological values on one (monolingual) or both (bilingual) sides of a bilingual pattern and (b) the generalization of these constraints. Two kinds of constraints can be built: value constraints and agreement/value constraints. A value constraint specifies which values are expected for the features on each side of a bilingual pattern. An agreement/value constraint, in turn, specifies which items on one or both sides have the same feature values (agreement constraint) and which are these values (value constraint).[3]

Thirdly, the induced rules are filtered to solve ambiguities. An ambiguous rule has the same sequence of PoS tags in the source side, and different PoS tag sequences in the target side. To decide, the filtering module looks for morphological or lexical values, which could distinguish them. For example, it could be possible to distinguish between two ambiguous rules with "`n adj`" (noun adjective) as their sequence of source PoS tags finding out that one rule was induced from examples with feminine nouns and the other, from masculine nouns as stated in their constraints.

Finally, the rule ordering specifies the order in which transfer rules should be applied. It is done implicitly by setting the occurrence frequency of each rule, each target side and each constraint set. The occurrence frequency of a rule is the number of times its source sequence of PoS tags was found in the training corpus. Then, for each rule, the occurrence frequency of a target side (constraint set) is the number of times this target side (constraint set) was found for this specific rule, in the training corpus. The frequencies are used to choose the "best suitable rule" as explained in the next section.

3.2 Translation in the ReTraTos Environment

As shown in Figure 1, the induced sets of transfer rules (transfer grammar) and bilingual entries (bilingual dictionary) are used by the MT module to translate an already analysed source sentence into a representation of a target sentence. The analysed source sentence and the representation of a target sentence are both sequences of *lexical forms*, each composed of a lemma, PoS tag, and morphological information.

The ReTraTos MT module, besides looking for translations in the bilingual dictionary, applies the best suitable transfer rules following a left-to-right longest-match procedure. The "best suitable rule" is the most frequent rule which: (a) matches the source pattern (sequence of source PoS tags), (b) matches one of the associated sets of source constraints (there can be more than one) and (c) the matching source constraint set is the most frequent. Therefore, the selected rule might not be the most frequent. Finally, if there is more than one "best suitable rule", the rule defined first is the one chosen.

A backtracking approach is used in rule application: if a source pattern *abcd* matches the input sentence, but cannot be applied, because it has no compatible

[3] Our *value* constraints are like in [16], but our *agreement/value* constraints are different from their *agreement* constraints since, here, the values are explicitly defined.

constraint, the system will try to apply the sub-pattern *abc*. This backtracking goes on until the sub-pattern has just one item and, in this case, word-by-word translation is applied to one word and the process restarts immediately after it.

4 Experiments and Results

The training and test/reference parallel corpora used in these experiments are described in section 4.1. The training corpora were used to induce the bilingual resources while the test/reference corpora were used to evaluate the performance of the translation based on the induced rules. The effect of the investigated factors in the induction of bilingual resources was measured by means of the BLEU [8] and NIST [9] measures, which give an indication of translation performance. Both take into account, in different ways, the number of n-grams common to the automatically translated sentence and the reference sentence, and estimate the similarity in terms of length, word choice and order.

A baseline configuration was defined as the one resulting from the ReTraTos project and explained in section 4.2. The investigated factors were confronted with the baseline configuration as explained in the following sections: (4.3) the coverage of the preprocessing resources and (4.4) the maximum length threshold.

4.1 Parallel Corpora

The experiments described in this paper were carried out using the training and the test/reference `pt–es` and `pt–en` parallel corpora. These corpora contain articles from a Brazilian scientific magazine, *Pesquisa FAPESP*.[4]

The training corpora were preprocessed as explained below. First, they were automatically sentence-aligned usingan implementation of the Translation Corpus Aligner [20]. Then, both corpora were PoS-tagged using the morphological analyser and the PoS tagger available in the `Apertium`[5] open-source machine translation platform. The morphological analysis provides one or more *lexical forms* or analyses (information on lemma, lexical category and morphological inflection) for each surface form using a monolingual morphological dictionary (MD). The PoS tagger chooses the best possible analysis based on a first-order hidden Markov model (HMM) [21].

To improve the coverage of the morphological analyser, the original MDs were enlarged with entries from `Unitex`[6] (`pt` and `en`) and from `interNOSTRUM`[7] (`es`).[8] The number of surface forms covered by the original and the extended versions

[4] `http://revistapesquisa.fapesp.br`.

[5] The open-source machine translation platform `Apertium`, including linguistic data for several language pairs and documentation, is available at `http://www.apertium.org`.

[6] `http://www-igm.univ-mlv.fr/~unitex/`.

[7] `http://www.internostrum.com/`.

[8] The original MDs are available as part of the `apertium-es-pt` (version 0.9) and the `apertium-en-ca` (version 0.8). The new `es` entries derived from `interNOSTRUM` were provided by the Transducens research group from the Universitat d'Alacant.

of the morphological dictionaries are: 128,772 vs. 1,136,536 for pt, 116,804 vs. 337,861 for es and 48,759 vs. 61,601 for en. This extension decreased the percentage of unknown words in the corpora used in our experiments from 11% to 3% for pt, from 10% to 5% for es and from 9% to 5% for en. The effect of this increase of coverage in the induction of bilingual resources is one of the factors under investigation in this paper (see section 4.3).

Finally, the translation examples were word-aligned using LIHLA [15] for pt–es and GIZA++ [14] for pt–en as explained in [10]. The translation examples were aligned in both directions and the union was obtained as in [22].

The pt–es training corpus consists of 18,236 pairs of parallel sentences with 503,596 tokens in pt and 545,866 in es. The pt–en training corpus, in turn, has 17,397 pairs of parallel sentences and 494,391 tokens in pt and 532,121 in en.

The *test corpus* consists of 649 parallel sentences with 16,801 tokens in pt, 17,731 in es and 18,543 in en (about 3.5% of the size of training corpora). The pt–es and pt–en *reference corpora* were created from the corresponding parallel sentences in the test corpus.

4.2 The Baseline

The baseline configuration used in our experiments is the automatic translation produced by ReTraTos (see section 3.2) based on the parallel corpora preprocessed with the *extended MDs* and using a *maximum length threshold* for rule induction set to *5*. Table 1 shows the values of BLEU and NIST for the translation performed in all possible directions and using the bilingual resources induced for both pairs of languages.

Table 1. Values of BLEU and NIST for ReTraTos translation

	pt–es		es–pt		pt–en		en–pt	
	BLEU	**NIST**	**BLEU**	**NIST**	**BLEU**	**NIST**	**BLEU**	**NIST**
baseline	65.13	10.85	66.66	10.98	28.32	7.09	24.00	6.11

4.3 The Coverage of Preprocessing Resources

The first factor investigated in this paper is related to the coverage of the MDs used in the morphological analysis of the training corpora. The effect of using the richer/extended MDs (the baseline) instead of the original ones (see section 4.1) was evaluated in terms of (1) the increase in the number of entries in the induced bilingual dictionary and (2) the effect on overall translation performance.

The number of bilingual entries induced from the training corpora preprocessed with the extended MDs was *twice* as large as using the original ones: 7,862 vs. 4,143 (pt–es) and 10,856 vs. 5,419 (pt–en). The effect on the overall translation was measured by means of BLEU and NIST as shown in Table 2.

From Table 2, it is possible to notice that, for the pt–es language pair, the extended MDs brought about an increase in the measures in relation to the original MDs: 1.96–3.23 points in BLEU and 0.22–0.36 points in NIST. About 23–24% of the sentences in pt–es test corpus were translated differently by the

Table 2. Values of BLEU and NIST according to the original and the extended MDs

MD	pt–es		es–pt		pt–en		en–pt	
	BLEU	**NIST**	**BLEU**	**NIST**	**BLEU**	**NIST**	**BLEU**	**NIST**
Extended	65.13	10.85	66.66	10.98	28.32	7.09	24.00	6.11
Original	63.17	10.63	63.43	10.62	28.12	7.00	23.72	6.08

resources induced from each version of MD. The improvement in `pt–es` language pair is mainly due to the new transfer rules generated by the extended MDs —293 `pt–es` and 258 `es–pt`— since the translations obtained using just the bilingual dictionaries (the word-by-word translation) give only a small improvement of 0.21–1.61 points in BLEU and 0.24–0.28 points in NIST.

On the other hand, the values of BLEU and NIST are roughly the same for `pt–en` increasing just 0.20–0.28 points in BLEU and 0.03–0.09 points in NIST. Although 17–21% of the sentences in `pt–en` test corpus were translated differently by the resources induced from each version of MD the values of BLEU and NIST do not change significantly. For `pt–en`, the new 167 `pt–en` and 159 `en–pt` transfer rules did not lead to better values of these measures.

From these values it is possible to conclude that more morphological information has a larger effect on related language pairs, but it does not seem to have influence on the translation for pairs of more distant languages. This fact shows that for more distant languages more than just an improvement in preprocessing linguistic resources is needed to achieve better values for BLEU and NIST.

4.4 The Maximum Length Threshold

An experiment was also carried out to investigate the effect of the maximum length threshold (maximum number of source items that a transfer rule can contain) used in the transfer rule induction process. Thus, the translations produced by ReTraTos based on rules induced using different maximum length thresholds (and the dictionaries generated by the baseline configuration) were analysed.

Table 3 shows the values of BLEU and NIST when the maximum length threshold was set to 3, 4, 5 (the baseline), 6, 7, 8 and 10. It is possible to see that the best maximum length threshold seems to be between 4 and 5 for both language pairs. The similar values of these metrics reflect the small percentage of sentences translated differently when the rules induced using threshold lengths of 4 and 5 were applied: 2–3% for `pt–es`, `es–pt` and `en–pt` and 12% for `pt–en`.

Table 3. Values of BLEU and NIST according to the maximum length threshold

Max. length	pt–es		es–pt		pt–en		en–pt	
	BLEU	**NIST**	**BLEU**	**NIST**	**BLEU**	**NIST**	**BLEU**	**NIST**
3	63.40	10.86	65.08	10.91	27.23	7.07	23.48	6.11
4	65.15	10.85	66.68	10.98	28.56	7.12	24.01	6.11
5	65.13	10.85	66.66	10.97	28.32	7.08	24.00	6.11
6–8	63.42	10.86	65.05	10.91	27.19	7.07	23.64	6.06
10	65.17	10.85	66.68	10.97	28.35	7.10	24.00	6.11

It is also important to say that although several new transfer rules have been generated when the length threshold increased from 4 to 5 —334 (pt–es), 337 (es–pt), 102 (pt–en) and 113 (en–pt)— less than a half of them were applied to translate the test corpora —45% (pt–es), 28% (es–pt), 30% (pt–en), and 13% (en–pt). Thus, these new rules seem not to be useful to improve translation performance as measured by BLEU and NIST.

5 Conclusions and Future Work

In this paper we described experiments carried out to investigate the effect of certain factors in the automatic induction of bilingual resources useful for RBMT considering the ReTraTos methodology. The factors under investigation were: (1) the coverage of preprocessing resources and (2) the maximum length threshold used in transfer rule induction process. From the experiments carried out on pt–es and pt–en language pairs it is possible to draw some conclusions.

First, a great coverage of the morphological dictionaries used in preprocessing the training corpora improved the translation generated based on the induced resources mainly for the pt–es pair: 1.96–3.23 points in BLEU and 0.22–0.36 points in NIST. For pt–en it seems that more than an improvement in lexical coverage is needed to improve the MT performance since the values increased just 0.20–0.28 points in BLEU and 0.03–0.09 points in NIST. Second, the maximum length threshold used in transfer rule induction process also proved to have a small influence in the translation performance and the best thresholds were 4 and 5. However, thresholds bigger than 5 did not bring measurable improvements.

Future work includes the design of new experiments to investigate the effect of other factors in the transfer rule induction process, such as the application (or not) of rule filtering and rule ordering. We also want to evaluate the translations by means of the word error rate (WER) using post-edited output as a reference. Finally, we are already carrying out experiments to compare the performance of the system presented here (using automatically induced resources) and that of a SMT system trained and tested on the same corpora.

Acknowledgements

We thank the financial support of the Brazilian agencies FAPESP, CAPES and CNPq, and of the Spanish Ministry of Education and Science.

References

1. Wu, D., Xia, X.: Learning an English-Chinese lexicon from parallel corpus. In: Proc. of AMTA 1994, Columbia, MD, pp. 206–213 (October 1994)
2. Fung, P.: A pattern matching method for finding noun and proper noun translations from noisy parallel corpora. In: Proc. of ACL 1995, pp. 236–243 (1995)
3. Koehn, P., Knight, K.: Learning a translation lexicon from monolingual corpora. In: Proc. of SIGLEX 2002, Philadelphia, pp. 9–16 (July 2002)

4. Schafer, C., Yarowsky, D.: Inducing translation lexicons via diverse similarity measures an bridge languages. In: Proc. of CoNLL 2002, pp. 1–7 (2002)
5. Kaji, H., Kida, Y., Morimoto, Y.: Learning translation templates from bilingual text. In: Proc. of COLING 1992, pp. 672–678 (1992)
6. McTait, K.: Translation patterns, linguistic knowledge and complexity in an approach to EBMT. In: Carl, M., Way, A. (eds.) Recent Advances in EBMT, pp. 1–28. Kluwer Academic Publishers, Netherlands (2003)
7. Menezes, A., Richardson, S.D.: A best-first alignment algorithm for automatic extraction of transfer mappings from bilingual corpora. In: Proc. of the Workshop on Data-driven Machine Translation at ACL 2001, Toulouse, France, pp. 39–46 (2001)
8. Papineni, K., Roukos, S., Ward, T., Zhu, W.: BLEU: a method for automatic evaluation of machine translation. In: Proc. of ACL 2002, pp. 311–318 (2002)
9. Doddington, G.: Automatic evaluation of machine translation quality using n-gram co-occurrence statistics. In: Proc. of ARPA Workshop on Human Language Technology, San Diego, pp. 128–132 (2002)
10. Caseli, H.M., Nunes, M.G.V., Forcada, M.L.: Automatic induction of bilingual resources from aligned parallel corpora: application to shallow-transfer machine translation. Machine Translation 20(4), 227–245 (2006)
11. Koehn, P., Och, F.J., Marcu, D.: Statistical phrase-based translation. In: Proc. of HLT/NAACL pp. 127–133 (2003)
12. Och, F.J., Ney, H.: The alignment template approach to statistical machine translation. Computational Linguistics 30(4), 417–449 (2004)
13. Brown, P., Della-Pietra, V., Della-Pietra, S., Mercer, R.: The mathematics of statistical machine translation: parameter estimation. Computational Linguistics 19(2), 263–312 (1993)
14. Och, F.J., Ney, H.: Improved statistical alignment models. In: Proc. of ACL 2000, Hong Kong, China, pp. 440–447 (October 2000)
15. Caseli, H.M., Nunes, M.G.V., Forcada, M.L.: Evaluating the LIHLA lexical aligner on Spanish, Brazilian Portuguese and Basque parallel texts. Procesamiento del Lenguaje Natural 35, 237–244 (2005)
16. Carbonell, J., Probst, K., Peterson, E., Monson, C., Lavie, A., Brown, R., Levin, L.: Automatic rule learning for resource-limited MT. In: Richardson, S.D. (ed.) AMTA 2002. LNCS (LNAI), vol. 2499, pp. 1–10. Springer, Heidelberg (2002)
17. Sánchez-Martínez, F., Forcada, M.L.: Automatic induction of shallow-transfer rules for open-source machine translation. In: Proc. of TMI 2007, pp. 181–190 (2007)
18. Caseli, H.M., Nunes, M.G.V.: Automatic induction of bilingual lexicons for machine translation. International Journal of Translation 19, 29–43 (2007)
19. Pei, J., Han, J., Mortazavi-Asl, B., Wang, J., Pinto, H., Chen, Q., Dayal, U., Hsu, M.: Mining sequential patterns by pattern-growth: the PrefixSpan approach. IEEE Transactions on Knowledge and Data Engineering 16(10), 1–17 (2004)
20. Hofland, K.: A program for aligning English and Norwegian sentences. In: Hockey, S., Ide, N., Perissinotto, G. (eds.) Research in Humanities Computing, pp. 165–178. Oxford University Press, Oxford (1996)
21. Armentano-Oller, C., Carrasco, R.C., Corbí-Bellot, A.M., Forcada, M.L., Ginestí-Rosell, M., Ortiz-Rojas, S., Pérez-Ortiz, J.A., Ramírez-Sánchez, G., Sánchez-Martínez, F., Scalco, M.A.: Open-source Portuguese-Spanish machine translation. In: Vieira, R., Quaresma, P., Nunes, M.d.G.V., Mamede, N.J., Oliveira, C., Dias, M.C. (eds.) PROPOR 2006. LNCS (LNAI), vol. 3960, pp. 50–59. Springer, Heidelberg (2006)
22. Och, F.J., Ney, H.: A systematic comparison of various statistical alignment models. Computational Linguistics 29(1), 19–51 (2003)

Experiments in the Coordination of Large Groups of Robots

Leandro Soriano Marcolino and Luiz Chaimowicz

VeRLab - Vision and Robotics Laboratory
Computer Science Department - UFMG - Brazil
{soriano,chaimo}@dcc.ufmg.br

Abstract. The use of large groups of robots, generally called *swarms*, has gained increased attention in recent years. In this paper, we present and experimentally validate an algorithm that allows a swarm of robots to navigate in an environment containing unknown obstacles. A coordination mechanism based on dynamic role assignment and local communication is used to help robots that may get stuck in regions of local minima. Experiments were performed using both a realistic simulator and a group of real robots and the obtained results showed the feasibility of the proposed approach.

1 Introduction

Cooperative robotics has become an important and active research field in the last couple of decades. Fundamentally, it consists of a group of robots working cooperatively to execute various types of tasks in order to increase the robustness and efficiency of task execution. The use of multi-robot teams brings several advantages over single robot approaches. Firstly, depending on the type of the task, multiple robots can execute it more efficiently by dividing the work among the team. More than that, groups of simpler and less expensive robots working cooperatively can be used instead of an expensive specialized robot. Robustness is also increased in certain tasks by having robots with redundant capabilities and dynamically reconfiguring the team in case of robot failures.

A natural evolution of this paradigm is the use of large groups of simpler robots, generally called swarms. Inspired by their biological counterparts, swarms of robots must perform in a decentralized fashion using limited communication. Normally these groups have to work in dynamic, partially-observable environments which increase the challenges in terms of coordination and control.

In [1] we proposed an algorithm that allows a large group of robots to overcome local minima regions while navigating to a specific goal in environments containing unknown obstacles. A coordination mechanism, based on dynamic role assignment and local communication was used to deal with robots stuck in local minima. In this paper, we experimentally validate this algorithm using both a realistic simulator and a group of real robots. We also describe and analyze the communication chain mechanism, one of the key features of the algorithm that is responsible for spreading feasible path information among team members.

G. Zaverucha and A. Loureiro da Costa (Eds.): SBIA 2008, LNAI 5249, pp. 268–277, 2008.

2 Related Work

The general area of motion planning for large groups of robots has been very active in the last few years. One of the first works to deal with the motion control of a large number of agents was proposed for generating realistic computer animations of flocks of birds (called *boids*) [2]. In the robotics community, the more classical approaches for planning the motion of groups of robots have generally been divided into centralized and decentralized. Centralized planning consists of planning for the entire group, considering a composite configuration space. It normally leads to complete solutions but becomes impractical as the number of robots increases due to the high dimensionality of the joint configuration space. On the other hand, decentralized approaches plan for each robot individually and later try to deal with the interactions among the trajectories. This reduces the dimensionality of the problem, but can result in a loss of completeness.

A common decentralized approach for motion planning is the use of potential fields [3], in which robots are individually attracted by the goal and repelled by obstacles and other robots. In swarms, attractive forces are generally modeled through the gradient descent of specific functions [4,5]. Unfortunately, as in regular potential field approaches, the presence of obstacles and local repulsion forces among the robots may cause convergence problems in general gradient descent approaches, mainly when robots are required to synthesize shapes. In this context, Hsieh and Kumar [6] are able to prove convergence properties and the absence of local minima for specific types of shapes and environments. Also, special types of navigation functions can be used to navigate swarms in cluttered environments [7]. But these approaches may be hard to compute in dynamic, partially-observable environments.

Another way of avoiding the dimensionality problem is to treat groups of robots as a single entity with a smaller number of degrees of freedom and then perform the motion planning for this entity. In the work presented in [8], for example, the robots can be dynamically grouped together in a hierarchical manner using a sphere tree structure. Belta et al. [9] show how groups of robots can be modeled as deformable ellipses, and presented complex decentralized controllers that allowed the control of the shape and position of the ellipses. This approach was extended in [10] with the development of a hierarchical framework for trajectory planning and control of swarms.

Certain types of tasks may require a greater level of coordination. For example, more sophisticated task allocation must be necessary for some tightly coupled tasks. When dealing with swarms, coordination mechanisms have to scale to tens or hundreds of robots. Scalable approaches for the coordination of large groups of agents (not necessarily robots) have been proposed in [11,12] among others. It is important to mention that most of the works that deal with swarms validate their approaches only through simulation. Few papers use a real robotic infrastructure and provide experimental results, for example [7,13,14].

In our approach, instead of restricting the environment or developing complex controllers and coordination mechanisms, we rely on the composition of a gradient descent controller and a simple coordination mechanism to navigate swarms in environments containing unknown obstacles. And differently from most of the papers that deal with robotic swarms, in this paper we perform real experiments to validate our approach.

3 Swarm Navigation

3.1 Controller

In this paper, the robots must move towards and spread along a goal region in an environment containing unknown obstacles. The goal region is specified by a 2D curve S given by implicit functions of the form $s(x, y) = 0$. This implicit function can be viewed as the zero isocontour of a 3D surface $f = s(x, y)$ whose value is less than zero for all points (x, y) that are inside the S boundary and is greater than zero for all points outside the S boundary. By descending the gradient of this function and applying local repulsion forces, robots are able to reach the goal and spread along the 2D curve. Details of this controller can be found in [4]. For obstacle avoidance, we augmented this controller using a regular potential field approach: if an obstacle is detected by a robot, this obstacle applies a repulsive force that is inversely proportional to the distance between them.

Thus, considering a fully actuated robot i with dynamic model given by $\dot{\mathbf{q}}_i = \mathbf{v}_i$ and $\dot{\mathbf{v}}_i = \mathbf{u}_i$, where $\mathbf{q}_i = [x_i, y_i]^T$ is the configuration of robot i, $\mathbf{u}_i$ is its control input and $\mathbf{v}_i$ is the velocity vector, the control law used by each robot is given by:

$$\mathbf{u}_i = -\alpha \nabla f^2(\mathbf{q}_i) - C\dot{\mathbf{q}}_i - \beta \sum_{k \in O_i} \frac{1}{\mathbf{d}_{ik}} - \gamma \sum_{j \in N_i} \frac{1}{\mathbf{q}_j - \mathbf{q}_i}. \quad (1)$$

Constants α, β, γ and C are positive. The first term is the inverse of the gradient used to guide the robots towards the specified shape. The second term is a damping force. The third term is the sum of repulsive forces applied by the obstacles ($\mathbf{d}_{ik}$ is the distance vector between robot i and obstacle k). Only the obstacles that are inside robot i sensing region, represented by the set O_i, are considered in the computation of forces. The fourth term computes the repulsive interaction of a robot with its neighbors, represented by the set N_i.

Unfortunately, the sum of these forces can lead to the appearance of local minima regions. Since robots are attracted by the goal and repelled by obstacles and other robots, they can be trapped in regions where the resultant force is zero or where the force profile leads to repetitive movements (for example, continuous circular movements in a specific region). Therefore, there are no formal guarantees that the robots will converge to the desired pattern. To overcome this, we rely on swarm coordination: robots may escape from local minima with the help of their teammates, as will be explained in the next subsection.

3.2 Coordination

Our coordination is based on a mode switching mechanism, generally known in robotics as dynamic role assignment [15]. A robot can switch between different

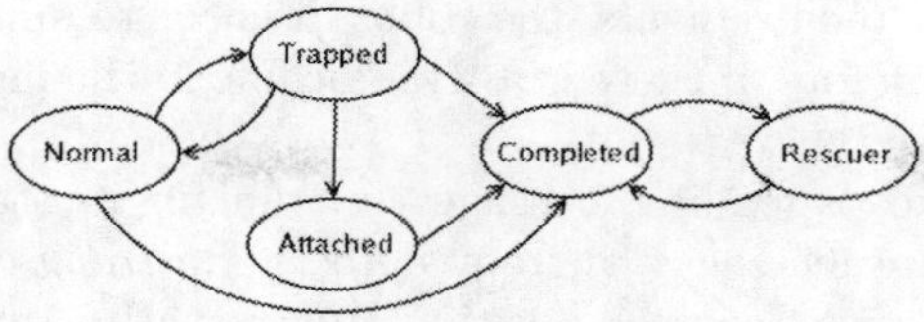

Fig. 1. Finite state machine showing the possible modes and transitions for one robot

modes (or roles) during the execution of the task. Each mode determines a different behavior for the robot and will be executed while certain internal and external conditions are satisfied.

These different modes can be better modelled by a finite state machine (FSM), in which the states and edges represent respectively the modes and the possible transitions between them. In the mechanism presented in this paper, the FSM for each robot is shown in Figure 1. It is composed of five different modes: *normal*, *trapped*, *rescuer*, *attached* and *completed*.

All robots start in the *normal* mode. A *normal* robot performs a gradient descent according to equation 1, trying to reach the goal while avoiding obstacles. It switches its mode to *trapped*, if it considers itself trapped in a local minima region. This transition is determined by the variation in the robot's position over time. If its position does not change significantly during a certain amount of time, it becomes *trapped*. A *trapped* robot may switch back to *normal* if its position start changing considerably again. It is important to note that sometimes, due to the resultant forces in the controller, robots may switch to *trapped* even if they are not in a local minima region. These false-positives do not compromise the performance of the algorithm since *trapped* robots are also controlled by equation 1, as will be explained below.

A *trapped* robot acts similarly to a *normal* one, except for the following facts: (i) a *trapped* robot strongly repels another *trapped* robot and this repulsion is stronger than the one between two *normal* robots. As a local minima region tends to attract many robots, the local interactions through these stronger repulsion forces will help some of the robots to escape this region; (ii) a *trapped* robot broadcasts messages announcing its state; (iii) *trapped* robots accept messages from *rescuers* (or *attached*) robots that will help them to escape from local minima and move towards the target. This will be better explained later in this section. We consider that all communication is local, *i.e.*, only messages received from robots within a certain distance are considered.

When a robot arrives at the target it may become a *rescuer*. Basically, when moving towards the goal, a robot saves a sequence of waypoints that is used to mark its path. If it becomes a *rescuer* it will retrace its path backwards looking for *trapped* robots. After retracing its path backwards, the robot moves again to the goal following the path in the correct direction. The number and frequency of *rescuer* robots are set empirically and may vary depending on the total number of robots and characteristics of the environment. The method used to determine which robots become rescuers is explained in [1]. In order to minimize memory requirements, the robot discards redundant information in the path stored. Therefore, if there is a straight line in the path, ideally only two waypoints will be used.

A *trapped* robot keeps sending messages announcing its state. When a *rescuer* listens to one of these messages, thereby detecting a *trapped* robot in its neighborhood, it broadcasts its current position and its path. Any *trapped* robot will consider the message if it is within a certain distance from the *rescuer* and there is a direct line of sight between them. This restriction enables the robot to reach the rescuer's position. After receiving the message, the *trapped* robot changes its mode to *attached*.

An *attached* robot will move to the received position and then follow the received path to the goal. An *attached* robot can also communicate with other *trapped* robots, spreading the information about the feasible path to the goal. Thus, for the *trapped* robots, an *attached* robot also works as a *rescuer*. Moreover, when an *attached* robot sends a message to a *trapped* robot, it adds its current position as a new waypoint in the path information. Therefore, robots that would not have a line of sight with any *rescuer* can easily escape the local minima region thanks to the extra waypoints created in this process. As will be shown in Section 5, the *attached* robots create a powerful communication chain that allows a large number of robots to be rescued from local minima.

Finally, a robot will change its mode to *completed* when it reaches the target. *Completed* robots will not switch to *trapped* again but may become *rescuers* as explained above.

4 Testbed

As mentioned, one of the contributions of this paper is the validation of the proposed algorithm using both a realistic simulator and a group of real robots. In this section we present the infrastructure used for experimentation.

In terms of simulation, we used *Gazebo* [16]. Gazebo is a multi-robot simulator for both indoor and outdoor environments. It is capable of simulating a population of robots, sensors and objects in a three-dimensional world. It generates both realistic sensor feedback and physically plausible interactions between objects (it includes an accurate simulation of rigid-body physics based on ODE – Open Dynamics Engine). Figure 2(a) shows a snapshot of a simulation running on Gazebo.

Gazebo is used in conjunction with the *Player* framework [17]. Player is a network server for robot control, that provides a clean and simple interface to the robot's sensors and actuators over an IP network. It is designed to be language and platform independent, allowing the same program to run on different robotic platforms. In fact, most of the time, the algorithms and controllers used in simulation do not need to be changed to run in real robots. This specific feature is very useful for experimentation in robotics.

In the real experiments, we used a group of 7 *scarab* robots, developed in the GRASP Lab. – University of Pennsylvania (Figure 2(b)). The Scarab is a small

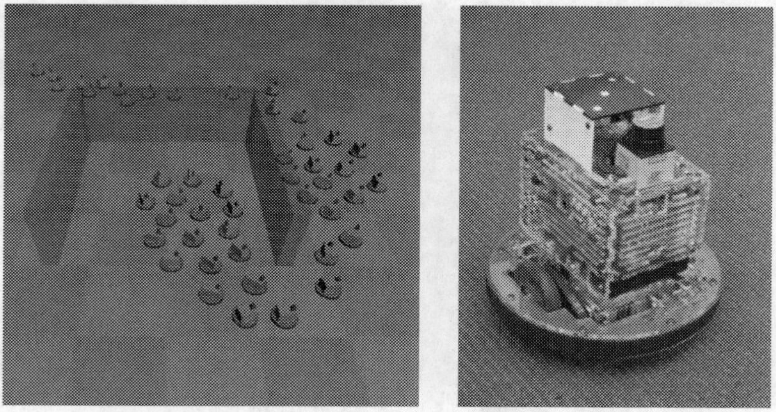

(a) (b)

Fig. 2. Testbed used in the experiments: (a) snapshot of the gazebo simulator with 48 robots and (b) one of the 7 scarab robots used in the experiments

differential drive robot equipped with an on-board computer, a Hokuyo URG laser range finder, wireless communication (802.11) and two stepper motors for actuation. The sensors, actuators, and controllers are modular and connected through the Robotics Bus (which is derived from the CAN bus protocol). A external localization system composed by a set of overhead cameras provides accurate pose information for the robots. More details about the *scarabs* and the localization system can be found in [18].

In the simulations and experiments performed in this paper, the only information provided a priori to the robots is the implicit function that attracts them to the target (function f in equation 1). Each robot knows its pose in a global reference frame, but does not have access to the pose of its teammates. This information, when needed, is explicitly transmitted by the robots using wireless communication. Also, there is no map of the environment and all obstacles are locally detected using lasers.

The laser is also used to check the existence of line of sight between a *trapped* and a *rescuer* robot. In the coordination mechanism presented in Section 3.2, *trapped* robots only accept messages from a *rescuer* if there is a line of sight between them, *i.e.*, if there is a free path connecting their positions (otherwise, it will not be able to move to the rescuer's position and then follow the feasible path to the goal). A simple algorithm, using the laser, enabled the robots to estimate the existence of line of sight. When a *trapped* robot receives a message from the *rescuer*, containing the location of the *rescuer* and a path to the target, it computes the euclidean distance (δ) and the bearing between them. Then, it turns in the direction of the *rescuer* and checks the distance returned by the laser. If this distance is smaller than δ, it means that the laser detected something (an obstacle or another robot) and there is no free path between them. In this case, the *trapped* robot will ignore the message and wait for another *rescuer*.

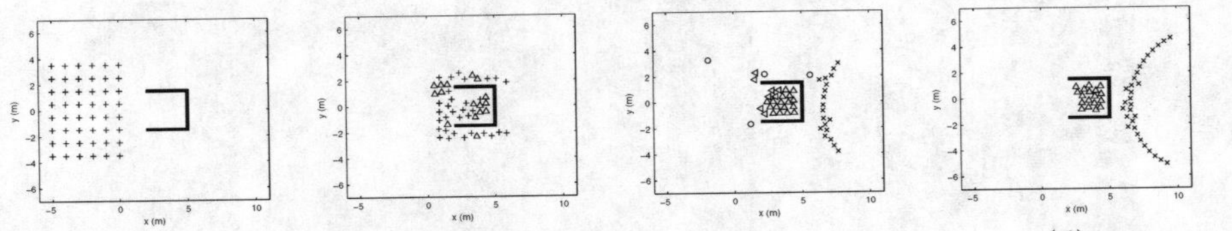

(a) (b) (c) (d)

Fig. 3. Simulation of the coordination algorithm without the communication chain. Robots are represented by different shapes according to their states.

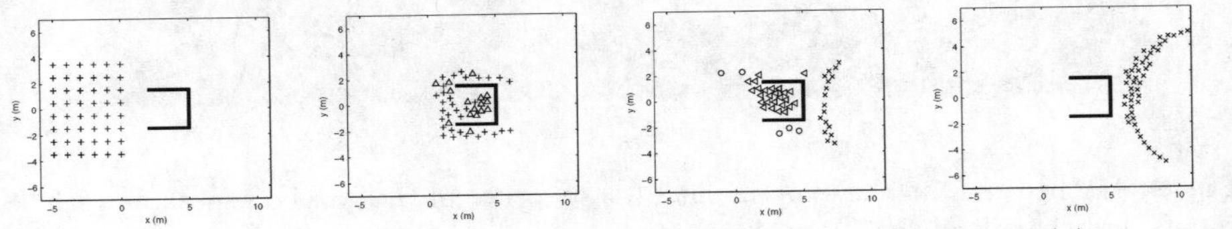

(a) (b) (c) (d)

Fig. 4. Simulation of the coordination algorithm using the communication chain

5 Simulations

We simulated forty-eight *scarabs* navigating in an environment containing an u-shaped obstacle. This is a typical local minima scenario in robotics. In these simulations, we focus our attention on the impact of the communication chain mechanism. Other experiments, using a simpler simulator, were executed to analyze the performance of the algorithm with a varying number of robots and communication parameters in different environments. Those were previously presented in [19].

Two versions of the algorithm were tested: one without using the communication chain and the other one with this mechanism enabled. Figures 3 and 4 respectively present graphs of these two variations. In both figures, robots are represented by different shapes according to their states: *normal* (+), *trapped* (△), *attached* (◁), *rescuer* (○), and *completed* (×). Robots start on the left and the target is on the right. The u-shaped obstacle is shown in black at the middle.

In the execution without the communication chain (Figure 3), *rescuer* robots successfully reached the region where many robots were *trapped*, but only the ones in the border of the obstacle are able to escape. The other *trapped* robots did not have a direct line of sight with the rescuers and remained stuck in the local minima region. The execution with the communication chain (Figure 4), on the other hand, was very effective, with all robots converging to the target. The information about the feasible path easily spread through the robots stuck in local minima, reaching even those that were closer to the bottom of the u-shaped obstacle.

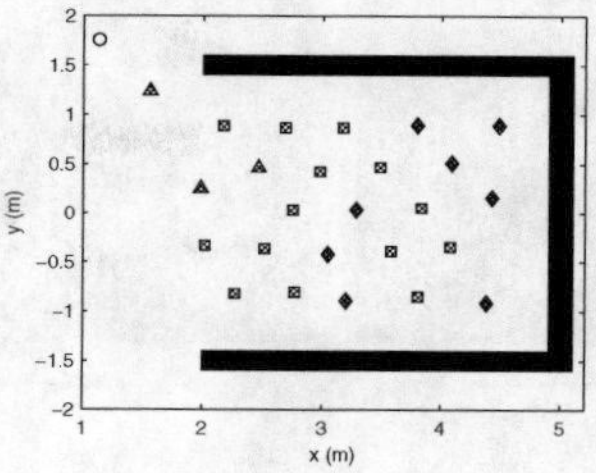

Fig. 5. Path information spread through the swarm. Shapes represent the number of message hops needed to receive the information.

In Figure 5, we take a closer look on how the information spread through the *trapped* robots when the *rescuer* (○) in the upper left sent the viable path to the goal. In this figure, shapes represent the number of message hops needed to receive the path information: triangle received with one hop, square with two hops and diamond with three hops. As can be seen, only a small number of robots, that were near the border of the obstacle, could be rescued with just one hop. Almost all robots were rescued with two hops, upon receiving a viable path from the *attached* robots. Robots that were near the bottom of the obstacle needed three hops, but could be rescued as well. As was explained, *attached* robots created new waypoints in the viable path, enabling these robots to effectively escape the local minima region.

6 Real Experiments

Real experiments are very important to show that an algorithm can effectively work in robots acting in the real world, with all the problems caused by uncertainties due to sensors and actuators errors, communication problems and real time constraints. In order to show the effectiveness of our proposed coordination mechanism, we performed real experiments with seven *scarabs* robots using a similar scenario with an u-shaped obstacle.

The sequence of snapshots of one execution can be seen in Figure 6, where the graphs on the bottom depict the robots' positions and states. The robots start in the left of the scenario and must converge to the target beyond the u-shaped obstacle in the middle. In Figure 6(a), three robots are able to move to the target, while four others are trapped in a local minima region, in front of the obstacle. The *trapped* robots are spread in this region, because of the local repulsion forces. Figure 6(b) shows a *rescuer* robot at the right of the obstacle, sending a viable path to the target. The robots that have a direct line of sight with the *rescuer* accept this message and change their status to *attached*. One of the *attached* robots retransmits the information to the other two *trapped* robots that did not have line of sight, allowing all of them to escape the local minima region, as can be seen in Figure 6(c). Soon the state in Figure 6(d) is achieved, where almost all robots reached the target.

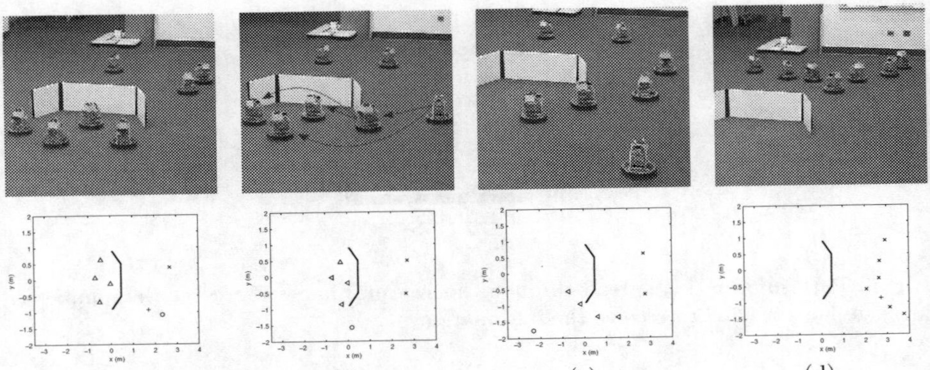

Fig. 6. Seven robots converge to the target with the proposed coordination mechanism. Exchange of messages is shown by the arrows.

Thus, using the proposed algorithm, all robots effectively escaped the local minima region. Only one *rescuer* was enough to save all four *trapped* robots, thanks to the communication chain mechanism: an *attached* robot was able to spread the information to the robots that did not receive it directly.

7 Conclusion

In this paper we experimentally validated a distributed coordination mechanism for navigating a swarm of robots in environments containing unknown obstacles. Realistic simulations and real experiments with seven robots showed the viability of the proposed technique and the benefits of the communication chain.

Our future work is directed towards the improvement of the mechanism, with the development of "congestion control" techniques for the swarm. We observed that many times, when a large number of robots tried to reach the same waypoint or robots navigated in opposite directions, congestion caused by the local repulsion forces increased the time needed to reach convergence, wasting time and resources.

Acknowledgments

This work is partially supported by PRPq–UFMG, Fapemig and CNPq.

References

1. Marcolino, L.S., Chaimowicz, L.: No robot left behind: Coordination to overcome local minima in swarm navigation. In: Proc. of the 2008 IEEE Int. Conf. on Robotics and Automation, pp. 1904–1909 (2008)

2. Reynolds, C.W.: Flocks, herds and schools: A distributed behavioral model. In: Proc. of the 14th Conf. on Computer Graphics (SIGGRAPH), pp. 25–34 (1987)
3. Khatib, O.: Real-time obstacle avoidance for manipulators and mobile robots. Int. Journal of Robotics Research 5(1), 90–98 (1986)
4. Chaimowicz, L., Michael, N., Kumar, V.: Controlling swarms of robots using interpolated implicit functions. In: Proc. of the 2005 IEEE Int. Conf. on Robotics and Automation, pp. 2498–2503 (2005)
5. Bachmayer, R., Leonard, N.E.: Vehicle networks for gradient descent in a sampled environment. In: Proc. of the IEEE Conf. on Decision and Control, pp. 112–117 (2002)
6. Hsieh, M.A., Kumar, V.: Pattern generation with multiple robots. In: Proc. of the 2006 Int. Conf. on Robotics and Automation, pp. 2442–2447 (2006)
7. Pimenta, L., Michael, N., Mesquita, R.C., Pereira, G.A.S., Kumar, V.: Control of swarms based on hydrodynamic models. In: Proc. of the 2008 IEEE Int. Conf. on Robotics and Automation, pp. 1948–1953 (2008)
8. Li, T.Y., Chou, H.C.: Motion planning for a crowd of robots. In: Proc. of the 2003 IEEE Int. Conf. on Robotics and Automation, pp. 4215–4221 (2003)
9. Belta, C., Kumar, V.: Abstraction and control for groups of robots. IEEE Transactions on Robotics 20(5), 865–875 (2004)
10. Kloetzer, M., Belta, C.: Hierarchical abstractions for robotic swarms. In: Proc. of the 2006 Int. Conf. on Robotics and Automation, pp. 952–957 (2006)
11. Jang, M.W., Agha, G.: Dynamic agent allocation for large-scale multi-agent applications. In: Proc. of the Workshop on Massively Multi-Agent Systems, pp. 19–33 (2004)
12. Scerri, P., Farinelli, A., Okamoto, S., Tambe, M.: Allocating tasks in extreme teams. In: Proc. of the Int. Joint Conf. on Autonomous Agents and Multiagent Systems, pp. 727–734 (2005)
13. Correll, N., Rutishauser, S., Martinoli, A.: Comparing coordination schemes for miniature robotic swarms: A case study in boundary coverage of regular structures. In: Proc. of the Int. Symposium on Experimental Robotics (ISER) (2006)
14. McLurkin, J., Smith, J.: Distributed algorithms for dispersion in indoor environments using a swarm of autonomous mobile robots. In: Proc. of the 7th Int. Symposium on Distributed Autonomous Robotic Systems (2004)
15. Stone, P., Veloso, M.: Task decomposition, dynamic role assignment, and low-bandwidth communication for real-time strategic teamwork. Artificial Intelligence 110(2), 241–273 (1999)
16. Gazebo Simulator (acessed, March 25, 2008), `http://playerstage.sourceforge.net/gazebo/gazebo.html`
17. Vaughan, R.T., Gerkey, B.P., Howard, A.: On device abstractions for portable, resuable robot code. In: Proc. of the IEEE/RSJ Intl. Conf. on Intelligent Robots and Systems (IROS 2003), pp. 2121–2427 (2003)
18. Michael, N., Fink, J., Kumar, V.: Experimental testbed for large multi-robot teams: Verification and validation. IEEE Robotics and Automation Magazine 15(1), 53–61 (2008)
19. Marcolino, L.S., Chaimowicz, L.: A coordination mechanism for swarm navigation: Experiments and analysis (short paper). In: Proc. of the Seventh Int. Conf. on Autonomous Agents and Multiagent Systems, pp. 1203–1206 (2008)

Evolving an Artificial Homeostatic System

Renan C. Moioli[1], Patricia A. Vargas[2], Fernando J. Von Zuben[1], and Phil Husbands[2]

[1] Laboratory of Bioinformatics and Bio-Inspired Computing - LBiC
School of Electrical and Computer Engineer - FEEC/Unicamp Campinas-SP, Brazil
[2] Centre for Computational Neuroscience and Robotics (CCNR)
Department of Informatics, University of Sussex, Falmer, Brighton, BN1 9QH, UK
{moioli,vonzuben}@dca.fee.unicamp.br,
{p.vargas,p.husbands}@sussex.ac.uk

Abstract. Theory presented by Ashby states that the process of homeostasis is directly related to intelligence and to the ability of an individual in successfully adapting to dynamic environments or disruptions. This paper presents an artificial homeostatic system under evolutionary control, composed of an extended model of the GasNet artificial neural network framework, named NSGasNet, and an artificial endocrine system. Mimicking properties of the neuro-endocrine interaction, the system is shown to be able to properly coordinate the behaviour of a simulated agent that presents internal dynamics and is devoted to explore the scenario without endangering its essential organization. Moreover, sensorimotor disruptions are applied, impelling the system to adapt in order to maintain some variables within limits, ensuring the agent survival. It is envisaged that the proposed framework is a step towards the design of a generic model for coordinating more complex behaviours, and potentially coping with further severe disruptions.

Keywords: Evolutionary Robotics, Homeostasis, Adaptation, Artificial Neural Networks.

1 Introduction

The term *homeostasis* has its origins in the work of the French physiologist Claude Bernard (1813-1878), who founded the principle of the internal environment, further expanded by Cannon in 1929 as the process of homeostasis [1]. Nonetheless, for Pfeifer & Scheier [2], homeostasis was completely defined by the English psychiatrist William Ross Ashby in 1952 [3]. For Ashby, the ability to adapt to a continuously changing and unpredictable environment (*adaptivity*) has a direct relation to intelligence. During the adaptive process, some variables need to be kept within predetermined limits, either by evolutionary changes, physiological reactions, sensory adjustment, or simply by learning novel behaviours. Therefore, with this regulatory task attributed to the homeostatic system, the organism or the artificial agent can operate and stay alive in a viability zone.

G. Zaverucha and A. Loureiro da Costa (Eds.): SBIA 2008, LNAI 5249, pp. 278–288, 2008.

Basically, homeostasis can be considered paramount for the successful adaptation of the individual to dynamic environments, hence essential for survival. Moreover, Dyke & Harvey [4][5] have pointed out that in order to understand real or artificial life it is necessary to first understand the conceptual framework and basic mechanisms of homeostasis. In the human body some particular sensory receptors trigger specific responses in the nervous, immune and endocrine systems, which are the main systems directly related to the process of homeostasis [6]. Therefore, one can say that it is a consensus that homeostatic processes are strictly connected to the balance of any real or artificial life.

The theory presented by Ashby has motivated applications of homeostasis in the synthesis of autonomous systems in mobile robotics [7][8][9][10][11][12]. The ideas presented by Alife researchers like Di Paolo [7] and Hoinville & Hnaff [10] encompass homeostasis within one unique structure, i.e. an artificial neural network (ANN) capable of dynamically changing their connections by means of plasticity rules. Our present work goes in a different direction and is an extension of a previous work by Vargas *et al.* [11] and Moioli *et al.* [12]. In the framework developed here the entire artificial homeostatic system (AHS) is under evolutionary control. An artificial endocrine system (AES) is synthesized by means of evolution and is responsible for controlling the coordination of two evolved spatially unconstrained GasNet models, named non-spatial GasNets (NSGasNets) [13].

This work is organized as follows: section 2 presents the basis of the approach adopted in this work, together with the details of our proposal of an evolutionary artificial homeostatic system. Section 3 depicts the suggested experiments and their implementation procedures. Section 4 encloses the simulation results of the artificial homeostatic system subject to some degrees of disruption. Finally, section 5 presents final remarks and suggests directions for future investigation.

2 Evolutionary Artificial Homeostatic System

The framework to be proposed is particularly concerned with neuro-endocrine interactions. The endocrine system employs chemical substances, called hormones, to promote homeostasis, metabolism and reproduction. The release of hormones can also affect the nervous system, which in turn can transmit nerve impulses affecting the production and secretion of hormones, thus establishing a control loop mechanism. There are positive and negative feedback processes represented by coupled difference equations, which are reminiscent of the biological neuro-endocrine interaction.

The new evolutionary artificial homeostatic system (EAHS) is composed of an artificial endocrine system (AES) and two NSGasNet models (Figure 1). Our work proposes not only the evolution of both NSGasNets, but also of the AES, aiming at developing a more specialized architecture.

The AES consists of three main modules: hormone level (HL), hormone production controller (HPC), and endocrine gland (EG). The hormone level has a record of the level of hormone in the organism; the hormone production controller is responsible for controlling the production of hormones in response to

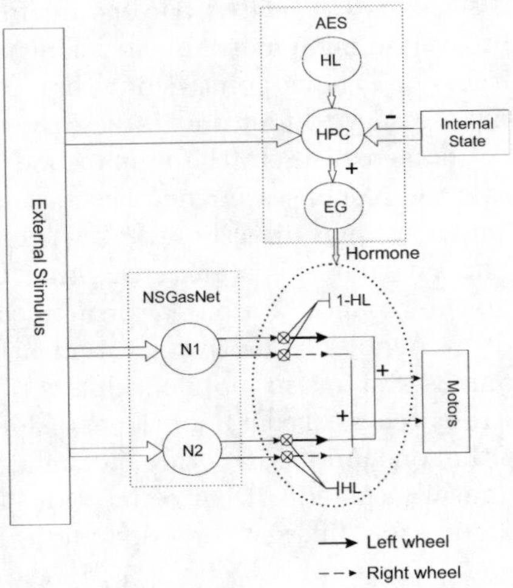

Fig. 1. The basic framework of the Evolutionary Artificial Homeostatic System

variations in the internal state of the organism and external stimulation; and the endocrine gland receives inputs from the HPC, being responsible for producing and secreting hormones when required. The hormone production HP is updated as follows:

$$HP(t+1) = \begin{cases} 0, & if\ IS < \theta \\ (100 - \%ES) \times \alpha(Max(HL) - HL(t)), & \text{otherwise} \end{cases} \quad (1)$$

where θ is the target threshold of the internal state IS; ES is the external stimulus; α is the scaling factor; HL is the hormone level; and t is the discrete time index. If the internal state IS is greater than or equal to a target threshold, then hormone will be produced at a rate that will depend upon the level of the external stimulus received and the level of hormone already present within the artificial organism. Otherwise, the hormone production will cease. The internal state IS is governed by:

$$IS(t+1) = \begin{cases} 0, & if\ (ES \geq \lambda)\ and\ (HL \geq \omega) \\ IS(t) + \beta(Max(IS) - IS(t)), & \text{otherwise} \end{cases} \quad (2)$$

where λ and ω are pre-determined thresholds associated with ES and HL, respectively, and β is the increasing rate of the internal state. The hormone level HL represents the amount of hormone stimulating the artificial neural network (ANN), and is a function of its current value and of the amount of hormone produced:

$$HL(t+1) = HL(t) \times e^{-1/T} + HP(t) \quad (3)$$

where T is the half-life constant.

The two NSGasNets (N1 and N2 in Figure 1) are previously and separately evolved to accomplish two distinct and possibly conflicting behaviours. The NS-GasNet is a spatially unconstrained model, which was proven to present superior performance when compared to the original GasNet model on a pattern generation task and on a delayed response robot task [13][14].

The outputs of the NSGasNets are modulated by the hormone level HL (Eq. 3), giving rise to the dynamical coordination of both behaviours. In particular, for N1 the modulation factor is given by $(1 - HL)$ and for N2 by (HL).

Given the original architecture of the AES, one might enquiry: "Why not evolving the rules that dictate the AES coupling behaviour?". Our work draws an analogy between the laws of physics, which dictate the nature of the physical coupling between a cell and its exterior environment [15], and the rules that govern the artificial coupling between the AES, the NSGasNets, the environment and the artificial agent (Eqs. 1, 2 and 3). In this sense, our new AES will evolve only the parameters of these rules.

3 Methods

A simulated mobile robotic agent equipped with an internal battery meter has to perform two coupled but distinct tasks: to explore the scenario while avoiding collisions and to search for a light source when its battery level is low (the light source indicates the location of the battery charger). This experiment was first proposed by Vargas *et al.* [11] to assess the performance of the artificial homeostatic system that has served as inspiration to the current proposal.

The robot is a Khepera II with two wheels which are responsible for its motion (each wheel has an independent electric motor). The robot has 8 infrared sensors that incorporate emitters and receptors. The sensors measure the environment luminosity (ranging from 50 to 500 - 50 being the highest luminosity that can be sensed) and the obstacle distance (ranging from 0 to 1023 - the latter value represents the closest distance to an object). The range of the obstacle sensors is roughly 10 cm. The simulations were carried out using a robot simulator named KiKS [16]. It reproduces both sensory behaviours of the real robot.

The evolution of the system was divided in two steps. First, the two NS-GasNets are evolved independently employing a distributed genetic algorithm [17][18](one NSGasNet evolved for each task). The AES is evolved thereafter as a module of coordination and it was responsible for the swapping of behaviour between both NSGasNets. No crossover is employed. A generation is defined as twenty five breeding events, and the evolutionary algorithm runs for a maximum of 50 generations. The fitness criteria are specific for each task. There are two mutation operators applied to 10% of the genes. The first operator is only for continuous variables. It produces a change at each locus by an amount within the $[-10, +10]$ range. For the second mutation operator, designed to deal with discrete variables, a randomly chosen gene locus is replaced with a new value that can be any value within the [0, 99] range, in a uniform distribution.

For further details about the application of the genetic algorithm to the evolution of NSGasNet models, the reader should refer to [19].

The first experiment only assesses the performance of the evolved EAHS without internal disruption. Thus, the parameters β and λ are not changed (Eq. 2). The second and third experiments focus on the analysis of the parameters of the artificial endocrine system under internal disruptions, a constant disruption for the second experiment and a variable disruption for the third experiment. This is an attempt to verify the performance of the EAHS in the homeostatic regulation process. It is expected that such an endeavour will promote the dynamic adjustment of variables under control, aiming at better coping with environmental changes and/or sensory and physical disruptions.

Evolution of the NSGasNets. The transfer function of the node i in the network is given by Eq. 4:

$$O_i(t) = tanh\left[K_i(t)\left(\sum_{j\in C_i} w_{ji}O_j(t-1) + I_i(t)\right) + b_i\right] \tag{4}$$

where C_i is the set of nodes with connections to node i, w_{ji} is the connection weight value (ranging from -1 to $+1$), $O_j(t-1)$ is the previous output of neuron j, $I_i(t)$ is the external input to neuron i at time t, if the node has external inputs, b_i is the bias of the neuron, and $K_i(t)$ represents the modulation of the transfer function caused by the gases.

The network genotype consists of an array of integer variables lying in the range [0, 99] (each variable occupies a gene locus). The decoding from genotype to phenotype adopted is the same as the original model [19]. The NSGasNet model has six variables associated with each node plus one modulator bias ($Mbias_{ij}$) for each node, plus task-dependent parameters. The modulator bias is responsible for dictating to what extent the node could be affected by the gases emitted by all the other nodes. Therefore, here, networks N1 and N2 (Figure 1) will have six Mbias for each node. For a more detailed explanation of the mechanisms of the GasNets and NSGasNets, the reader should refer to [13][19].

For the straight motion with obstacle avoidance behaviour, the network N1 (see Figure 1) had four inputs: the most stimulated left, right, front and back distance sensors. Two additional neurons were considered to be output neurons, so the network consisted of six neurons. The output neurons correspond to the motor neurons. The fitness function (Eq. 5) and the training scenario were inspired by the work of Nolfi & Floreano [20]:

$$\phi = V(1 - \sqrt{\Delta v})(1 - i) \tag{5}$$

where V is the sum of the rotation speed of the wheels (stimulating high speeds), Δv the absolute value of the algebraic difference between the speeds of the wheels (stimulating forward movement), and i is the normalized value of the distance sensor of highest activation (stimulating obstacle avoidance). A trial is considered to be 2,000 iterations of the control algorithm. At the end of each trial, the robot is randomly repositioned in the scenario.

The structure of network N2 for the phototaxis behaviour was similar to the obstacle avoidance network. Only the distance sensors were replaced by the luminosity sensors. The training scenario consisted of a squared arena, where the robot has an initial fixed position at the beginning of each trial. Each trial corresponds to 2,000 simulation steps. The fitness function is given by Eq. 6:

$$\phi = V(1 - i) \tag{6}$$

Parameter i (refers to sensory activation) is maximized when the robot is near the light, due to the sensory structure of the robot.

Evolution of the Artificial Endocrine System. The genotype consisted of four parameters: ω, θ, α and T (Eqs. 1-3). A trial is considered to be 800 iterations of the control algorithm. The parameters λ and ω stand for minimum light intensity and hormone level, respectively. β is the internal state (*IS*) growing rate. In our model, as on [12], the internal state (*IS*) of the artificial agent (Eq. 2) stands for the inverse of the battery meter reading. It implies that the lower the battery level, the higher the *IS*. θ is the *IS* level threshold, above which the hormone production starts at a rate influenced by α. T is simply the half-life of the hormone. As the battery is always discharging (given that the robot is turned on), β should have a predefined value associated with it. Similarly, the minimum light intensity above which the robot could recharge should also be predefined. The presence of light here indicates a recharging area.

The experiment starts with a robot exploring the arena, controlled by the obstacle avoidance network. The AES was designed to sense the internal state of the robot. If the internal state grows above 90, in a 0 to 100 scale, the robot is considered to be dead. To obtain a successful performance the robot should be able to efficiently switch between exploration behaviour and phototaxis behaviour. This switching is expected to be due to the production of the hormone related to the decrease of battery level. After the recharge of the battery (associated with being close to the light), and consequently the decrease in the related hormone level, the robot should return to its original exploratory behaviour. Eq. 7 shows the fitness function adopted for this task:

$$\phi = V(1 - i)t/M \tag{7}$$

where V is the absolute value of the sum of the rotation speed of the wheels (stimulating forward movement), i is the normalized value of the distance sensor of highest activation (stimulating obstacle avoidance), t is the number of iterations in which the robot remains alive and M is the maximum number of iterations a trial can have. Thus, a good performance would consist of adjusting the hormone production thresholds and growing rate in order to allow maximum exploration interspersed with recharging steps. Due to the environment set-up, the robot could not stay closer to the light sources when performing exploration of the environment, as the light sources are themselves located close to the wall.

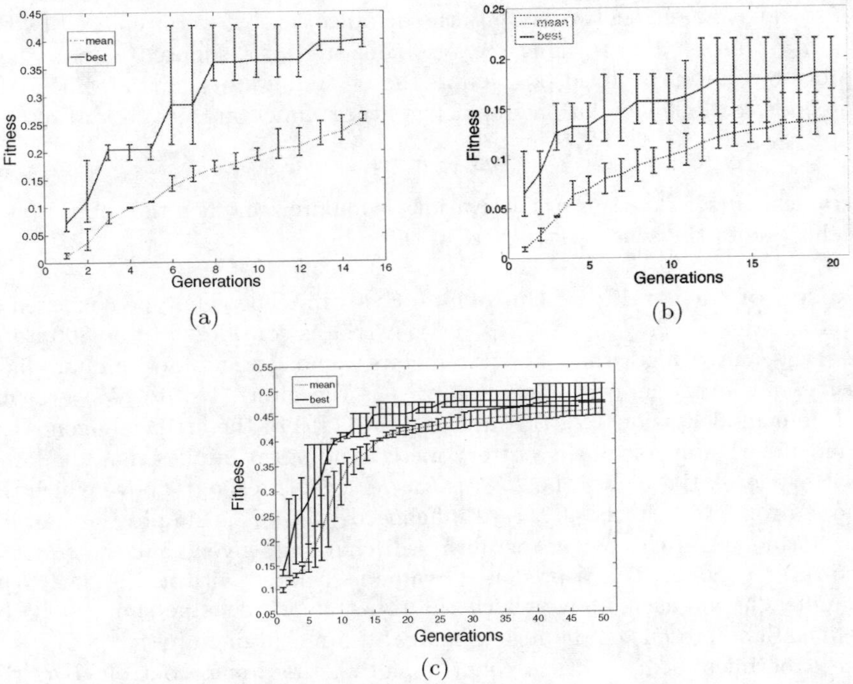

Fig. 2. Fitness evolution: obstacle avoidance (a), phototaxis (b) and EAHS (c). Mean values refer to a set of 3 experiments.

4 Results

Experiment 1 - EAHS Performance under Normal Conditions. Figure 2 depicts the fitness value along the evolutionary process, for the NSGasNets ((a) and (b)) and the whole system (c). The final network configuration indicated that network N1 (for obstacle avoidance) has evolved a more symmetrical architecture in terms of connections and gas emission.

The performance of the evolved EAHS was evaluated under the same conditions described above. The phenotype of the best individual is: $\alpha = 0.0099$; $T = 11.1$; $\omega = 50.5$; and $\theta = 52.0$. Parameters $\beta = 0.01$ and $\lambda = 103$ were defined empirically and kept fixed. When the hormone level increases above ω, the robot stops exploring the scenario and starts chasing the light. This confirms the influence of the hormone level over the robot's autonomous behaviour. During the experiment, the robot traverses the arena in maximum speed, only adjusting its speed when avoiding collision courses or when changing the behaviour to phototaxis.

Experiment 2 - EAHS Performance under Constant Disruptions. This experiment aims at submitting the evolved artificial homeostatic system to perturbations in order to analyse its performance in dealing with situations not

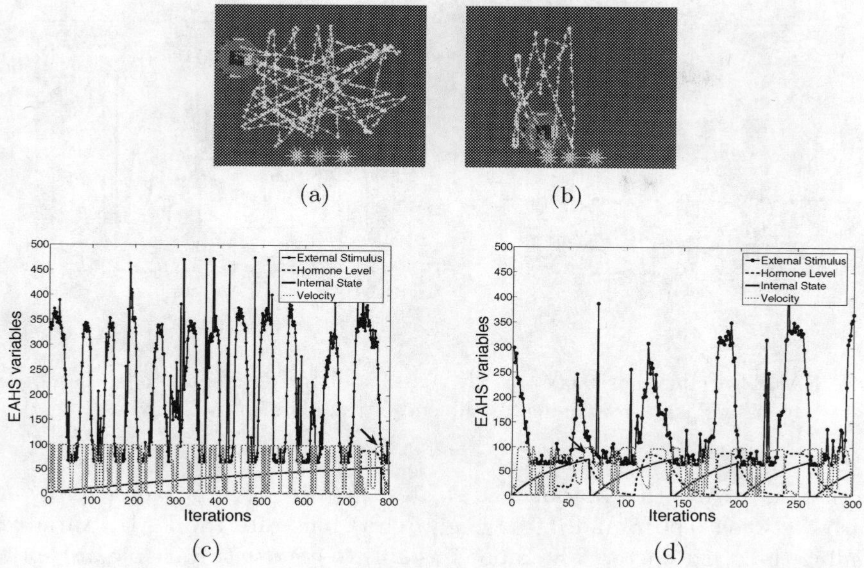

Fig. 3. Trajectories and EAHS variables for $\beta = 0.001$ ((a) and (c)) and for $\beta = 0.02$ ((b) and (d)). The black arrow indicates the end of the phototaxis behaviour.

presented during evolution. The task consists of changing the value of the parameter β, simulating a faster or slower battery discharge. Figure 3 presents the trajectory and the parameter analysis for two different values of β. Remember that, during evolution, β was kept fixed in the value 0.01. Figures 3(a) and 3(c) shows the results for $\beta = 0.001$, i.e. a slower discharge rate of the battery, and Figures 3(b) and 3(d) shows the results for $\beta = 0.02$, simulating a faster discharge rate.

In the first case (Figure 3(a)), the robot's internal state grows slowly enough to allow the robot to explore the whole arena before recharging is needed (around iteration 800). However, when the discharge rate is increased, the robot has a smaller period of time to explore the arena between consecutive recharging (Figure 3(b)). The robot is very good in timing the switch of behaviour, avoiding to get too far from the battery charger and eventually "dying". The behaviour of the velocity curve also indicates a quantitative change of behaviour. With a lower β, the robot only needs to turn (and consequently reduce its velocity) when facing a wall and sometimes when adjusting its direction. At higher values, β forces the robot to turn and seek the light more frequently, which requires operation at a lower velocity. Thus, when the exploratory behaviour is enforced, the robot tends to move much faster, on average.

Experiment 3 - EAHS Performance under Variable Disruptions. In this experiment, parameter β varies within a single trial. The absolute rotation speed of the wheels of the robot is linked to a greater energy consumption. In

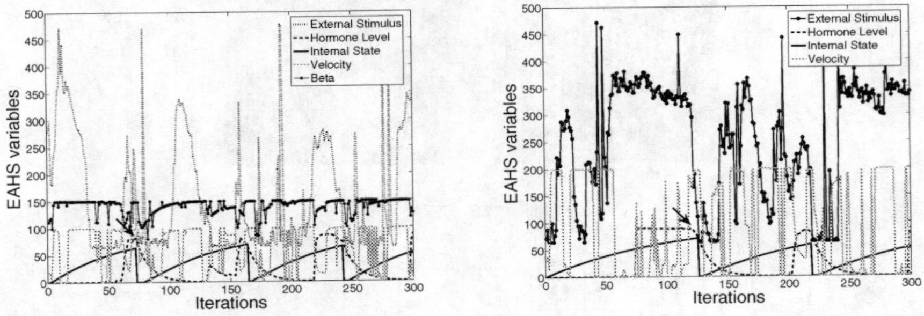

(a)

(b)

Fig. 4. EAHS variables for $0.005 \leq \beta \leq 0.015$ and for double velocity disruption, respectively. The black arrow indicates the end of the phototaxis behaviour.

this way, the greater the rotation, the greater the value of β. β is allowed to vary linearly between 0.0015 and 0.015. Figure 4(a) illustrate the EAHS variables, including the β parameter. The robot was able to cope with variable disruption by successfully navigating and exploring the scenario, chasing the light when recharge is necessary, and consequently maintaining its integrity. The next experiment simulates a disruption in the speed of the robot: the velocity value command is multiplied by 2 before reaching the robot wheels, but this information is not directly available to the robot or the control system. Figure 4(b) depicts the experiment results. Once more the system was able to self-adapt to this unpredicted and unknown variable disruption, being robust enough to present a good performance.

5 Discussion and Future Work

This work is a step forward in the design of an evolutionary artificial homeostatic system (EAHS). Towards the goal of creating an even more robust system, this work has introduced an artificial homeostatic system whose parameters are defined by means of an evolutionary process. It consists of two evolved ANNs coordinated by an artificial endocrine system. The ANNs followed the model proposed by Vargas *et al.* [13], drawing inspirations from gaseous neuro-modulation in the human brain. The objective was to design a more biologically-plausible system inspired by homeostatic regulations pervasive in nature, which could be able to tackle key issues in the context of behaviour adaptation and coordination.

This work deals with a flexible behaviour-based approach in the sense that coordination between modules of behaviour is evolved, not predesigned. A series of experiments were performed in order to investigate the performance of the system and its robustness to internal sensory disruptions.

When β increases, the need for precise coordination also increases on account of the environment and physical conditions of the robot becoming more severe.

Recall from the experiments that, when $\beta = 0.02$, the EAHS was able to adapt to the disruption, albeit reducing the exploration ratio. When employing only the exploration behaviour, if $\beta = 0.02$, the robot could eventually gets its internal state above 90, meaning its "death". This aspect reinforces the homeostatic regulatory behaviour of the proposed system.

We aimed at designing a system that would be robust enough to self-adapt to a wider variety of disruptions and yet perform well. Future work would include deeper analysis of the neuro-endocrine interactions, eventually proposing mechanisms for the coordination of more complex sensorimotor behaviours. Furthermore, the interactions of more than two behaviours and/or the implementation of more than one hormone, including hormone receptors, are going to be investigated. Also, future work would include transferring the evolved controllers from simulated agents to a real robot, analysing their robustness when crossing the *"reality gap"*.

Acknowledgments

Moioli and Von Zuben would like to thank CAPES and CNPq for their financial support. Vargas and Husbands were supported by the Spatially Embedded Complex Systems Engineering (SECSE) Project, EPSRC grant no. EP/C51632X/1.

References

1. Cannon, W.B.: Organization for physiological homeostasis. Physiological Review 9, 399–431 (1929)
2. Pfeifer, R., Scheier, C.: Understanding Intelligence. MIT Press, Cambridge (1999)
3. Ashby, W.R.: Design for a Brain: The Origin of Adaptive Behaviour. Chapman and Hall, London (1952)
4. Dyke, J., Harvey, I.: Hysteresis and the limits of homeostasis: From daisyworld to phototaxis. In: Capcarrère, M.S., Freitas, A.A., Bentley, P.J., Johnson, C.G., Timmis, J. (eds.) ECAL 2005. LNCS (LNAI), vol. 3630, pp. 332–342. Springer, Heidelberg (2005)
5. Dyke, J.G., Harvey, I.R.: Pushing up the daisies. In: Proc. of Tenth Int. Conf. on the Sim. and Synthesis of Living Systems, pp. 426–431. MIT Press, Cambridge (2006)
6. Besendovsky, H.O., Del Rey, A.: Immune-neuro-endocrine interactions: Facts and hypotheses. Endocrine Reviews 17, 64–102 (1996)
7. Di Paolo, E.A.: Homeostatic adaptation to inversion of the visual field and other sensorimotor disruptions. In: From Animals to Animals, Proc. of the 6th Int. Conf. on the Simulation of Adaptive Behavior, pp. 440–449. MIT Press, Cambridge (2000)
8. Harvey, I.: Homeostasis and rein control: From daisyworld to active perception. In: Proc. of the 9th Int. Conf. on the Simulation and Synthesis of Living Systems, ALIFE9, pp. 309–314. MIT Press, Cambridge (2004)
9. Neal, M., Timmis, J.: Timidity: A useful mechanism for robot control. Informatica 7, 197–203 (2003)

10. Hoinville, T., Henaff, P.: Comparative study of two homeostatic mechanisms in evolved neural controllers for legged locomotion. In: Proccedings of 2004 IEEE/RSJ Int. Conf. on Intelligent Robots and Systems, vol. 3, pp. 2624–2629 (2004)
11. Vargas, P.A., Moioli, R.C., Castro, L.N., Timmis, J., Neal, M., Von Zuben, F.J.: Artificial homeostatic system: a novel approach. In: Proc. of the VIIIth European Conf. on Artificial Life, pp. 754–764 (2005)
12. Moioli, R.C., Vargas, P.A., Von Zuben, F.J., Husbands, P.: Towards the evolution of an artificial homeostatic system. In: 2008 IEEE Congress on Evolutionary Computation (CEC 2008), pp. 4024–4031 (2008)
13. Vargas, P.A., Di Paolo, E.A., Husbands, P.: Preliminary investigations on the evolvability of a non-spatial GasNet model. In: Almeida e Costa, F., Rocha, L.M., Costa, E., Harvey, I., Coutinho, A. (eds.) ECAL 2007. LNCS (LNAI), vol. 4648, pp. 966–975. Springer, Heidelberg (2007)
14. Vargas, P.A., Di Paolo, E.A., Husbands, P.: A study of gasnet spatial embedding in a delayed-response task. In: Proc. of the XIth Int. Conf. on the Sim. and Synthesis of Living Systems, ALIFE-XI, Winchester, UK, August 5-8 (to appear, 2008)
15. Di Paolo, E.A.: Autopoiesis, adaptivity, teleology, agency. Phenomenology and the Cognitive Sciences 4(4), 429–452 (2005)
16. Storm, T.: KiKS, a Khepera simulator for Matlab 5.3 and 6.0, `http://theodor.zoomin.se/index/2866.html`
17. Collins, R., Jefferson, D.: Selection in massively parallel genetic algorithms. In: Proc. of the 4th Intl. Conf. on Genetic Algorithms, ICGA 1991, pp. 249–256. Morgan Kaufmann, San Francisco (1991)
18. Hillis, W.D.: Co-evolving parasites improve simulated evolution as an optimization procedure. Physica D 42, 228–234 (1990)
19. Husbands, P., Smith, T., Jakobi, N., Shea, M.O.: Better living through chemistry: Evolving GasNets for robot control. Connection Science 10, 185–210 (1998)
20. Nolfi, S., Floreano, D.: Evolutionary Robotics: The Biology, Intelligence, and Technology of Self-Organizing Machines. MIT Press, Cambridge (2004)

Author Index